History, Philosophy and Theory of the Life Sciences

Volume 31

History, Philosophy and Theory of the Life Sciences is a space for dialogue between life scientists, philosophers and historians – welcoming both essays about the principles and domains of cutting-edge research in the life sciences, novel ways of tackling philosophical issues raised by the life sciences, as well as original research about the history of methods, ideas and tools, which constitute the genealogy of our current ways of understanding living phenomena.

The series is interested in receiving book proposals that • are aimed at academic audience of graduate level and up • combine historical and/or philosophical and/or theoretical studies with work from disciplines within the life sciences broadly conceived, including (but not limited to) the following areas: • Anatomy & Physiology • Behavioral Biology • Biochemistry • Bioscience and Society • Cell Biology • Conservation Biology • Developmental Biology • Ecology • Evolution & Diversity of Life • Genetics, Genomics & Disease • Genetics & Molecular Biology • Immunology & Medicine • Microbiology • Neuroscience • Plant Science • Psychiatry & Psychology • Structural Biology • Systems Biology • Systematic Biology, Phylogeny Reconstruction & Classification • Virology The series editors aim to make a first decision within 1 month of submission. In case of a positive first decision the work will be provisionally contracted: the final decision about publication will depend upon the result of the anonymous peer review of the complete manuscript. The series editors aim to have the work peer-reviewed within 3 months after submission of the complete manuscript. The series editors discourage the submission of manuscripts that contain reprints of previously published material and of manuscripts that are below 150 printed pages (75,000 words). For inquiries and submission of proposals prospective authors can contact one of the editors: Charles T. Wolfe: ctwolfe1@gmail.com Philippe Huneman: huneman@wanadoo.fr Thomas A.C. Reydon: reydon@ww.uni-hannover.de

Giuseppe Bianco • Charles T. Wolfe
Gertrudis Van de Vijver
Editors

Canguilhem and Continental Philosophy of Biology

Springer

Editors
Giuseppe Bianco
Department of Philosophy
and Cultural Heritage (DFBC)
Ca'Foscari University
Venice, Italy

Charles T. Wolfe
Département de Philosophie & ERRAPHIS
Université de Toulouse Jean-Jaurès
Toulouse, France

Gertrudis Van de Vijver
Department of Philosophy
and Moral Sciences
Ghent University
Ghent, Belgium

ISSN 2211-1948 ISSN 2211-1956 (electronic)
History, Philosophy and Theory of the Life Sciences
ISBN 978-3-031-20528-6 ISBN 978-3-031-20529-3 (eBook)
https://doi.org/10.1007/978-3-031-20529-3

This Springer imprint is published by the registered company Springer Nature Switzerland AG
The registered company address is: Gewerbestrasse 11, 6330 Cham, Switzerland

Contents

Introduction

Giuseppe Bianco, Charles T. Wolfe, and Gertrudis Van de Vijver

Abstract In this Introduction we lay out the context of a 'Continental philosophy of biology' and suggest why Georges Canguilhem's place in such a philosophy is important. There is not one single program for Continental philosophy of biology, but Canguilhem's vision, which he referred to at one stage as 'biological philosophy', is a significant one, located in between the classic holism-reductionism tensions, significantly overlapping with philosophy of medicine, philosophy of technology and other themes moving away from the more common existential and phenomenological motifs of post-war European thought. Chapters examine (among other themes) his relation to *Lebensphilosophie*, to authors such as Kant, Nietzsche and Marjorie Grene, and to current theoretical biology

Keywords Canguilhem · Philosophy of biology

G. Bianco (✉)
Department of Philosophy and Cultural Heritage (DFBC), Ca'Foscari University, Venice, Italy
e-mail: giuseppe.bianco@unive.it

C. T. Wolfe
Département de Philosophie & ERRAPHIS, Université de Toulouse Jean-Jaurès, Toulouse, France
e-mail: ctwolfe1@gmail.com

G. Van de Vijver
Department of Philosophy and Moral Sciences, Ghent University, Ghent, Belgium
e-mail: gertrudis.vandevijver@ugent.be

G. Bianco et al. (eds.), *Canguilhem and Continental Philosophy of Biology*, History, Philosophy and Theory of the Life Sciences 31,
https://doi.org/10.1007/978-3-031-20529-3_1

1 Introduction

For a variety of historical and theoretical reasons, little of the work that is now done under the heading of "philosophy of biology" deals with the core questions that the life sciences have traditionally posed to philosophers. Those questions primarily pertain to the concepts of "life" and "organism," which can be seen as constitutive of the proper domain of the biological sciences, and which are inhabited by metaphysical, epistemological, and moral issues that directly impact one's approach within the life sciences.

Although contemporary philosophy of biology, as developed in the Anglo-American world since the 1960s, has turned its attention away from the fundamental issues proper to these sciences, the very mention of which seems to summon the ghosts of vitalism and teleology, philosophers of the life sciences on the continent have never ceased to be preoccupied by them. Despite François Jacob's (1973) famous claim that life is no longer an object of inquiry in laboratories, and despite Michel Foucault's (1994) idea that the very concept of life is doomed to disappear, since it belongs to an old "*epistemé*," the fundamental notions proper to the life sciences still provoke important debates which resonate with discussions that emerged during the long nineteenth century, and are often tied to what went under the name of "life-philosophies" or *Lebensphilosophie* (see Bianco, 2019). Refusing to leave behind concepts such as "life" and "organism", but on the contrary insisting upon addressing them, time and again, along and in dialogue with the developments of the life sciences – that is perhaps what best characterizes the various continental approaches in philosophy of biology. The figure of Georges Canguilhem is paramount here.

The challenge is thereby not so much to attempt to comply with specific scientific standards paradigmatically identified with those of physics and chemistry (which can be seen as reductionist, as Canguilhem did), nor to conclude, in the absence thereof, that there are concepts, such as life, which irremediably escape scientific treatment. The challenge is, rather, to pursue the question of what it can mean to consider a living being, i.e. an organism, as an object of science, and to make room, through the living, for a more generous, a more subtle, an "extended" conception of science and scientific objectivity.

At times there were competitor terms for the Continental version of philosophy of biology, such as "biophilosophy" or "biological philosophy" (Gayon, 2009). Some of these versions had a pronounced anti-reductionist or anti-naturalist focus, which will be discussed and evaluated in this volume (for a study that looks beyond thinkers like Marjorie Grene and Canguilhem, to the Cambridge Theoretical Biology Club in the early twentieth century, see Peterson, 2017). Sometimes these projects have a pronounced Kantian focus (Van de Vijver & Demarest, 2013; Huneman, 2017), which is also present in this volume, to which we return below.

Rather than anti-naturalism or anti-reductionism, perhaps the term "heterodox naturalism" can best capture this specific focus we aim at here: it is a naturalism indeed, in as far as the aim is not to be opposed to what is being discovered and articulated in the sciences, but in addition, heterodox, because the conception of scientific objectivity that is strived for does not primarily have to obey standards as identified in orthodox naturalism.[1] We suggest that there are two main directions along which these questions have been approached in continental philosophy of biology, two directions that also operate in the works of Canguilhem that we put at center stage here, and that continue to make themselves felt up to this very day.

Firstly, it is possible to start from the qualitative, phenomenal differences that we can observe between living and dead bodies. A "*Körper*" is not a "*Leib*" – the former being subject to reduction to elementary parts, the latter resisting any such form of reduction. A living being, a *Leib*, cannot be touched by reductive science, so it is said. But it should, and it can, be touched differently, for instance in more experiential ways, in ways corresponding to our lived experience, in ways that we, living, sentient beings have more or less faithfully access to. Here we find, roughly, much of the phenomenologically inspired continental philosophy of biology, of which Canguilhem seems to be implicitly rather critical (for instance, see Canguilhem, 2012), in as far as it carries with it the risk of installing a kingdom alongside the "orthodox" kingdom of the natural sciences, a kingdom namely of lived experience, a kingdom sometimes even bordering on things qualified as "non-material"[2] (on the difference between these approaches see Wolfe in this volume).

Secondly, there are critical, transcendental voices focusing mainly on the question whether, and how, the fact of being alive impacts on what it can mean to know and on what it can mean to be driven to know (see Kolen & Van de Vijver, 2007; Van de Vijver et al., 2005; Van de Vijver & Haeck, in press). The epistemological implications of taking the concepts of "life" and "organism" seriously are in this case potentially more revolutionary, because the upshot is no longer to find a more adequate scientific account of those systems that we call the living beings, but to address the question of the possibility (i.e. the meaning) of scientific knowledge as a product of a peculiar living being (the human), confronted with other living beings. It is in the heart of knowledge that we should search for the living, in the heart of its structure, in the heart of its judgments. Georges Canguilhem sides very much with this

---

[1] The term naturalism lends itself to various interpretations. In speaking here of heterodox naturalism, we take it in its broad and open-ended sense of striving for theories that are not in contradiction with the data and insights produced in the various sciences. There is, however, a much narrower sense that has played an important role in Anglo-American philosophy of biology, namely a more dogmatic or ideological form of naturalism, here called orthodox, in as far as it takes for granted that there is only one truth possible in and on science and scientific objectivity. In this regard, see Kitcher (1992) and Callebaut (1993).

[2] Canguilhem (2008, 70) writes: "the classical vitalist accepts the insertion of the living organism into a physical milieu to whose laws it constitutes an exception. Therein lies, in our opinion, the philosophically inexcusable fault. There cannot be an empire within an empire without there being no longer any empire, neither as container nor as contents. There can be only one philosophy of empire, that which refuses any division: imperialism." We use 'kingdom' instead of 'empire' to render Canguilhem's deliberate echo of Spinoza's *imperium in imperio*, translated in English as 'kingdom within a kingdom'. On vitalism in Canguilhem's thought see Wolfe and Wong (2014), and in relation to contemporary biology such as Jacob, see Etxeberria and Wolfe (2018).

approach, even if he does not always seem to pursue this track in its utmost consequences (see, in this volume, the chapters by Gandolfi, Van de Vijver & Haeck).

We are aware that the term 'transcendental' is used in quite different ways in the philosophical literature. In Anglo-American contexts, it quite often refers to what is outside of our material grasping, to what transcends material processes at large. When things are considered to be "transcendentalized", they are said to belong to another realm, to have acquired another status. From within critical philosophy as initiated by Kant, however, transcendentality does concern still another realm, the one namely of ideality, but it is a realm that, however much it takes precedence over materiality, is also grafted upon it by articulating its conditions of possibility. We consider the difference between these two conceptions of transcendentality to be symptomatic of the "analytical-continental divide" in philosophy of biology. As it stands now, the transcendental is a fault line, a "*ligne de fracture*", and it seems that there is still a long way to go to develop a genuinely critical viewpoint on objectivity in which the positions of both are being estimated in an adequate way.

There is no doubt that the work of Georges Canguilhem is highly relevant for the philosophy of the life sciences, and this for both theoretical and historical reasons, as can be seen by the recent turn to his work in English-language scholarship, including several monographs (Elden, 2019; Talcott, 2019; Méthot, ed. 2020). His theories on life are located at the convergence of many antinomies such as objectivism versus subjectivism, realism versus relativism, or the cumulative idea of science versus the discontinuous one. And these antinomies, which Canguilhem alternately challenges and employs, can also be 'transcended' in a current critical analysis. Because indeed, Canguilhem is not so much bothered by the idea that he might occupy a certain position – as he writes in the preface to his study of the history of reflex action, "I'm not worried about being called a vitalist" (Canguilhem, 1955/1977, 1) – rather, he is bothered by what it involves and implies to adopt a certain position, in relation to other positions.

From an historical standpoint, Canguilhem – whose work on the history and philosophy of the life-sciences spans a period running from the 1940s until the 1990s – played an important role as a mediator between different scientific traditions: starting from the 1940s, his essays discussed German authors like Uexküll or Goldstein, that were almost unknown to French philosophers; during the 1960s, his works influenced Anglo-American philosophers of biology, like Marjorie Grene, even if this impact on the field did not last long; starting from the 1990s, Canguilhem's approach to the history of the sciences, often labeled as "historical epistemology," had a huge impact on generations of German and Anglo-American historians and philosophers, many of whom worked in the history and philosophy of the life-sciences, such as, for instance Hans-Jörg Rheinberger (2010) and his group at the Max Planck Institute for the History of Science in Berlin (who had devoted a research project to Canguilhem in the late 1990s, later published in Borck et al., 2005).

This volume presents studies focusing – in different ways – on this alternative philosophy of biology here termed "continental philosophy of biology," with special focus on Canguilhem. In doing so, it contributes to debates in the history and philosophy of science and the history of philosophy of science, as well as to the craving for 'history' and/or 'theory' in the theoretical biological disciplines (cf. Noble, 2008; Nicholson & Gawne, 2015). This is a different approach, although hopefully complementary, to that which seeks to relate (and contrast) Canguilhem to current philosophy of medicine (Giroux, 2010; Méthot, 2020). In addition, however, it also aims to provide inspiration for a broader conception of philosophy of biology, in which these traditional issues may have a place. This aspect of a broadening or widening of the narrative of philosophy of biology also complements more historical projects which present a broadened historical narrative of the constitution of biology (see e.g. Bognon-Küss & Wolfe, 2019), in a more contemporary context.

We have arranged the present volume around Canguilhem's work in the field of philosophy and history of the life sciences, but also includes work on the genesis and outcomes of Continental philosophy of biology as a whole. A first part is composed by texts situating Canguilhem more generally in the field of philosophy of biology, Anglo-American as well as continental, and opens the horizon of what prevails in the field. The second part situates his works historically and theoretically, and the third part studies their influence, departing from his concepts and theories and examining further cases.

In the first chapter Pierre-Olivier Méthot shows how, during the 60s and the 70s, philosophy of biology emerged as a distinguishable field traversed by conflicts; he particularly focuses on the debates between Marjorie Grene – a reader of Canguilhem – and David Hull. He raises a question about the historical conditions that make fields such as the philosophy of biology possible, and calls attention to the exclusions that permeated the philosophy of biology from its inception. In the second chapter, Philip Honenberger studies the intellectual trajectory of Grene, showing how she developed an original philosophical position, "ecological epistemology," which situates organic life at the center of the interpretation of reality and human affairs. The chapter reviews Grene's work in the main research areas for which she is best known with an eye to how these tensions were ultimately resolved in her account. In the third chapter, Thomas Ebke adopts a comparative approach in taking into account the relation between Canguilhem and German authors belonging to the tradition of modern German philosophical anthropology. All these scholars used, in different ways, the expression "philosophical biology," they all shared some common theoretical roots and put, at the center of their attention, a strong concept of life and human life.

The studies belonging to the second section deal with Canguilhem's work in historical-conceptual terms, locating forgotten sources, implicit and explicit debates between him and his contemporaries. In the fourth chapter Giuseppe Bianco tries to explain why, while being a philosophy professor, during the mid-1930s Canguilhem decided to embark on a long medical training. According to the author, who examines unpublished manuscripts, such as a series of lectures given in Toulouse between 1933 and 1935, the motivation of his turn has to be related to the readings of works in psychology and ethology undertaken during this period. The fifth and sixth chapters focus on Canguilhem's complicated relation with Kant and the heritage of transcendental philosophy. In "Life, concept and purpose," Giulia Gandolfi shows the close connection between concept, life and knowledge in Canguilhem's and Kant's philosophical projects; in both these two authors the organism appears as the condition of possibility for knowledge, as a material *a priori.* In Canguilhem, this idea will go through a change after the discovery of DNA which complexified the idea of organic totality. In the sixth chapter, "Canguilhem's Divided Subject," Levi Haeck and Gertrudis Van de Vijver deepen this Kantian analysis to Canguilhem's reappropriation of the vitalist tradition. The authors develop the idea that Canguilhem's late writings on logic and life, if we tackle them from a Kantian point of view, bring to the fore a theory of the subject. If the human subject appears as a living instance, it simultaneously appears as a knowing instance, the needs of which are however at least to a certain extent contingent on the subject's participation in the realm of logic. The seventh chapter deals with Canguilhem's partially hidden Nietzschean legacy and like the previous two, seeks to connect Canguilhem's works from the 1940s and the 1950s, with the turn his thought undergoes during the 1960s. In "Knowledge, Life, and Error," Henning Schmidgen argues that the philosophy of Nietzsche inspired his ideas of the relationship between health and disease, of the influence of language on perception and knowledge, and of philosophy as a philosophy of values. Canguilhem has a deep relation to Nietzsche's idea of a "General Physiology," investigating the relation of the organism to its environment; this attempt remains intact even with the emergence of genetics and molecular biology, and even if, starting from the 1960s, the life sciences reconfigure themselves with respect to the concept of information. The eighth and ninth chapters deal, in different ways, with Canguilhem's complicated relation to phenomenology, and more particularly, with the work of Maurice Merleau-Ponty. In "Neither Angel nor Beast," Sebastjan Vörös addresses the problem of the relationship between life (vitality) and mind (thought) by drawing on the resources available in Canguilhem's and Merleau-Ponty's philosophies. In "The promise of the flesh," Charles T. Wolfe discusses the phenomenological distinction between body and flesh, *Körper* and *Leib* and compares it critically to Canguilhem's approach. According to the author, his critique of mechanism and mechanisms is not done in the name of a wholescale organicism and/or an unproblematized notion of embodiment and privacy. The authors try, in contrasting ways, to explain how the dynamism between life and mind also holds true for the researcher investigating these topics, and hint at the broader philosophical implications of such a view for the practice of science and philosophy.

In the tenth chapter, "What is biological normativity?," Paul-Antoine Miquel discusses two philosophical assumptions: according to the first one, biological normativity is not an irreducible property of the living, but rather the living is the historical result of its normative activity, according to the second, the concept of "operational closure" is used in order to naturalize biological normativity. The author proposes at looking at the problem, implying the idea of the architectural constraints of a biological system, which enable us predict the presence of specific propulsive and repulsive devices in every organism, by which its organization can be constantly rebuilt, and through which biological disruption can also be amplified. In the eleventh chapter, "Self-Organizing Life," Massimiliano Simons comes back to the problem of organization, and analyses the work of one of Canguilhem's pupils, Michel Serres. Along with Jacques Monod and Henri Atlan, Serres was one of the first scholars who started to think about biology in terms of second-order cybernetics and information theory. Simons maps the relation between Serres and Canguilhem, fleshes out Serres' peculiar "bio-philosophy" and explores the consequences through a brief examination of two authors whose work lies within this tradition: René Girard and Bruno Latour. The twelfth chapter, "French Philosophy of Technology and Technoscience," focuses on the heritage of the work of another of Canguilhem's pupils, Gilbert Simondon. Jessica Lombard focuses on the concept of "technoscience," used by French and Belgian philosophers of technology such as Bernadette Bensaude-Vincent, Xavier Guchet and Gilbert Hottois. The chapter reflects on an ontology of the existence proper to technoscientific objects. The author focuses on a submicroscopic technoscientific object, the *bio-object*, and put forward an original analysis about the being of (bio)technoscientific objects in their own milieu.

Finally, in the closing chapter, "A Bergsonian Perspective on Causality and Evolution," Mathilde Tahar focuses on Henri Bergson, one of the sources of Canguilhem's biological philosophy, not, however, in historical but in theoretical fashion, insofar as Bergsonian philosophy of biology can, in her view, contribute to current debates in philosophy of biology. In *Creative Evolution* Bergson was able to understand the processual nature of evolution and of its consequences to conceive a causality that considers the efficacy of time in evolution. The chapter reassesses Bergson's philosophy of evolution not only to show that the Bergsonian philosophy of life is not completely outdated, but also to reevaluate its importance in the history of biology and the philosophy of biology.

Acknowledgments This publication was made possible by the generous support of FWO, within the frame of the project G05081N, Vitalism: A Counter-History of Biology.

References

Bianco, G. (2019). Philosophies of Life. In P. Gordon & W. Breckman (Eds.), *The Cambridge History of Modern European Thought* (pp. 153–175). Cambridge University Press.

Bognon-Küss, C., & Wolfe, C. T. (2019). The Idea of 'Philosophy of Biology Before Biology': A Methodological Provocation. In C. Bognon-Küss & C. T. Wolfe (Eds.), *Philosophy of Biology Before Biology* (pp. 4–23). Routledge.

Borck, C., Hess, V., & Schmidgen, H. (Eds.). (2005). *Mass und Eigensinn. Studien im Anschluss an Georges Canguilhem*. Wilhelm Fink.

Callebaut, W. (1993). *Taking the Naturalistic Turn, or How Real Philosophy of Science is Done*. University of Chicago Press.

Canguilhem, G. (1977). *La formation du concept de réflexe aux XVIIe et XVIIIe siècles*, 2nd rev ed. Vrin (first published 1955).

Canguilhem, G. (2008). Aspects of Vitalism. In *The Knowledge of Life*. (S. Geroulanos & D. Ginsburg, Trans.) (pp. 59–74). Fordham University Press.

Canguilhem, G. (2012). Health: Popular Concept and Philosophical Question. In G. Canguilhem, *Writings on Medicine* (S. Geroulanos & T. Meyers, Trans.) (pp. 43–52). Fordham University Press

Elden, S. (2019). *Canguilhem*. Polity Press.
Etxeberria, A., & Wolfe, C. T. (2018). Canguilhem and the Logic of Life. *Transversal: International Journal for the Historiography of Science, 4*, 47–63. https://doi.org/10.24117/2526-2270.2018.i4.06
Foucault, M. (1994). *The Order of Things: An Archaeology of the Human Sciences* (A. Sheridan, Trans.). Vintage Books.
Gayon, J. (2009). Philosophy of Biology: An Historico-Critical Characterization. In A. Brenner & J. Gayon (Eds.), *French Studies in Philosophy of Science: Contemporary Research in France* (pp. 201–212). Springer.
Giroux, É. (2010). *Après Canguilhem: définir la santé et la maladie*. PUF.
Huneman, P. (2017). Kant's Concept of Organism Revisited: A Framework For a Possible Synthesis Between Developmentalism and Adaptationism? *The Monist, 100*(3), 373–390.
Jacob, P. (1973). *The Logic of Life: A History of Heredity* (B. E. Spillmann, Trans.). Pantheon Books.
Kitcher, P. (1992). The Naturalists Return. *Philosophical Review, 101*, 53–114.
Kolen, F., & Van de Vijver, G. (2007). Philosophy of Biology: Naturalistic or Transcendental? *Acta Biotheoretica, 55*, 35–46.
Méthot, P.-O. (Ed.). (2020). *Vital Norms. Canguilhem's* The Normal and the Pathological *in the Twenty-First Century*. Editions Hermann.
Nicholson, D., & Gawne, R. (2015). Neither Logical Empiricism nor Vitalism, but Organicism: What the Philosophy of Biology Was. *HPLS, 37*(4), 345–381.
Noble, D. (2008). Claude Bernard, the First Systems Biologist, and the Future of Physiology. *Experimental Physiology, 93*(1), 16–26.
Peterson, E. L. (2017). *The Life Organic: The Theoretical Biology Club and the Roots of Epigenetics*. University of Pittsburgh Press.
Rheinberger, H.-J. (2010). *An Epistemology of the Concrete: Twentieth-Century Histories of Life* (G. M. Goshgarian, Trans.). Duke University Press.
Talcott, S. (2019). *Georges Canguilhem and the Problem of Error*. Palgrave Macmillan.
Van de Vijver, G., & Demarest, B. (Eds.). (2013). *Objectivity After Kant. Its Meaning, its Limitations, its Fateful Omissions*. Olms.
Van de Vijver, G., & Haeck, L. (In press). Judging Organization. A Plea for Transcendental Logic in Philosophy of Biology. In M. Mossio (Ed.), *Biological Organization*. Springer.
Van de Vijver, G., Van Speybroeck, L., De Waele, D., De Preester, H., & Kolen, F. (2005). Current Philosophy of Biology: Outline of A Transcendental Project. *Acta Biotheoretica, 53*(2), 57–75.
Wolfe, C. T., & Wong, A. (2014). The Return of Vitalism: Canguilhem, Bergson and the Project of Biophilosophy. In G. Bianco, M. de Beistegui, & M. Gracieuse (Eds.), *The Care of Life: Transdisciplinary Perspectives in Bioethics and Biopolitics* (pp. 63–75). Rowman & Littlefield International.

Giuseppe Bianco is a senior researcher, Cà Foscari University, Venice has published on the history of philosophy, the humanities and the social sciences (nineteenth and twentieth centuries). He is the author of *Après Bergson* (PUF, 2015) and edited several books on French philosophy. He is developing three intertwined research projects: a book on Gilles Deleuze's intellectual trajectory (*Gilles Deleuze and Philosophy*), a history of biological philosophy and *Lebensphilosophie*, and a sociological study about the emergence of a transnational European space of circulation of philosophical books during the twentieth century.

Charles T. Wolfe is a Professor in the Department of Philosophy, Université de Toulouse-2 Jean-Jaurès. He works primarily in history and philosophy of the early modern life sciences, with a particular interest in materialism and vitalism. He is the author of *Materialism: A Historico-Philosophical Introduction* (2016), *La philosophie de la biologie avant la biologie: une histoire du vitalisme* (2019) and *Lire le matérialisme* (2020), and recently coedited (w. D. Jalobeanu) the *Encyclopedia of Early Modern Philosophy and the Sciences*. He is co-editor of the book series 'History, Philosophy and Theory of the Life Sciences' (Springer). Papers available at [https://univ-tlse2.academia.edu/CharlesWolfe]

Gertrudis Van de Vijver is full professor in philosophy at Ghent University. Her research deals with the issue of complexity, self-organization and teleology in the life sciences and with issues around the system of the psyche from within the psychoanalytical theories of Freud and Lacan. For both, she contributed to the articulation of a transcendental epistemology, centring on the idea of co-constitution. From 2016 till 2018 she was the director of the Centre for History of Philosophy and Continental Philosophy (HICO) at Ghent University. Between 2005 and 2010, she led an interdisciplinary research project on Complexity and Systems Biology that involved philosophers, biologists, bio-engineers and communication scientists.

Part I
Biophilosophical Backgrounds: Faultlines and Forerunners

Analytic and Continental Approaches to Biology and Philosophy: David Hull and Marjorie Grene on 'What Philosophy of Biology Is Not'

Pierre-Olivier Méthot

Abstract Gaining momentum during the last third of the twentieth century, the philosophy of biology is now a distinct field with its own debates, journals, audiences, and professional societies. This professionalization came along with the forging of an intellectual identity based on the existence of disciplinary frontiers that demarcated philosophy of biology from neighboring disciplines such as philosophy of medicine, history of biology, or general philosophy of science. Here, I argue that the identity of this emerging philosophy of biology also excluded Continental traditions often called "biological philosophy" or "historical epistemology of the life sciences". Going back to the 60s and 70s, I explore the emergence of the philosophy of biology at a time when its identity was still in flux and its analytic orientation debated. To do so, I focus primarily on the works of David Hull and Marjorie Grene, and I draw on their unpublished correspondence. Although Grene's intellectual contribution to the philosophy of biology has been widely acknowledged, her coming from a different philosophical universe created tensions with the identity Hull and others sought to establish. Overall, this chapter raises a question about the historical conditions that make fields such as the philosophy of biology possible, and calls attention to the exclusions that permeated the philosophy of biology from its inception and what this involves in terms of the proper relation between philosophy and science, especially biology.

Keywords Philosophy of biology · Biological philosophy · Canguilhem · Identity · Disciplines · Life sciences

P.-O. Méthot (✉)
Université Laval, Québec City, QC, Canada

G. Bianco et al. (eds.), *Canguilhem and Continental Philosophy of Biology*, History, Philosophy and Theory of the Life Sciences 31,
https://doi.org/10.1007/978-3-031-20529-3_2

1 Introduction

There has not always been an autonomous philosophical domain uniquely devoted to the life sciences, but questions about the nature of life have historically been at the center of major philosophical debates. Gaining momentum during the last third of the twentieth century, the philosophy of biology is now a distinct field with its own debates, journals, audiences, and professional societies. Taking as its starting point Charles Darwin (instead of Immanuel Kant or Claude Bernard, for instance), the philosophy of biology has for 30 years mostly focused on issues arising in, or related to, evolutionary theory, and remained a long way from considering the full range of biological sciences (Pradeu, 2017).[1] Elliott Sober spoke for many when he declared that "[t]he philosophy of biology does not end with evolutionary issues, but that is where I think it begins" (1993, xix).[2] In the past two decades, however, the field diversified and became more inclusive. Philosophers of biology are now addressing issues in systems biology, development, immunology, biomedicine, agriculture, microbiology, ecology, synthetic biology, and ethnobiology (among others) as well as the relation between biology and the wider society (Odenbaugh & Griffiths, 2020).

Tracing the formation of a discipline matters to understand its relation to other fields and to characterize its own identity. Since its foundation in the United States in the 60s and 70s, the philosophy of biology has professionalized. This professionalization came along with the forging of an "intellectual identity" based on the existence of "disciplinary frontiers" that distinguished philosophy of biology from neighboring disciplines such as philosophy of medicine, history of biology, or general philosophy of science (Gayon, 2009a). Any claim to identity, indeed, "involves deciding what to include or exclude, what will be in and what will be out" (Hopwood, 2019, 1). As we will see, the identity of the philosophy of biology also excluded Continental traditions often called "biological philosophy" or "historical epistemology of the life sciences".[3]

Going back to the 60s and 70s, this chapter explores the emergence of the philosophy of biology at a time when its identity was still in flux and its analytic orientation debated. To do so, I focus primarily on the works of Marjorie Grene and

[1] In Werner Callebaut's view, "philosophy of biology started not with Aristotle, or Spinoza, or Kant, or even Claude Bernard, but with Darwin […]" (2005, 93). Callebaut saw no reason to "invok[e] a pre-Darwinian philosopher – transcendental stance included – to shed light on contemporary debates" (Ibid., 114). For a response from a Kantian perspective: Kolen and Van de Vijver (2007).

[2] While evolutionary theory raises interesting philosophical issues, this growing interest in evolutionary theory in the 80s and 90s could be connected to the fact that several philosophers of biology came through "Lewontin's lab" at Harvard (Smocovitis, 1996, 41).

[3] The term "biophilosophy" is often used in the German context ("*Biophilosophie*") rather than the French "*philosophie biologique*" (Gayon, 2009a; Köchy, 2008). In the strict sense, one should also distinguish between "*biological* philosophy" and "*philosophical* biology". On this: Ebke (this volume). On Continental philosophy of science: Gutting (2005), Méthot (2020).

David Hull, and I draw on their unpublished correspondence.[4] Grene and Hull are often taken to be the respective founding "grandmother" and the founding "father" of philosophy of biology thanks to their roles in the professionalization of the field.[5] Although Grene's intellectual contribution on the philosophy of biology has been widely acknowledged (Burian, 2009; Smocovitis, 2009; Gayon, 2009b), her coming from a different philosophical universe created tensions with the identity Hull and others sought to establish. In this chapter, I also examine the relations of Grene and Hull to other influential figures such as Ernst Mayr, Everett Mendelsohn, and Georges Canguilhem. After a biographical section contrasting Grene's and Hull's trajectories, I critically review the standard historiography according to which the decline of logical positivism led to the philosophy of biology. Building on recent scholarship, I problematize the claim that Hull is the sole figure responsible for the emergence of the field. Then, zooming in on his famous article "What Philosophy of Biology is Not" (WPOBIN) (1969), I show how it originated at the request of Mendelsohn, then editor of the new *Journal of the History of Biology*. Finally, in the last section of the paper, I present Hull's perspective on the state of philosophical works in biology in the 60s and the critical response it elicited from Grene, who argued that the Continental tradition should not be ignored. In the following decades, she went on to defend the works of Canguilhem against the direction taken by Hull and the philosophy of biology community.

A note on terminology is in order before proceeding further. The term "biological philosophy" (*philosophie biologique*) is often loosely used in French to denote the biologically informed view of a scientist, a historian, a philosopher, etc. Thus, it was (and is) common to speak of the "biological philosophy" of Canguilhem or of Jacques Monod, for instance (Gayon, 1998a, b). As Jean Gayon observed, the term did not refer to a separate field, a subfield, or a specialty. The philosophy of biology, in contrast, self-identifies as a *field* or a *discipline* that arose in English-speaking countries (first in the United States, the United Kingdom, and Australia). Another observable difference between these traditions is that philosophers such as Canguilhem or biologists like Kurt Goldstein – closely associated with biological philosophy – did not limit their philosophical inquiries to biology (or to some parts of it – like evolution) but included medicine as well as physiology and other biological disciplines (Canguilhem, 1996 [1991]; Goldstein, 1934). These two intellectual traditions also differ markedly in terms of classical figures worthy of study: for instance, and still according to Gayon: "Nietzsche, together with, say, […] Bergson, provided a perfect example of the old fashioned 'biological philosophy' that the

[4]The unpublished correspondence of Hull is in the David L. Hull Papers, 1965–2004, ASP 2005.01 in the Archives of Scientific Philosophy, Archives & Special Collections at the University of Pittsburgh Library System. Other unpublished sources in this paper include materials from the Fonds Canguilhem at the Centre d'Archives en Philosophie, Histoire et Éditions des sciences (CAPHÉS) at the École Normale Supérieure; the Indiana University Department of History and Philosophy of Science Records; and the Archives of the Illinois Wesleyan University.

[5]https://believermag.com/an-interview-with-marjore-grene/; on Hull: Sect. 2.

new 'philosophy of biology' has wanted to avoid at all costs" (Gayon, 1999, 155–56).

Importantly, biological philosophy and philosophy of biology articulate the connection between biology and philosophy in opposite ways: briefly put, while the former draws on biological sciences to answer questions within philosophy – that is, biology provides them with "sources of inspiration" or "themes of reflexions"[6] – the latter is concerned with the clarification of methodological or conceptual puzzles arising within the biological sciences themselves. As philosophers of biology often address issues considered relevant to biology, collaborations with scientists are strongly encouraged and valued (Laplane et al., 2019). In this respect, the philosophy of biology differs not just from biological philosophy but also from an earlier view in philosophy of science where biology provides examples to test philosophical claims about, for instance, the nature of scientific change or scientific explanation. In contrast, Hull's view is that biology is distinct from general models drawn from physics, on the one hand, and that philosophy of biology must make itself relevant or at least useful and comprehensible to biologists, on the other. In the late 90s, following "the agenda of science" (Sterelny & Griffiths, 1999, xi), and not that of the general philosophy of science, became the rallying call of philosophers of biology, eager to gain autonomy and to make themselves useful to the actual scientific enterprise.

Finally, while both traditions encourage philosophers to become knowledgeable about the science they study, each of them nevertheless conceptualizes science and philosophy as different kinds of inquiry: "The position of the epistemologist", Canguilhem writes, "is set back [*en retrait*]"; "one should not intervene to participate in scientific research" (2018, 1002 [1984] [my translation]). Why shouldn't philosophers contribute to scientific knowledge? Because, for Canguilhem at least, the object of science is distinct from the object of the history of science; and the role of philosophy is to analyze the historical conditions under which this "cultural object" is being formed (2019, 265 [1968]). This statement will certainly sound more like history than philosophy to philosophers reading this and, indeed, Canguilhem was considered primarily as a historian of biology and medicine by his American colleagues (Méthot 2020). But as Gayon has put it "this is precisely the issue: historical epistemology claims that conceptual history is a major task for philosophers" (2009a, 206).[7] In sum, whereas biological philosophy characterizes philosophy and science as two distinct endeavors – each with their own objects and methods – in philosophy of biology there is no difference in kind, only in degree,

[6] As Canguilhem put it: "*Par philosophie de type biologique nous entendons une philosophie qui tire d'une vision du phénomène vital pris dans son originalité des sources d'inspiration et des thèmes de réflexions*" (1946–1948; CAPHÉS, GC. 12.1.8, fol. 15). Canguilhem's papers are located in the CAPHÉS (ENS, Paris). References to Canguilhem's notes and manuscripts (GC) are followed by box and page numbers (fol.). On Canguilhem's biological philosophy: Limoges (2015, 2018), Sholl (2016), and Méthot (2020).

[7] The relation between history and philosophy of science has developed differently in English-speaking and French or German traditions but I cannot address this complex issue here.

between (the theoretical parts of) science and the philosophy of science. I will return to these differences in outlook, styles, and methods throughout the paper (especially in Sects. 2 and 5) when I will contrast the biographies and the views of Grene and Hull about what the philosophy of biology was, is, or should be.

2 Marjorie Grene and David Hull Contrasted

Marjorie Grene was born and educated in the United States. She first trained in zoology (major) before switching over to philosophy. In the early 30s, she went to Germany to study with Martin Heidegger and Karl Jaspers in Freiburg and Heidelberg. After that, she returned to Radcliffe College, then the women's college at Harvard, to attend lectures by Alfred N. Whitehead and C.I. Lewis. In 1935, she obtained her PhD in philosophy with a dissertation on the philosophy of existence. Following a research stay in Denmark to study Kierkegaard, she moved to the University of Chicago in 1937 to participate in Rudolph Carnap's seminar where she was hired as a teaching assistant. Retrospectively, she admitted this was her only "commitment to logical positivism" (Grene, 2000, 48, 2002, 9). Grene stayed in Chicago until 1944 (when she lost her position) and, from then until 1957, she was out of academic life, except to serve as Michael Polanyi's research assistant while living on a farm with her husband (David Grene) and her children in Ireland (Nye, 2011).

She returned to teaching at the University of California at Davis in the mid-60s. Given her long interest in the life sciences, she took part in the new philosophy of biology that was developing across the United States. At Davis, Grene was in contact with the leading evolutionary biologist G. Ledyard Stebbins, with whom she even co-taught a course in Philosophy of the Biological Sciences. She was also in correspondence with another "architect" of the modern synthesis, Theodosius Dobzhansky, who came to Davis from the Rockefeller University in the early 70s along with his *protégé*, the population geneticist Francisco Ayala. At that time, the biologist and philosopher Michael Ghiselin was also at Davis, which made Davis, along with Chicago and later Harvard, one the first places where the philosophy of biology developed in the United States. Other places included the University of Minnesota and the University of Pittsburgh.

At Davis, Cornell, and later at Viriginia Tech Grene mentored young historians (Betty Smocovitis), scientists (Stuart Kauffman) and philosophers (William Wimsatt, Richard Burian, Jean Gayon) interested in biology, but her being outside academia for a long period of time made it difficult to have many graduate students (Smocovitis, 2009). Like everyone working in the history or philosophy of biology in the 60s, Grene was in contact with Ernst Mayr, who was quick to direct her attention towards Hull's papers, even suggesting she write to him. Her relationship with Mayr was mixed, however, because her unorthodox views of evolution were in disagreement with some tenets of the modern synthesis (Honenberger, 2015).[8] While her early work had been mostly in the history of philosophy (Aristotle, Descartes, Spinoza), with a particular focus on existentialism (Martin Heidegger, Jean-Paul Sartre), Grene started publishing on Darwinism, reductionism, and the structure of evolutionary biology; she taught the philosophy of biology in the late 60s and 70s and co-edited one of the first anthologies (Grene & Mendelsohn, 1976), a collection of "Essays in the Philosophy of Biology" (Grene, 1974), and reviews of developments of the field (Grene, 1983, 1997).

Next to this, Grene continued working within the Continental tradition and in philosophical anthropology. In 1968, she published *Approaches to Philosophical Biology*. This book introduced European scientists and philosophers such as Adolph Portmann, Helmuth Plessner, Arnold Gehlen, Maurice Merleau-Ponty, and Kurt Goldstein to an English-speaking readership. These authors also informed her philosophical work as Grene developed an "ecological epistemology", that is, a view of "situated" or "embodied" knowledge that insisted on the role of perception in shaping science and knowledge (Gayon, 2009b). Her approach was thoroughly historical and contextualist: rejecting both social-constructivist approaches to knowledge and a-historical positivism as untenable, she defended a view of epistemology "turned toward the life sciences" (Grene, 2006, 27 [translation mine]) where "all knowledge is orientation" (Grene, 2002, 25). In so doing, she drew on recent theories of perception (Gibson, 1979). As Honenberger has put it (this volume), the historical character of her position "involved a commitment to realism about the objects of perception and knowledge" while recognizing the "partiality, and historical contingency" of any of those scientific models and theories.

In the 70s and 80s, Grene took on a more institutional role in the philosophy of biology as she actively co-organized several key meetings that helped structure the field. Worthy of notice is the Summer Institute on "Biological and Social Perspectives on Human Nature" (Colorado College) she co-organized with William Wimsatt in 1977 that brought biologists, neuroscientists, philosophers of biology, and social scientists together for 5 weeks. Five years later the Summer Institute in the "Philosophy of Biology" (Cornell University) she co-organized with Richard Burian was – at least in retrospect – an intellectual landmark in the emergence of the philosophy of biology. Lecturers included Stephen J. Gould and Richard Lewontin among the scientists, but also Richard Levins and others. Importanly, this meeting was intended to produce material in order to teach philosophy of biology and to build curricula, with an emphasis on interdisciplinarity.[9] As these and other meetings (1983, 1987, 1989) continued to attract more people and fostered interactions between philosophers of biology and biologists, Grene's efforts led to the International Society for the History, Philosophy, and Social Studies of Biology

[8] Her "Two Evolutionary Theories" (reprinted in Grene, 1974), led her to be subjected some harsh criticisms by leading evolutionary biologists, to which she responded (Grene, 1963). This paper was written when she was working with Polanyi, many years before she moved to Davis.

[9] For a list of regular participants and lecturers: Grene and Burian (1983). For a list of the other meetings that led to the foundation of ISHPSSB: Burian (1986).

(ISHPSSB) in the early 90s.[10] With Mayr, she was named the first Honorary President of ISHPSSB, and a Marjorie Grene Prize was soon created in her honor. In addition to numerous Festschrifts that celebrate her works (Donovan et al., 1986; Burian & Pitt, 1992; Gayon & Burian, 2006), Grene was also the first woman to be selected for the *Library of Living Philosophers*. Her unique contributions have been widely recognized by the philosophy of biology community but she did not see eye-to-eye with Hull about several issues, including how to define the field.

Twenty-five years younger, David Hull's trajectory was different than Grene's although he, too, initially trained in biology before turning to philosophy. Following his military service, he attended classes in comparative anatomy, microbiology, and zoology taught by Dorothea Franzen, an expert in invertebrate zoology at Illinois Wesleyan University, where he graduated as a biology major in 1960. Following his premed education, Hull enrolled in one of the first History and Philosophy of Science programs in the United States at the University of Indiana (Grau, 1999). According to him, one reason he was admitted was because "he knew some biology": "In a way, my writing papers emphasizing biology was a help because none of my professors knew much biology. If they faulted me, it would have to be on the philosophy, not the biology", he said.[11] At Indiana, Hull took history of science classes with Edward Grant and attended seminars by invited lecturers such as Alexandre Koyré (1961) and Karl Popper (1963) (on Hull and Popper: Winsor, 2006).

The "driving forces" behind the new department, however, were philosophers Russel N. Hanson and Michael Scriven (Horder, 2013), who were part of the group of historically-inclined philosophers of science, influenced by Thomas Kuhn and Paul Feyerabend, who challenged the positivist view of scientific explanation (Honenberger, 2018). With Morton Beckner (1959) and Thomas Goudge (1961), Scriven (1959) was, according to Hull, one of the few philosophers interested in biology in the 50s. Hanson and Scriven possibly made Hull aware of how looking at biology instead of physics might challenge the positivist framework – especially since Hull was initially working under Hanson. But Hull considered them "the most arrogant men [he] has ever met".[12] So, when Hanson moved to Yale, and since Scriven was often away for research, he transitioned to philosopher of science Roger Buck, who provided him with the "emotional stability" to go through graduate school. Reading drafts of his dissertation, Buck was also "willing to sit down with a beginning student and teach the basics".[13] As an undergraduate, Hull took only one course in the philosophy of science and, although he studied German, his philosophical training was very different from Grene's in terms of contents and world-view.[14] In 1964, he obtained his PhD with a dissertation on "The Logic of

[10] Richard Burian and Jane Maienschein were closely involved in the writing of the by-laws of ISHPSSB, which were finalized in London, Ontario, in 1989 (Burian, 2009, 186).

[11] "Buck", Hull Papers, ASP 2005.1, box 7, folder 4, undated.

[12] "Buck", Hull Papers, ASP 2005.1, box 7, folder 4, undated.

[13] "Buck", Hull Papers, ASP 2005.1, box 7, folder 4, undated.

[14] David L. Hull, Transcript of Record, Illinois Wesleyan University, The Registrar's Office.

Phylogenetic Taxonomy" (Hull, 1964) in which he analyzed the term "species" using Scriven's notion (inspired by Wittgenstein) of a "cluster concept". Opposing the view that species can be defined by necessary and sufficient conditions, he suggested that the clustering approach was more promising.

Hull started corresponding with Ernst Mayr at the end of 1965, who, by then, had read his thesis and was already advancing the population/typological thinking dichotomy.[15] Using his institutional power, he was also pushing the philosophy of biology away from general philosophy of science inspired by the physical sciences (Mayr, 1959, 1969). This "cooptation" (Hull, 1994, 378) was a good working relation. While Mayr was "moving more and more in the direction of the history and philosophy of biological science" in the late 60s,[16] Hull enjoyed the possibility of "communicating with a biologist who [he felt] confident will understand what [he is] saying".[17] Mayr also borrowed the term "essentialism" from Hull (Winsor, 2006) and used it to articulate his argument about the static view of species that Darwin allegedly dislodged, an approach historians of biology have now called into question (Müller-Wille, 2011). In addition to his attempt to correct the ills of the past and to demarcate the field, Hull was involved in debates over the nature of species, biological classification, reductionism, presentism, the concept of human nature, and the dynamics of conceptual change. He often urged philosophers to use real, biological examples instead of thought-experiments (1982).

During his career, Hull published on problems arising in the biological sciences and authored and edited textbooks and anthologies (Hull, 1973, 1974; Hull & Ruse, 1998, 2007) in addition to serving for 21 years as the editor of the Chicago book series on the Conceptual Foundations of Science. Like Grene, he had few doctoral students, but he mentored and advised several younger philosophers who took an interest in biological science; he was generous with his time and involved in the organization of meetings that brought together biologists and philosophers (Ruse, 1989a, b) in addition to fighting for several social causes, including gay rights (Overman, 2000). Hull taught at the University of Milwaukee from 1965 to 1984, when he moved to Northwestern, where he stayed until his retirement in 2000. In the mid-1980s, he became president of the Philosophy of Science Association and the Systematic Zoology Association almost simultaneously. Overall, Hull was less directly involved with the creation of ISHPSSB than Grene, though he is often considered the intellectual founder of the field and became its first president. In 2011, after his passing, the David L. Hull Prize was established by the officers of ISHPSSB to honor his life and his legacy.

[15] On the history of this distinction in Mayr's thought, see Witteveen (2015).

[16] Mayr to Hull, June 2nd 1966, Hull Papers, ASP 2005.1, box 35, folder 2. Hull was also in contact with Michael Ghiselin from the late 60s in the context of the debate about the species concept.

[17] Hull to Mayr, June 4th 1966, Hull Papers, ASP 2005.1, box 35, folder 2.

3 Contested Origin Stories About the Philosophy of Biology

During his address at the annual meeting of the American Association for the Advancement of Science in December 1965, Mayr declared that "five or six volumes on [his] bookshelves […] include the misleading words 'philosophy of science'". In fact, "each of these volumes is a philosophy of physics" (1969, 197). This *philosophy-of-science-as-philosophy-of-physics* view was largely promoted by the first generation of philosophers of biology (see Ruse, 1973, 9; Hull, 1974, 6). A companion claim to this view – also hammered home by Mayr – was that "there was hardly any mention of biology in the writings of the logical positivists" (Mayr, 1996, vii). Worse, when the logical positivists *did* take an interest in it, we are told, they were misled by an ideological framework infused with vitalism. Their sharp criticisms of such thinking, in favour of versions of "physicalism", lead them to be blind to biological thought or fascinated by formalist approaches to biology (Wolters, 2018; 1999; for critiques: Nicholson & Gawne, 2014, 2015; Byron's (2007) bibliometric survey). Thus, the rejection of a general model of scientific explanation based on physical sciences and, with it, the decline of the logical positivist tradition, came to be regarded as what "triggered" the emergence of philosophy of biology (Ducheseneau, 1997, xii; Sober in Callebaut, 1993, 74). According to this narrative, the positivist approach had first to be abandoned for the philosophy of biology to flourish and take its rightful place within the philosophy of science.

Scholars have recently pointed out that there was something wrong with this received view: on the one hand, contrary to Mayr's claims, biology was not neglected by the logical positivist tradition whose interests included much more than vitalism and its critique including (second-generation) Ernst Nagel on organicism, which is still critical but nevertheless a sustained engagement with biology (Byron, 2007; Hoffer, 2013); on the other, there was a "discourse" and perhaps even an "international community" of philosophy of biology in the 30s and 40s that flourished *outside* the positivist tradition (Nicholson & Gawne, 2015, 372).[18] Furthermore, it would be misleading to assume that the decline of the logical positivist tradition was necessary for philosophy of biology to emerge as a subfield of the philosophy of science: several contemporary philosophers such as Michael Ruse, Alex Rosenberg, and Ken Schaffner saw themselves as *extending* the positivist program towards the life sciences or, at least, to be working in its philosophical spirit (but without the verificationist approach).[19] The "*high standards of clarity and rigor* introduced by logical empiricists", Callebaut wrote, "deserve further emulation" (2005, 107 [emphasis in original]). While Hull's work on theory reduction was compatible with

[18] Like Honenberger (2018, 296), however, I incline to think that the philosophy of biology as an academic discipline emerged in the sociological sense only after 1970.

[19] Ruse (1973) is a direct application of positivist approach to biology; Schaffner's work on theory reduction in the sciences during the 60s and 70s has followed closely the path opened by Ernest Nagel, his thesis advisor, and Hempel; as to Rosenberg, he defined himself as a "positivist *manqué*" for keeping-up with the discovery/justification distinction (Rosenberg in Callebaut, 1993, 85).

this tradition, Ruse's and Schaffner's attempts to reduce biology to chemistry and to physics were openly working within it (Honenberger, 2018). Others, like Grene, rejected this framework altogether (Gayon, 1998a).

Another aspect of this origin story is that the same Hull is largely responsible for the emergence of the philosophy of biology as a field in the last third of the past century. This *Hull-as-founder* narrative is found almost everywhere in the literature.[20] Its roots are in the late 80s, shortly after Hull's *Science as a Process* came out (1998a), when a second generation of philosophers of biology started producing papers and books at a high rate and the field began to look like a discipline. Around the late 80s, in a discipline-building move (Bensaude-Vincent, 2013), Ruse did two things to bring the field forward: firstly, he launched a new journal, *Biology and Philosophy* (1986) to welcome these contributions and institutionalize the discipline. Secondly, he made Hull – his long-time friend, colleague, and early mentor – *the* founding figure of the field.[21] Ruse published a "target article" by Hull in *Biology and Philosophy* (Hull, 1998b) along with 18 responses, showcasing the significance of his work for philosophy of biology and beyond. The next year, he published a collection of essays in his honor titled *What the Philosophy of Biology Is* (1989a). In the preface, Ruse emphasized that "the success of this subject [philosophy of biology] is due above all to the work and influence of one man: David Hull" (1989b, ix). Finally, he effectively declared Hull "the father of modern studies of biology from a philosophical point of view" (2008, 3). Mayr provided support when he declared that "so far as I can see" [...], "David Hull [...] was the first philosopher to dedicate himself to philosophy of biology" (Mayr, 1996, viii). Of course, no one has suggested that Hull *alone* created the philosophy of biology. The first post-50s generation included Michael Ruse, Ken Schaffner, Bill Wimsatt, and Marjorie Grene. Philosophically inclined biologists such as Ernst Mayr, Richard Lewontin, Richard Levins, Stephen J. Gould, Michael Ghiselin, Francisco Ayala, John Maynard Smith, and Edward O. Wilson were instrumental at the institutional level as well as in setting the agenda (core set of problems, etc.) for philosophy of biology. The problem of the units of selection, for instance, was taken up and discussed with vigor by both scientists and philosophers in the 70s and 80s, and continues to be an important area of research in the philosophy of biology community.[22]

Despite this collective dimension, Hull's reputation as *the* intellectual founder of the field went uninterrupted in the Anglophone world from the late 80s

[20] Smocovitis credits Lewontin with the view that Grene was the most influential figure and gives credit to Mayr in playing a key in shaping the philosophy of biology. But others would cite Lewontin as having played "the greatest role by permitting numerous workers to train in his lab" (Smocovitis, 1996, 105).

[21] As Honenberger points out, "nearly every text that either [Hull or Ruse] published from 1969 to 1975 appears to have been shared, and commented on at least partially, in correspondence" (2018, 292). Hull also commented on and shared drafts with William Wimsatt, Ken Schaffner, and many others.

[22] According to Callebaut, the levels of selection is "certainly the biggest [issue] in current philosophy of biology in sheer terms of investment in man – and woman power" (1993, 265).

onward. When he retired as Dressler Professor of the Humanities in 2000, Paul Griffiths published a series of essays in his honor in *Biology and Philosophy*. "In the forty years since David Hull began to study the history and philosophy of science", he wrote, "biology has moved from the periphery of the discipline to its center. No one has played a larger role in producing this shift than David Hull" (2000, 299). When Hull died, Peter Godfrey-Smith declared him "the person most responsible for the philosophy of biology achieving the status within philosophy that it has today. He is the single figure most responsible for its flourishing" (2010, 749).[23]

The characterization of Hull as the first philosopher to take biology seriously or as the most responsible for the emergence of the field has been criticized (Byron, 2007; Nicholson & Gawne, 2015). From a historiographical point of view, Nicholson and Gawne imagined how a different assessment of the past could have resulted in a different future for the philosophy of biology: "we cannot help but wonder that, had Hull made a greater effort to 'study the literature of the subject' [...] he almost certainly would have come away with a decidedly different impression of the field's accomplishments" (2015, 375). Hull not only left out much of the past, but he also missed two recent contributions titled "Biology and Philosophy" published in international philosophical encyclopedias (edited by Raymond Klibansky) and covering the period under consideration, namely: from 1948 to 1958 and from 1958 to 1968 (Van der Klaauw, 1958; Chiaraviglio, 1968). Instead of faulting Hull for his neglect of early or recent works, I seek to understand why he considered that only some of this literature "*might legitimately* be called philosophy of biology" (1969, 241 [emphasis mine]). A greater attention to the philosophical and intellectual context in which his paper was written, I argue, can shed some light on the vision he started articulating in the late 60s and where it came from.

4 Writing 'What Philosophy of Biology Is Not' (WPOBIN)

In the context of this (doubly problematic) received view, Hull's WPOBIN, presented in the first stand-alone Philosophy of Science Association meeting on October 12 1968, in Pittsburgh, stands out.[24] WPOBIN appeared the next year both

[23] See the testimonies by Jane Maienschein, Robert J. Richards, and Christopher D. Horvath following Hull's passing that were posted on the ISHPSSB website (Newsletter 2011).

[24] David Hull, Ken Schaffner, Michael Ruse, and William Wimsatt met in this PSA meeting and started corresponding with each other afterwards. Though each of them addressed topics in the philosophy of biology, they were not in the same session. The four men found themselves together again, and this time in a same session (on reductionism), at the 1974 PSA meeting. For reminiscences: Ruse (1993).

in the *Journal of the History of Biology* and *Synthese*.[25] Despite its critical outlook, "the paper was very well received in Pittsburgh",[26] Hull said, though he did receive "an irate letter from Mario Bunge" shortly after it appeared in print.[27] In his talk and in the paper, Hull defended the possibility of a more significant role for philosophy in biology: philosophers could "uncover, explicate, and possibly solve problems in biological theory and methodology" (1969, 268). Hull noted that so far, philosophers have been "misinformed" and that their analyses of biology "have not been motivated […] by any concern with issues *currently of interest to biologists*" (1969, 241, 249 [emphasis mine]). Thus, he railed against Bunge, who confused moths with flies, and criticized Grene's allegedly poor understanding of evolutionary theory. He made a point that philosophers must have sufficient understanding of the science and get their biology right. At least this is what philosophers of biology took to be the central message of this paper. As Robert Brandon put it: "I really see [Hull] as the person who more than anyone else made it *unacceptable* to work in this area without knowing biology" (in Callebaut, 1993, 239 [emphasis in original]). Rephasing this point, Callebaut contended that Hull's message was that philosophers should "do justice to the facts and the theories of biological sciences" (2005, 97; a critique: Van Speybroeck, 2007).

The actual writing of WPOBIN started in the Spring 1968, 6–8 months before the PSA congress in Pittsburgh. On November 27th 1967, Mendelsohn invited Hull to submit a paper to *JHB* asking "whether we might be able to induce [him] to write an essay-length review and commentary on one or two recent attempts at dealing with the philosophy of biology". Mendelsohn was thinking of Ludwig von Bertalanffy's *Robots, Men, and Minds* (1967) as well as Hans Jonas's *The Phenomenon of Life* (1966). Since "working biologists" were to be part of the readership, he noted, "there might be provocative things that you would like to present in the form of an essay review".[28] Although he was already at work on a paper dealing with philosophers of science who were contemporaries of Darwin (Herschel, Whewell, Mill), Hull said he would be interested in reviewing "Morton Beckner's new book (which will be out shortly) in conjunction with his earlier book *The Biological Way of Thought* (which is to be issued in paperback by the University of California Press)". Hull considered Beckner's (1959) book as the "single major work of a philosopher of biology in over a decade" (Hull, 1969, 267). There is no

[25] In 1968, Hull also agreed to participate in a conference organized by Jaakko Hintikka, who planned on publishing the proceedings in *Synthese*. However, he was already in conversation with Mendelsohn about publishing WPOBIN in the *Journal of the History of Biology*. Since the audience of the two journals were different, Mendelsohn gave his permission that the paper be published in *Synthese* as well.

[26] Hull to Mendelsohn, November 19th 1968, Hull Papers, ASP 2005.1, box 36, folder 6.

[27] Hull to Mendelsohn, November 12th 1968, Hull Papers, ASP 2005.1, box 36, folder 6. Incidentally, Bunge did not adopt the term "philosophy of biology" and used "biophilosophy" instead, perhaps because of this clash with Hull, not because of the German use of the word. See Bunge (1979).

[28] Mendelsohn to Hull, November 27th, 1967, Hull Papers, ASP 2005.1, box 36, folder 6.

mention in his response of a critical review of the *field* – only of two *books* by Beckner, who was one of the major figures in emerging philosophy of biology.[29] If Beckner "doesn't meet with approval", Hull said, he would be "willing to try the [Hans] Jonas volume" or Hans Selye's *In Vivo* (1967) but he "would like to beg off [Ludwig] von Bertalanffy".[30]

Mendelsohn "liked the idea of a fairly lengthy review of the Beckner books" and invited Hull to structure his essay in the following way: "it might be that you will want to take several books dealing with biology in a fairly extensive review in which you comment on the state of studies in the philosophy of biology in general, even as you refer to the several books at hand".[31] One month later, he wrote to Hull with bad news: "Although it is true that his *Biological Way of Thought* is to be brought out in paperback, the rumor about a second book in the philosophy of biology is unfortunately not true". In the same letter, he continued wearing his editor's hat: "It strikes me that one type of article which might be of great interest to the readers of a journal like ours would be an essay on the recent history of the philosophy of biology and the various controversies and major points of view that have arisen".[32] Following that letter, Hull finally agreed to go on with Mendelsohn's strategy: "You seem intent on embarking me on a major paper". "OK, I'll see if I can work up a paper on a 'recent history of the philosophy of biology'".

The main emphasis would be on Beckner and William S. Beck since "these are the only two men [*sic*] who have produced what I consider a substantial work on the philosophy of biology which is *actually relevant to biology*" [emphasis mine]. In addition, he would comment on "[Gavin] de Beer, [Hans] Selye, [Thomas] Goudge, [Felix] Mainx, Block, the [John] Gregg volume on Woodger, the [John V.] Canfield book, two paperbacks edited by Vincent Smith on the philosophy of biology, perhaps reference to the [Michael] Polanyi, [Marjorie] Grene, [Gertrude] Himmelfarb school and recent development in taxonomy".[33] Not all these authors found themselves in the published review. Importantly here, Hull made clear he did not want to talk about Bertalanffy, not because of the holistic outlook of his works, but because of its formalist language: "I'd like to avoid the general system boys [*sic*]. It's just too cracker barrel for me".

In the same letter, he opened some more about what the philosophy of biology was for him and went on identifying two groups of philosophers, the "formalists" and the "mystics", neither of which, he claimed, were of any relevance to biologists.

---

[29] Beckner, a student of Nagel, published several papers on the philosophy of biology but left the field before it took off the ground. As Hull's letter to Mendelsohn indicates, he thought that Beckner "might have served as a catalyst for an unpsurge in the philosophy of biology" (Hull, 1994, 37). Hull tried to engage him in correspondence but his attempts "were met with no encouragement" (*Ibid.*).

[30] Hull to Mendelsohn, December 5th 1967, Hull Papers, ASP 2005.1, box 36, folder 6.

[31] Mendelsohn to Hull, December 15th 1967, Hull Papers, ASP 2005.1, box 36, folder 6.

[32] Mendelsohn to Hull, January 9th 1968, Hull Papers, ASP 2005.1, box 36, folder 6.

[33] Hull to Mendelsohn, January 17th 1968, Hull Papers, ASP 2005.1, box 36, folder 6.

> The major contributions to the philosophy of biology in the past have been either formalists (Woodger, Gregg, Mainx and to some extent Beckner) or mystics with an axe to grind (the contributors to the Smith paperbacks, Grene and a host of others). The formalists have been rigorous, hard-nosed and the like but *irrelevant*. Biologists don't read them, and I honestly can see no reason to encourage any extensive study of the literature. In spite of all this, there is some work in the philosophy which is both respectable and *relevant*." [emphasis mine]

Hull ended his letter with a caveat: "Needless to say I'll make myself very unpopular if I express myself this bluntly in the paper. I'll try to tone it down". In the end he did not tone it down but, instead of being unpopular, his essay turned retrospectively into one of the landmarks of the emergence of the philosophy of biology.

What morals can be extracted from these letters? First, it was Mendelsohn, a historian, who initially set Hull on the path of an extensive critical review of the philosophy of biology. It is also he who first wrote that "there is a lot of dead wood to be cut away so that some of the basic questions can be asked", not Hull.[34] Mendelsohn even went to great lengths to suggest the general structure of the paper, which would appeal to the audience of his journal. Hull, thus, was not led to write this paper "out of exasperation" as a "brash young man with a mission", as he later put it (Hull, 1998c, 77), but at Mendelsohn's request. The letters also illustrate how Hull characterized some of the works done in the philosophy of biology based on their relevance to actual biological problems. He included Beckner and Beck because they are "the only two men" who have produced work "which is actually relevant to biology" but he came down hard on the "formalists" (Woodger, Gregg, Mainx) and the "mystics" (Grene, Smiths), because they appeared "irrelevant" to biology (Beckner excepted). Prior to delivering his paper in Pittsburgh, Hull had sent a copy to Mayr in which he explained that his criticisms against philosophers "are complaints which biologists have frequently felt like making but have been too polite". Mayr agreed: "I am tremendously pleased that you keep on showing the flimsiness of the argument of the so-called philosophers", he said.[35]

No doubt Hull's view of the relation between science and philosophy in WPOBIN benefited from the rich discussions that took place during the Asilomar Conference on explanation in biology organized by Mendelsohn in June 1968, an event that brought together nearly 40 scholars in history and philosophy of science and biology (Smocovitis, 1996, 185–86). In the 60s, biology was still generally treated as a source of case-studies to test general philosophical claims. For example Hanson, who wanted to hire a philosopher of biology at the University of Indiana, told Scriven in 1962 he wanted to find someone who could "draw on biological examples" to derive "logical or historical morals".[36] Scriven, who was a participant at the Asilomar conference, sketched a different view of philosophy of science whereby he encouraged philosophers "to think of philosophy of science as an activity defined

---

[34] Mendelsohn to Hull, February 15th, 1968, Hull Papers, ASP 2005.1, box 36, folder 6.

[35] Hull to Mayr, September 30th; Mayr to Hull, October 11th 1968, Hull Papers, ASP 2005.1, box 35, folder 2.

[36] Hanson to Scriven, March 12th 1962, Indiana University Department of History and Philosophy of Science Records, box 4.

by its relation to [...] science as it is usually conceived, rather than as what people called philosophers of science do. For what we do is not distinguishable from what some scientists do some of the time and very properly" (1969, 188). Finding common ground between science and philosophy in the "logical analysis of concepts", Scriven regarded the question as to whether "philosophers of science ever say anything that leads their listeners to improve their practice as scientists" to be a "legitimate challenge" (Ibid., 189). Hull, who listened to Scriven's talk, went on to invite philosophers to clarify some of the puzzles internal to biological concepts and theories. According to him, biology yields methodological and conceptual problems to which both scientists and philosophers could contribute.

In three other reviews of the field (1979, 1982, 1998c), he kept on defining the identity of the philosophy of biology based on the criterion of relevance and the critique of formalist methods in philosophy. Most (in)famously, Woodger's approach based on Boolean logic, Hull claimed, was "incomprehensible" to most biologists, and thus useless for the philosophy of biology (Hull, 1998c, 79).[37] As to Grene, he admitted her work on Otto Schindewolf and evolutionary theory "did have an influence on biologists", but it was "entirely negative". "This was philosophy of biology?" (Hull, 1998c, 79). In those reviews, Hull made room for philosophy in biology and, in so doing, he tried to bring down the disciplinary boundaries between science and philosophy. Thus, he encouraged philosophers to respond to Scriven's challenge and improve scientific practice. For example, Hull invited philosophers of science to "enter into [science] and propose improvements" (1982, 281–82). "The distinction between science and philosophy", he told Callebaut in an interview, "is not all that good" (1993, 280). While he rejected the idea that scientists "should be the arbiters of philosophical disputes",[38] Hull agreed that "because one's degree is in science does not mean that one cannot contribute to philosophy and vice versa".[39] Thus, "if there is any really difference between scientists and philosophers, it is in the sort of problems that we are trained to handle, but these differences are in degree, not kind" (1998c, 86).

Coming back to the letters, it is more difficult to make sense of the direction Mendelsohn suggested Hull should take in his critical review. It is well-known that he considered Canguilhem to have been "[his] teacher for many years" (Mendelsohn, 1985, 15). When *The Normal and the Pathological* was published in English translation, his hope was that it "would rupture the wall that has kept the works of this

[37] As Smocovitis pointed out, the fact that contemporary philosophers of biology or biologists could not understand Woodger's work, "does not in any manner prove that Woodger was not an influential source for biologists and philosophers in the 1930s" (1996, 105; Nicholson & Gawne, 2014). Even Hull later admitted that Woodger was "the most influential philosopher of biology at the time [...]" (1998c, 79).

[38] As Godfrey-Smith (2010) observed in his obituary, Hull would have rejected the idea that all philosophical problems are scientific problems, contrary to Quine.

[39] Hull alluded to this point when he described how biologists themselves have succeeded in solving the philosophical problem of the distinction between classes and individuals (1969; Van Speybroeck, 2007).

imaginative French scholar from English readers" (Mendelsohn, 1977; xxiii–iv; Méthot, 2020). Yet as we will see in the next section, Hull's review was implicitly based on the outright rejection of Continental approaches to biology and philosophy.

5 Philosophy of Biology Versus Biological Philosophy, or Hull Against Grene

From their personal trajectories and published works, the differences between Hull and Grene regarding the nature of philosophy and its relation to history and biology are clear. Their correspondence, starting in May 1967, brings it out more sharply still. In 1968, when Hull was writing WPOBIN, Grene was completing *Approaches to a Philosophical Biology*, a book going in the opposite direction to that of her younger colleague. In dense chapters, Grene looked at the work of Continental scientists/philosophers from Europe who favored a holistic approach to biology such as Helmuth Plessner, Adolph Portman, Kurt Goldstein, Frederik Buytendijk, and Kurt Goldstein. Grene's aim was to show how their reflections "have bearing on current problems in philosophy itself as well as in the methodological and metaphysical problems that cluster around the foundations of biology" (1968, 282). Her aim was thus to "bring to the attention of English-speaking readers a number of European scientists – or scientists-philosophers – whose reflections on the conceptual foundations of biology deserve more attention than they have so far received" (Ibid., v). But importantly, she was not a discipline builder and she did not seek "to produce a new philosophy of biology" (Ibid., 282).

Before going to press with his manuscript, Hull sent her a copy of WPOBIN to ensure he provided a fair account of her views on evolutionary theory. He warned her that "the whole paper had a bitchy tone" and that she and the people she admires are coming in "for the worst drubbing".[40] Hull's letter arrived on September 30th, a few months after Grene had written the preface of *Approaches* (July 1968). Having trained with Jaspers and Heidegger and worked with Polanyi, Grene was ready to respond to Hull: "it seems to me [...] that in a review of development in the philosophy of biology, you ought to take account of the very considerable literature of the continent of Europe".[41] To make her point, she cited the work of German philosophical anthropologist Arnold Gehlen as worth looking at from a philosophy of biology point of view. Grene ended up accusing Hull and "Anglo-American philosophers" in general of being "on the whole [...] as narrow-minded about this aspect of their subject as about most other things".[42] Hull, however, remained skep-

[40] Hull to Grene, September 30th 1968, Hull Papers, ASP 2005.1, box 21, folder 1–2.

[41] In her "Intellectual biography", Grene commented that she was asked to prepare a bibliography in the philosophy of biology and how this led her to discover "a number of writers, chiefly from the continent of Europe, whose work gave [her] a broader perspective on philosophical questions connected with biology" (2002, 18).

[42] Grene to Hull, October 3rd, 1968, Hull Papers, ASP 2005.1, box 21, folder 1–2.

tical and, in this response, insisted on the importance of "clarity" in philosophical writing, while also admitting his incapacity to make sense of Continental philosophy in general:

> You mention the Continental literature in the philosophy of biology. I knew that it was there and even tried reading Hartmann. Psychologically, I just can't. Such stuff, quite literally, runs me up the walls. I don't mind working out extremely intricate systems as long as there is some payoff. But just where there should be clarity there is the darkest murk. All I get is that they're unhappy about something but it's hard to say just what. Your piece on Plessner has even deepened my conviction on this point […] Is what they're trying to say so difficult and so radically new that *no* term can be found *ever* which means literally what they intend, and that after half a century there has not been any increase in clarity?[43]

He closed his letter with the following positive and polite remark, which was typical of most of Hull's personal writings:

> Since we have profoundly different philosophical prejudices, we'll certainly never be reconciled on our major area of disputes, but we should still be able to talk to each other with some profit. I suspect from our correspondence that if we ever get to know each other, we could really have whopping big arguments but probably nevertheless like each other.

Hull and Grene were never reconciled on their philosophical positions (Honenberger, 2015), but they indeed became good friends.

A few years later, in 1973, in Jyväskylä, Finland, an international conference jointly organized by the International Union of History of Science and the International Union of Philosophy of Science would allow for an unprecedented encounter between biological philosophy and philosophy of biology in the form of a debate between Canguilhem, Schaffner, and Grene (Méthot, 2018, 2020). Canguilhem's paper on "The Question of 'Normality' in the History of Biological Thought" was one of the two keynote papers of the session on the "Development and Use of Norms in History of Biology". Schaffner's paper, "Normalité and Teleology in Modern Biology", was a direct response to Canguilhem.[44] In his paper, Canguilhem argued that contrary to a popular misconception, Darwin "is far from having eliminated all consideration of normality when determining the biological object". On the contrary, he "introduced into biology a criterion of normality based on the relation of the living to life and to death" (Canguilhem, 1988, 137 [1977]). In his turn, Schaffner started by recalling the importance of the question of normality in the work of Canguilhem and stressed his "surprising" conclusions. Detailing his disagreement with his interpretation of Darwin, he maintained that Canguilhem's "quasi-teleological" concept of normality conflates several notions. Without dissimulating his astonishment in the face of the Canguilhemian interpretation of Darwin, Schaffner attributed this in part "to the intricacies of the French intellectual milieu" in which he is "not too well versed" (1973, 1, 3). Grene had been assigned to respond to Schaffner's paper, as is made clear in the notes on the conference

---

[43] Hull to Grene, October 10th, 1968, Hull Papers, ASP 2005.1, box 21, folder 1–2. Hull most likely is referring to Karl R.E. von Hartmann.

[44] Schaffner's unpublished paper is found in the Canguilhem Papers, CAPHÉS, G.C.22.5.

proceedings (Méthot, 2018). Her exact words have today been lost, but Schaffner summarized her contribution in the following terms:

> I have only hazy memories of the 1973 conference where I commented on Canguilhem. It is possible that he may have replied in French, and I could not follow it. I do recall that Marjorie Grene, who was at that conference, felt I had been inappropriately critical of Canguilhem, and gave a 10-minute defence of Canguilhem. I do not know if any of this was ever published, or if Marjorie ever wrote her comments down.[45]

Like Schaffner, Canguilhem would likely have been astonished at the vigorous defence offered by Grene of his arguments against Schaffner's interpretation. In the next few years, however, Grene published texts in philosophy of medicine in which the work of Canguilhem on norms and the normal presented at Jyväskylä is cited approvingly and used to support the indispensability of concepts of norms and normal in biology. Indeed, during the first PSA session in 1976 on the philosophy of medicine, she argued that philosophers of biology should "follow [Canguilhem's] example" for, "if we don't look at medicine, we are very likey to miss, in our addition to a particular model of physics as model for science, a characteristic, if not of science in general, certainly of the biological sciences [...]" (1977, 91). Elsewehere, she commented:

> As Georges Canguilhem (1976) has recently pointed out, evolutionary theory made the avoidance of death the only goal of life; indeed, strictly speaking, in terms of Darwinian theory, that is, the theory of natural selection, teleological discourse should have been eliminated altogether from biological, or at least from evolutionary explanation. But the normative character of biology, as distinct from the teleological nature of some of its process and some of its etiologies, cannot be thus eliminated. (1976, 186)[46]

In 1976, at the time she was organizing the Summer Institute on Human Nature, Grene told Hull about her project about "the role of 'norms' and 'normal' in biological thinking" she was developing (drawing on Canguilhem). Knowing this topic would make her colleague cringe, she wrote in parentheses: "'what philo. of biol. is not' again?"[47] Hull "enjoyed her aside" about 'what philosophy of biology is not again'" and thought "she had been too good about [his] jibes in the article for too long. Now the real Marjorie Grene shows herself!", he said in response.[48]

In a paper on Canguilhem published almost 25 years later, Grene talked about the difference between philosophy of biology and biological philosophy in terms of Ludwik Fleck's "thought styles" to emphasize the incompatibility of the worldviews (2000, 49). As this chapter illustrates, her observation might have stemmed from her own personal experience following several decades of trying to bridge these traditions: linguistically, conceptually as well as personally. Three decades

---

[45] Email from Schaffner to Méthot 2010/12/20.

[46] Grene's reference to Canguilhem (1976) is a reference to his paper presented at the conference in Finland, but the proceedings were never published (Méthot, 2018, 45). The paper, however, was included in *Ideology and Rationality in the History of the Life Sciences* (Canguilhem, 1988 [1977]).

[47] Grene to Hull, February 10th 1976, Hull Papers, ASP 2005.1, box 21, folder 1–2.

[48] Hull to Grene, March 8th 1976, Hull Papers, ASP 2005.1, box 21, folder 1–2. On Hull and Grene about types and norms in biology: Honenberger (2015); Grene (2002).

after her first clash over what the philosophy of biology is not, this paper on Canguilhem in the *Revue d'histoire des sciences* (commissioned by Gayon) once again "set [Hull's] teeth on edge". In his letter, Hull looked down with condescension on Canguilhem and Selye: "If there is more to such writers", he said, "I have yet to tumble to it. Yes, they can look good when compared with the worst possible positivists, but anyone looks good when compared to them. Of course, I suspect you think I am a positivist".[49] The legacy and meaning of "positivism" and the question of the relation between history, philosophy, and science were sensitive issues between Hull and Grene and for the identity of the new philosophy of biology more generally (Honenberger, 2018; Smocovitis, 1996; Callebaut, 1993).

In terms of topics, methods, and core problems, Hull was not as liberated from positivism as he thought he was, as he later admitted.[50] In contrast, Grene acknowledged only a "brief commitment" to logical positivism in the 30s (2000, 48). Her philosophical work also stayed closer to historical cases whereas Hull kept historical and philosophical work separated, despite having been educated in an HPS department. In addition, the two colleagues and friends held profoundly divergent views about the proper connection between science and philosophy, and thus about what philosophy of biology is or should be. As Grene explained to Hull in correspondence: "I do agree with Canguilhem that history of science needs methods different from those of the sciences [...] and that philosophy is one move away from the practices of the sciences themselves".[51] In *The Understanding of Nature*, she noted: "Scientists can't tell us how to solve our problems – which are always (almost always?) meta-problems". "But on the other hand", she continued "we certainly cannot tell them (except when they try to philosophize) how to solve theirs" (1974, viii). This way of demarcating science and philosophy undercuts Hull's model, but whether philosophical problems are "meta-problems" goes beyond the continental/analytic divide and the Hull/Grene opposition. Lastly, in contrast with Hull and Ruse, who excluded philosophical topics related to medicine (Gayon, 2004), Grene took part in the early days of the philosophy of medicine in the mid-70s, arguing the point that philosophers of biology would be well-advised to follow the example set by Canguilhem in *The Normal and the Pathological* (1991) and *Knowledge of Life* (2008) (Grene, 1977, 91; see Aranova, 2009). Despite her contribution to the field, the "grandmother" of the new philosophy of biology, in fact, remained somewhat of an outsider and much closer to the older biological philosophy.

[49] Hull to Grene, September 14th 1999, Hull Papers, ASP 2005.1, box 21, folder 1–2.

[50] "[...] looking back at my introductory text [1974], it is no more radical than Ruse's was. Perhaps we were not positivists, but we were close" (Hull, 2008, 24).

[51] Grene to Hull, June 8th, 1999, Hull Papers, ASP 2005.1, box 21, folder 1–2.

6 Conclusion

> One would have hoped that the example set by the works of Woodger in the last 40 years would have motivated other researchers into full-time activity in this field. One would have hoped, too, that the surprising transformation of biology in the last 20 years would have attracted the attention of philosophers of science or of scientists interested in the philosophy of science. But neither hope has materialized. Perhaps the coming decade will be more fruitful (Chiaraviglio, 1968, 384).

In 1982, Hull wrote the entry on "Biology and Philosophy" for the latest edition of *Contemporary Philosophy: A New Survey*. In his introduction, he cited Luca Chiaraviglio's conclusion (above) in the earlier edition (1968) approvingly but also selectively. Hull agreed with him about the recent developments of the field but ignored Chiaraviglio's hope to see Woodger's formal program extended. Right after Chiaraviglio's chapter in 1968 was Canguilhem's on "Biologie et Philosophie", which Hull ignored, too. In this chapter concerned with European works, Canguilhem argued that "biological philosophy [...] could no longer ignore how the discovery of the genetic code laid the groundwork for a new knowledge of life" (2018, 317 [1968] [translation mine]). Whereas Canguilhem went on in the late 60s to re-examine the foundations of his biological philosophy (2008 [1952]) against this new knowledge of life, Hull's discussion of Schaffner's work on theory reduction (1982) signals that the crucial issue was whether Mendelian genetics could be reduced to molecular biology, a problem Grene considered to be perhaps not solved but at least no longer debated (Grene, 2002, 16). This difference in terms of philosophical problems illustrates Grene's point that philosophy of biology and biological philosophy continued to develop as distinct styles.

Like any field of knowledge, philosophy is organized into smaller units called disciplines – of which the philosophy of biology is among the most recent ones. But "disciplines are made, not found", as Hopwood put it. "Making one is a political project of carving out questions, approaches and scope, and recruiting patrons and audiences, in relation to what went before and to other sciences" (Hopwood, 2019, 1). And like any discipline, this one includes and excludes people, and traditions; it involves the making, remaking, and unmaking of disciplinary boundaries as well as the choice of cognitive styles over others, such as Canguilhem's, but also the use of biology as merely case-studies for general philosophy of science. The Hull-Grene-Mendelsohn correspondence sheds new light on the intellectual origins of WPOBIN and illustrates how it laid the conceptual foundation on which the identity of an analytic philosophy of biology was to be erected in the ensuing decades. In this respect, this enterprise was immensely successful, much more so, perhaps, than Continental approaches to the life sciences in the second half of the twentieth century.

As Philip Sloan has observed: "[...] Bachelard, Foucault, and Canguilhem, do not seem to have been able to create anything like the disciplinary following generated by David Hull, Philip Kitcher, Alex Rosenberg, Elliot Sober, Bill Wimsatt, and the larger community of biological scientists who have found these general empiricist perspectives the most congenial to their research agendas" (2002, 227). But, in (re)shaping the relation between philosophy and biology and narrowing down the identity of the philosophy of biology, Hull and his colleagues also had to negotiate the intellectual relations of the newly emerging field with neighboring disciplines such as the history of biology, general philosophy of science, the philosophy of medicine, and all that went before the 60s and 70s in terms of the relation between biology and philosophy – including the Continental tradition. As it turned out, most of this was excluded from the new philosophy of biology that developed in close contact with logical positivism. Toward the late 90s and early 2000s, however, Hull became increasingly concerned with the specialization and professional isolation of philosophy, history, and sociology of science; he claimed that ISHPSSB should serve as a model for more interdisciplinary work (Hull, 2000).

This paper has raised a question about the historical conditions that makes fields such as the philosophy of biology possible and has called attention to the exclusions that this involves in terms of the proper relation between philosophy and science, especially biology. Scholars are now taking issue with the origin stories told by the philosophy of biology community and revisiting its connection to past intellectual traditions. While Hull is not the only intellectual founding figure of the philosophy of biology, it was possibly through him that the demand for a greater role for philosophy *in* actual biological science was most strongly felt from the late 60s and early 70s onward in the Anglophone world.[52] Despite claims to the contrary, the field gained traction not only because of the new direction taken by philosophy of science (the so-called "revolt" against logical positivism), but because of conscious, collective, and interdisciplinary efforts to construct a new discipline with its own scholarly venues (*Biology and Philosophy*, the Conceptual Foundations of Science series, ISHPSSB, etc.), styles, and values, such as the emphasis on the usefulness of "proper" philosophy of biology for biologists. Paying attention to *where* and *how* philosophy of biology was done in practice could offer critical insights into the factors that shaped the ethos of the discipline. In particular, the role of biologists in the emergence of analytic philosophy of biology in the 70s and 80s remains insufficiently acknowledged.[53]

Thus, to trace the history of the philosophy of biology, one must focus closely on the intellectual *and* biographical context that includes the places, peoples, and institutions where philosophical ideas developed. The foundation of the field reveals a profound tension between two of the most important actors over a critical point: how to articulate science, history, and philosophy, which remains an open question

[52] Thomas Pradeu and Maël Lemoine have taken the relation between philosophy and biology a step further with the idea of a philosophy *in* science (or "PinS"). Starting from scientific (and not philosophical) problems, they argue that PinS uses philosophical methods not only to clarify scientific issues but to solve scientific problems and thus advance scientific knowledge (Pradeu et al., 2021).

[53] The interdisciplinary reading group in biology and philosophy at Chicago University in the late 60s-early 70s that brought together Hull, Schaffner, Wimsatt, Levins Lewontin and other scientists (Wimsatt, 1999), or Lewontin's lab at Harvard in the 80s, where a number of philosophers stayed as postdocs, are obvious places to consider to trace the history of the philosophy of biology.

today (Huneman, 2019). Only by looking at the assumptions about the nature of philosophical (and historical) work and the values that were built into the crafting of the professional identity of the field, can one hope to grasp the logic of exclusion and inclusion that permeated the philosophy of biology in the Anglophone world from its inception.

Acknowledgments I would like to thank Charles Wolfe, Giuseppe Bianco, and Gertrudis Van de Vijver for their invitation to speak at the workshop on "Canguilhem and Continental Philosophy of Biology" held (online) on September 16-17 2021. I am especially grateful to Staffan Müller-Wille, Nick Hopwood, Jonathan Sholl, Betty Smocovitis, and Charles Wolfe for constructive feedback on an earlier draft of this chapter that significantly improved it. I also wish to thank Dick Burian, Philip Honenberger, Philippe Huneman, Katie Kendig, Camille Limoges, Laurent Loison, Ken Schaffner, Neeraja Sankaran, Philip Sloan, and Bill Wimsatt for helpful conversations about the topic of this paper. Lastly, I am happy to acknowledge the assistance of the following archivists: Jon Klosinski, at the Archives of Scientific Philosophy, Archives & Special Collections at the University of Pittsburgh; Dina Kellams and Isbell Mariah Sade, at Indiana University; Nathalie Queyroux at the Centre d'Archives en Philosophie, Histoire et Éditions des sciences (CAPHÉS) in Paris; and Meg Milner at The Tate Archives and Special Collections, the Ames Library, Illinois Wesleyan University, Bloomington, Illinois. Lastly, financial support from the Social Sciences and Humanities Research Council in Canada is gratefully acknowledged.

References

Aronova, E. (2009). In Search of The Soul in Science: Medical Ethics' Appropriation of Philosophy of Science in the 1970s. *History and Philosophy of the Life Sciences, 31*(1), 5–33.

Beckner, M. (1959). *The Biological Way of Thought*. Columbia University Press.

Bunge, M. (1979). Some Topical Problems in Biophilosophy. *Journal of Social and Biological Structures, 2*, 155–172.

Burian, R. (1986). The 'Internal Politics' of Biology and the Justification of Biological Theories. In A. Donagan & M. V. Wedin (Eds.), *Human Nature and Natural Knowledge. Essays Presented to Marjorie Grene on the Occasion of her Seventy-Fifth Birthday* (pp. 23–45). Reidel.

Burian, R. (2009). Marjorie Grene: A Remembrance With Special Attention to Her Importance for ISHPSSB. *Biological Theory, 4*, 183–187.

Burian, R., & Pitt, J. C. (Eds.). (1992). *Synthese, 92*(1), 3–166: special issue on *The Thought of Marjorie Grene*.

Byron, J. M. (2007). Whence Philosophy of Biology? *British Journal for the Philosophy of Science, 58*, 409–422.

Callebaut, W. (1993). *Taking the Naturalistic Turn: Or How Real Philosophy of Science is Done*. University of Chicago Press.

Callebaut, W. (2005). Again, What The Philosophy of Biology is Not. *Acta Biotheoretica, 53*, 93–122.

Canguilhem, G. (1988). *Ideology and Rationality in the History of the Life Sciences* (A. Goldhammer, Trans.). MIT Press.

Canguilhem, G. (1996 [1991]). *The Normal and the Pathological* (C. Fawcett & R.S. Cohen, Trans.). Zone Books.

Canguilhem, G. (2008 [1965/1952]). *Knowledge of Life* (S. Geroulanos & D. Ginsburg, Trans.). Fordham University Press.

Canguilhem, G. (2018 [1968]). Biologie et philosophie: publications européennes. In G. Canguilhem, *Œuvres complètes*, tome 5: *Histoire des sciences, épistémologie, commémorations 1966–1995*, ed. C. Limoges (pp. 315–328). Vrin.

Canguilhem, G. (2019 [1968]). L'objet de l'histoire des sciences. In G. Canguilhem, *Œuvres complètes*, tome 3: *Écrits d'histoire des sciences et d'épistémologie*, ed. C. Limoges (pp. 253–273). Vrin.

Chiaraviglio, L. (1968). Biology and Philosophy. In R. Klibansky (Ed.), *Contemporary Philosophy: A Survey* (pp. 376–386). La Nuova Italia Editrice.

Donovan, A., Perovich, A. N., Jr., & Wedin, M. V. (Eds.). (1986). *Human Nature and Natural Knowledge. Essays Presented to Marjorie Grene on the Occasion of her Seventy-Fifth Birthday*. Reidel.

Duchesneau, F. (1997). *Philosophie de la biologie*. Presses universitaires de France.

Gayon, J. (1998a). Philosophie et biologie. In Jean-François Mattei (Ed.), *Encyclopédie philosophique universelle,* vol. IV (Le Discours philosophique) (pp. 2152–2171). Presses universitaires de France.

Gayon, J. (1998b). The Concept of Individuality in Canguilhem's Philosophy of Biology. *Journal of the History of Biology, 31*, 305–325.

Gayon, J. (1999). Nietzsche and Darwin. In J. Maienschein & M. Ruse (Eds.), *Biology and the Foundation of Ethics* (pp. 154–197). Cambridge University Press.

Gayon, J. (2004). À quoi servent les philosophes… de la biologie ? Interview with Michel Imbert and Jean-Jacques Perrier. *Le vivant*, 2.

Gayon, J. (2009a). The Philosophy of Biology: An Historico-Critical Characterization. In A. Brenner & J. Gayon (Eds.), *French Studies in the Philosophy of Science. Contemporary Research in France* (pp. 201–212). Springer.

Gayon, J. (2009b). Marjorie Grene: Personal Memories. *Biological Theory, 4*, 188–190.

Gayon, J., & Burian, R. M. (Eds.) (2006). *Conceptions de la science: hier, aujourd'hui, demain. Hommage à Marjorie Grene*. Ousia.

Gibson, J. J. (1979). *Ecological Approach to Visual Perception*. Houghton.

Godfrey-Smith, P. (2010). David Hull. *Biology and Philosophy, 25*, 749–753.

Goldstein, K. (1934/1995). *The Organism: A Holistic Approach to Biology Derived From Pathological Data in Man*. American Book Company; reprint, Zone Books.

Goudge, T. A. (1961). *The Ascent of Life. A Philosophical Study of the Theory of Evolution*. University of Toronto Press.

Grau, K. T. (1999). Force and Nature. The Department of History and Philosophy of Science at Indiana University, 1960–1988. *Isis, 90*, 295–318.

Grene, M. (1963). Two Evolutionary Theories: A Reply. *The British Journal for the Philosophy of Science, 14*(54), 152–154.

Grene, M. (1968). *Approaches to a Philosophical Biology*. Basic Books.

Grene, M. (1974). *The Understanding of Nature. Essays in the Philosophy of Biology*. Reidel.

Grene, M. (1976). To Have A Mind…. *The Journal of Medicine and Philosophy, 1*(2), 177–199.

Grene, M. (1977). Philosophy of Medicine: Prolegomena to a Philosophy of Science. *Proceedings of the Biennial Meeting of the Philosophy of Science Association* 2 [Symposia and Invited Papers] (pp. 77–93).

Grene, M. (1983). Philosophy of biology 1983: Problems and prospects. In R. B. Marcus, G. J. W. Dorn, & P. Weingartner (Eds.), *Logic, Methodology and Philosophy of Science VII. Proceedings of the Seventh International Congress of Logic, Methodology and Philosophy of Science, Salzburg 1983* (pp. 433–452). Elsevier.

Grene, M. (1997). Current Issues in the Philosophy of Biology. *Perspectives on Science, 5*, 255–281.

Grene, M. (2000). The Philosophy of Science of Georges Canguilhem: A Transatlantic View. *Revue d'histoire des sciences, 53*, 47–64.

Grene, M. (2002). Intellectual Biography. In R. E. Auxier & L. E. Hahn (Eds.), *The Library of Living Philosophers, Vol. XXIV: The Philosophy of Marjorie Grene* (pp. 4–28). Open Court.

Grene, M. (2006). La vie des sciences et les sciences de la vie. In J. Gayon & R. M. Burian (Eds.), *Conceptions de la science : hier, aujourd'hui, demain. Hommage à Marjorie Grene* (pp. 23–36). Ousia.
Grene, M., & Burian, R. M. (Eds.). (1983). *Philosophy of Biology in the Philosophy Curriculum*. The Council for Philosophical Studies.
Grene, M., & Everett, M. (Eds.). (1976). *Topics in the Philosophy of Biology*. Reidel.
Griffiths, P. E. (2000). Dedication. *Biology and Philosophy 15*, 299.
Gutting, G. (2005). Introduction: What is Continental Philosophy of Science? In G. Gutting (Ed.), *Continental Philosophy of Science* (pp. 1–16). Blackwell.
Hofer, V. (2013). Philosophy of Biology in Early Logical Empiricism. In H. Andersen, D. Dieks, W. J. Gonzalez, T. Uebel, & G. Wheeler (Eds.), *New Challenges to Philosophy of Science. The Philosophy of Science in a European Perspective* (pp. 351–363). Springer.
Honenberger, P. (2015). Grene and Hull on Types and Typological Thinking in Biology. *Studies in History and Philosophy of the Biological and Biomedical Sciences, 50*, 13–25.
Honenberger, P. (2018). Darwin Among the Philosophers: Hull and Ruse on Darwin, Herschel, and Whewell. *HOPOS: The Journal of the International Society for the History of Philosophy of Science, 8*, 278–309.
Hopwood, N. (2019). Inclusion and Exclusion in The History of Developmental Biology. *Development, 146*, 1–7.
Horder, T. J. (2013). David Hull's Philosophical Contribution to Biology. In O. Harman & M. Dietrich (Eds.), *Outsider Scientists. Routes to Innovation in Biology* (pp. 238–255). Chicago University Press.
Hull, D. L. (1964). *The Logic of Phylogenic Taxonomy* (PhD. Dissertation). Indiana University.
Hull, D. L. (1969). What the Philosophy of Biology is Not. *Journal of the History of Biology, 2*, 241–268.
Hull, D.L. (Ed.) (1973). *Darwin and his Critics. The Reception of Darwin's Theory of Evolution by the Scientific Commnuity*. Harvard University Press.
Hull, D. L. (1974). *Philosophy of Biological Science*. Prentice-Hall.
Hull, D. L. (1979). Philosophy of Biology. In P. D. Asquith & H. E. Kyburg (Eds.), *Current Research in Philosophy of Science. Proceedings of the P.S.a. Critical Research Problems Conference* (pp. 421–435). East Lansing.
Hull, D. L. (1982). Philosophy and Biology. In G. Fløistad (Ed.), *Contemporary Philosophy: A New Survey* (Philosophy of science) (Vol. 2, pp. 281–316). Nijhoff.
Hull, D. L. (1994). Ernst Mayr's Influence on the History and Philosophy of Biology: A Personal Memoir. *Biology and Philosophy, 9*, 375–386.
Hull, D. L. (1998a). *Science as a Process. An Evolutionary Account of the Social and Conceptual Development of Science*. Chicago University Press.
Hull, D. L. (1998b). A Mechanism and its Metaphysics: An Evolutionary Account of the Social and Conceptual Development of Science. *Biology and Philosophy, 3*, 123–155.
Hull, D. L. (1998c). The Role of Biology in Philosophy of Biology. In J. T. Suárez (Ed.), *1st International Conference on Philosophy of Science: Philosophy of Biology Today, Universiade de Vigo* (pp. 75–94).
Hull, D. L. (2000). The Professionalization of Science Studies: Cutting Some Slack. *Biology and Philosophy, 15*, 61–91.
Hull, D. L. (2008). The History of the Philosophy of Biology. In M. Ruse (Ed.), *The Oxford Handbook of Philosophy of Biology* (pp. 11–33). Oxford University Press.
Hull, D. L., & Ruse, M. (Eds.). (1998). *The Philosophy of Biology*. Oxford University Press.
Hull, D. L., & Ruse, M. (Eds.). (2007). *The Cambridge Companion to the Philosophy of Biology*. Cambridge University Press.
Huneman, P. (2019). Philosophy after Philosophy of Biology Before Biology. In C. Bognon-Küss & C. T. Wolfe (Eds.), *Philosophy of Biology Before Biology* (pp. 198–212). Routledge.
Jonas, H. (1966). *The Phenomenon of Life. Toward a Philosophical Biology*. Northwestern University Press.

Köchy, K. (2008). *Biophilosophie. Zur Einführung*. Junius Verlag.
Kolen, F., & Van de Vijver, G. (2007). Philosophy of Biology: Naturalistic or Transcendental? *Acta Biotheoretica, 55*, 35–46.
Laplane, L., Mantovani, P., Adolphs, R., et al. (2019). Opinion: Why Science Needs Philosophy. *Proceedings of the National Academy of Sciences, 116*(10), 3848–3952.
Limoges, C. (2015). Introduction. Philosophie biologique, histoire des sciences et interventions philosophiques. Georges Canguilhem 1940–1965. In G. Canguilhem, *Œuvres complètes*, tome 4: *Résistance, philosophie biologique et histoire des sciences, 1940–1965*, ed. C. Limoges (pp. 7–48). Vrin.
Limoges, C. (2018). Introduction. La confirmation de l'historien des sciences et la mise à l'épreuve de sa philosophie biologique: Georges Canguilhem 1966–1995. In G. Canguilhem, *Œuvres complètes*, tome 5: *Histoire des sciences, épistémologie, commémorations 1966–1995*, ed. C. Limoges (pp. 7–60). Vrin.
Mayr, E. (1959). Where Are We? *Cold Spring Harbor Symposia on Quantitative Biology, 24*, 1–14.
Mayr, E. (1969). Footnotes on the Philosophy of Biology. *Philosophy of Science, 36*, 197–202.
Mayr, E. (1996). Foreword. In G. Wolters & J. G. Lennox (Eds.), *Concepts, Theories, and Rationality in the Biological Sciences* (pp. vii–xi). University of Pittsburgh Press.
Mendelsohn, E. (1977). Editorial Note. In Canguilhem, *The Normal and the Pathological* (trans. C. R. Fawcett). Reidel.
Mendelsohn, E. (1985). The Origin of Life and the Materialism Problem. *Revue de Métaphysique et de Morale, 90*(1), 15–28.
Méthot, P.-O. (2018). De l'épistémologie historique à la philosophie de la biologie: le double héritage philosophique de Jean Gayon. In F. Merlin & P. Huneman (Eds.), *Philosophie, histoire, biologie : mélanges offerts à Jean Gayon* (pp. 13–50). Éditions Matériologiques.
Méthot, P.-O. (2020). Introduction. Georges Canguilhem's Critique of Biological Thought and Historical Epistemology. In P.-O. Méthot (Ed.), with the collaboration of J. Sholl, *Vital Norms: Canguilhem's* The Normal and the Pathological *in the Twentieth-First Century* (pp. 13–51). Hermann.
Müller-Wille, S. (2011). Making Sense of Essentialism. *Critical Quarterly, 53*(4), 61–77.
Nicholson, D. J., & Gawne, R. (2014). Rethinking Woodger's Legacy in the Philosophy of Biology. *Journal of the History of Biology, 47*, 243–292.
Nicholson, D. J., & Gawne, R. (2015). Neither Logical Empiricism nor Vitalism, but Organicism: What the Philosophy of Biology Was. *History and Philosophy of the Life Sciences, 37*, 345–381.
Nye, M. J. (2011). *Michael Polanyi and his Generation. Origins of the Social Construction of Science*. University of Chicago Press.
Odenbaugh, J., & Griffiths, P. (2020). Philosophy of Biology. *The Stanford Encyclopedia of Philosophy (Summer 2022 edition)*, Edward N. Zalta (Ed.). https://plato.stanford.edu/archives/sum2022/entries/biology-philosophy/.
Overmann, R. J. (2000). David Hull, Hod Carrier. *Biology and Philosophy, 15*, 311–320.
Pradeu, T. (2017). Thirty Years of *Biology & Philosophy*: Philosophy of Which Biology? *Biology and Philosophy, 32*, 149–167.
Pradeu, T., Lemoine, M., Khelfaoui, M., & Gingras, Y. (2021). Philosophy in Science: Can Philosophers of Science Permeate Through Science and Produce Scientific Knowledge? *The British Journal for Philosophy of Science*.
Ruse, M. (1973). *The Philosophy of Biology*. Hutchinson.
Ruse, M. (1989a). David Hull Through Two Decades. In M. Ruse (Ed.), *What the Philosophy of Biology is: Essays Dedicated to David Hull* (pp. 1–15). Kluwer.
Ruse, M. (Ed.). (1989b). *What the Philosophy of Biology is: Essays Dedicated to David Hull*. Kluwer.
Ruse, M. (1993). Booknotes. *Biology and Philosophy, 8*, 477–483.
Schaffner, K. (1973). 'Normalité' and Teleology in Modern Biology. Unpublished.
Scriven, M. (1959). Explanation and Prediction in Evolutionary Theory. *Science, 130*(3374), 477–482.

Scriven, M. (1969). Explanation in the Biological Sciences. *Journal of the History of Biology, 2*, 187–198.
Selye, H. (1967). *In Vivo. The Case for Supramolecular Biology*. Liveright Publishing.
Sholl, J. (2016). Escaping the Conceptual Analysis Straightjacket. Pathological Mechanisms and Canguilhem's Biological Philosophy. *Perspectives in Biology and Medicine, 58*(4), 395–418.
Sloan, P. (2002). Reflections on the Species Problem: What Marjorie Grene Can Teach Us About a Perennial Issue. In R. E. Auxier & L. E. Hahn (Eds.), *The Library of Living Philosophers, Vol. XXIV: The Philosophy of Marjorie Grene* (pp. 225–255). Open Court.
Smocovitis, V. B. (1996). *Unifying Biology: The Evolutionary Synthesis and Evolutionary Biology*. Princeton University Press.
Smocovitis, V. B. (2009). Marjorie, Matriarchy, and 'Wretched Reflection': A Personal Remembrance of Marjorie Grene. *Biological Theory, 4*, 191–195.
Sober, E. (1993). *Philosophy of Biology*. Westview Press.
Sterelny, K., & Griffiths, P. E. (1999). *Sex and Death. An Introduction to Philosophy of Biology*. University of Chicago Press.
Van der Klaauw, C. J. (1958). Biology and Philosophy. In R. Klibansky (Ed.), *Philosophy in the Mid-Century* (pp. 315–328). La Nuova Italia Editrice.
Van Speybroeck, L. (2007). Philosophy of Biology: About the Fossilization of Disciplines and other Embryonic Thoughts. *Acta Biotheoretica, 55*, 47–71.
Vincent, B. B. (2013). Discipline-Building in Synthetic Biology. *Studies in History and Philosophy of the Biological and Biomedical Sciences, 44*, 122–129.
Von Bertalanffy, L. (1967). *Robots, Men, and Minds. Psychology in the Modern World*. George Braziller.
Wimsatt, W. C. (1999). Appendix: In The Lab of a 'Natural Philosopher'. Unpublished version.
Winsor, P. M. (2006). The Creation of The Essentialism Story: An Exercise in Metahistory. *History and Philosophy of the Life Sciences, 28*, 149–174.
Witteveen, J. (2015). 'A Temporary Oversimplification': Mayr, Simpson, Dobzhansky, and the Origins of the Typological/Population Dichotomy (part 1 of 2). *Studies in History and Philosophy of the Biological and Biomedical Sciences, 54*, 20–33.
Wolters, G. (2018). 'Wrongful Life' Reloaded: Logical Empiricism's Philosophy of Biology 1934-1936 (Prague/Paris/Copenhagen). *Philosophia Scientiae, 22*(3), 233–255.

Pierre-Olivier Méthot is Professor of History and Philosophy of Science in the Faculty of Philosophy at Laval University, Quebec City, where he holds the Canada Research Chair in Medical Humanities and History of Biological Thought. His research is in the philosophy and history of medicine and biology and in the history of historical epistemology. He is the coeditor of a recent volume in the philosophy of medicine on *Le normal et le pathologique: des catégories périmées?* (Éditions Matériologiques, 2022) and has edited a new collection of essays on Georges Canguilhem, *Vital Norms: Canguilhem's* The Normal and the Pathological *in the Twenty-First Century* (Hermann, 2020).

All Knowledge Is Orientation: Marjorie Grene's Ecological Epistemology

Phillip Honenberger

Abstract In the course of a more than 70-year philosophical career and over 100 publications, Marjorie Grene (1910–2009) developed an original and coherent philosophical position that placed situated organic life at the center of the interpretation of reality and human affairs. Grene sometimes described this position as an "ecological epistemology" and summarized its central thrust in the expression "all knowledge is orientation." However, Grene's view incorporated a set of apparently or potentially opposed commitments such as naturalism and anti-reductionism, pluralism and realism, and both a critique and affirmation of Darwinian evolutionary theory. This raises questions about precisely where Grene stood on the issues over which she argued and the coherence of her "ecological epistemology" as a whole. Here I review Grene's work in the main research areas for which she is best known – history of philosophy, philosophy of biology, epistemology, and philosophical anthropology – with an eye to how these tensions were ultimately resolved in her account.

Keywords Marjorie Grene · Life · Organisms · History of philosophy · Darwinism · Philosophy of biology · Perception · Reduction · Epistemology · René Descartes · Immanuel Kant · Michael Polanyi · Maurice Merleau-Ponty · Helmuth Plessner · J.J. Gibson

P. Honenberger (✉)
University of Nevada, NV, Las Vegas, USA

G. Bianco et al. (eds.), *Canguilhem and Continental Philosophy of Biology*, History, Philosophy and Theory of the Life Sciences 31,
https://doi.org/10.1007/978-3-031-20529-3_3

1 Introduction

In the course of a more than 70-year philosophical career and over 100 publications,[1] Marjorie Grene (1910–2009) developed an original and coherent philosophical position that placed *situated organic life*[2] at the center of the interpretation of reality and human affairs. Grene sometimes described this position as an "ecological epistemology" (2002, 25) and summarized its central thrust in the expression "all knowledge is orientation" (1969, 1995, 108, 2002, 25). The position involved a commitment to realism about the objects of perception and knowledge while recognizing the aspectivity, partiality, and historical contingency of any particular perception, model, or set of epistemic commitments. Grene thus asserted a pluralism about knowledge's character, contents, and objects.[3] Grene also denied the reducibility of organic life to chemistry or physics, as well as human affairs to biological accounts such as those of sociobiology. Yet she insisted on the inclusion of both life and human affairs within the category of "nature." She charged contemporary Darwinians such as R.A. Fischer and G.G. Simpson with theoretical inconsistency, yet also affirmed the basic correctness and significance of modern evolutionary theory. The apparent tensions between these commitments raise the question of whether and how Grene's ecological epistemology can be coherent.[4]

In what follows I do my best to articulate a point of view from which the grounds for, and overall coherence of, this set of commitments can be appreciated. A crucial component of Grene's overall philosophical position, I will argue, is the figure of *situated organic life*. This figure ties together Grene's work in the three main research areas for which she is best known – (1) history of philosophy, (2) philosophy of biology, and (3) a mixture of epistemology, metaphysics, and views about human nature, action, and behavior sometimes called "philosophical anthropology"[5] – and provides a crucial reference point for Grene's combination of realism, pluralism, naturalism, and anti-reductionism.

After a brief summary of Grene's biography (Sect. 2), I recount the development of Grene's own views by considering, in turn, her work in these three areas (Sects. 3 and 5). In each case, I will argue, the figure of "life" appears in a central role, together with qualifications and complications arising from life's "situated" (ecological) character. By the end of this recounting, the grounds for and consistency of

---

[1] For a full bibliography of Grene's publications, see Auxier & Hahn, 2002.

[2] I use the term "*organic* life" to clarify that I mean "life" in the sense of "living things," those things that become the objects of biological science, rather than "life" in the less specific sense of "everyday life" or "life experience."

[3] Grene's pluralism might instructively be compared to others (e.g. Dupré, 1993; Mitchell, 2002), but for shortage of space I won't pursue this comparison here.

[4] The apparent tension between pluralism and realism in Grene's position has been addressed before, for instance by Longino, 2002 and Brandon, 1984.

[5] As exemplified by the three main sections of the Library of Living Philosophers volume devoted to her work: Auxier & Hahn, 2002.

Grene's apparently paradoxical suite of commitments will, I hope, be more readily understandable.

2 Grene's Intellectual Biography

Grene was born in Wisconsin in 1910. She completed a BA in Zoology from Wellesley College, then studied with Heidegger and Jaspers in Germany in 1931–2 and with Whitehead and C. I. Lewis at Harvard in 1933–4. She earned a Ph.D from Radcliffe in 1935, "the nearest a woman could get to Harvard in those days" (2002, 7). In 1937 she moved to Chicago to work with Carnap and Hempel. Yet Grene did not become a devotee of anyone with whom she studied in this period. She published a largely negative review of Heidegger's *Sein und Zeit* in 1938 and her first two published monographs were highly critical commentaries on the existentialist tradition, particularly Kierkegaard, Heidegger, and Sartre (1948, 1957). And, regarding Carnap: "I didn't understand how anyone could construct a philosophy of science with a purely extensional logic… I hadn't anything like [Carnap's] patience in slowly modifying whatever proved untenable in one's overall enterprise" (2002, 9).

In 1944, Grene and her husband, the classicist David Grene, lost their positions at the University of Chicago and for the next sixteen years, first in Illinois and then in Ireland, ran a farm and raised two children before Grene's eventual return to teaching in the late 1950s. Grene later described this period as an "exile," but also suggested that close contact with animals and children influenced her subsequent philosophy: "a close acquaintance with infants, as well as members of other species, does make a difference" (2002, 12).[6] Yet even during this period she published several books and more than ten essays in philosophy journals.

In 1950 Grene heard a lecture by a chemist with philosophical aspirations, Michael Polanyi, and, by the exchange that ensued, became Polanyi's "research assistant… editor and …advisor in the history of philosophy" (2002, 13). Grene's contributions in the years leading to publication of Polanyi's *Personal Knowledge* (Polanyi, 1958) led him to credit her in unequivocal terms: "[t]here is hardly a page that has not benefited from her criticism. She has a share in anything that I may have achieved here."[7] Soon thereafter and partly through Polanyi's influence, Grene

[6] She also recounted that she found it hard to fathom Kant in this period: "My first ten years farming I found I had lost any ear for the sacred text [i.e. the Transcendental Analytic in the first *Critique*]. We had a great gray Percheron mare named Kitty; I couldn't look at her and ask, was she an appearance or a thing in herself. Of course the question is equally absurd for a gnat or a mouse; but somehow a ton work horse seems more absolutely real, out of all relation, or relativity, to our mode of perception, than smaller critters. … [B]abies are not just phenomenal either (or not in the Kantian sense!) Whatever the reason, there it was: agricultural duties and critical philosophy didn't mix. … [W]hen I could read Kant again, later on, it was perhaps the immersion in farm life that made my rereading … more radically realistic than it had been" (1995, 35).

[7] Polanyi, 1958: ix. For more on the Grene-Polanyi relationship, see Mullins, 2002 and Nye, 2011.

began teaching philosophy again at the universities of Belfast and Leeds in the UK, eventually securing a continuing position at the University of California, Davis from 1965 to 1978.

In the late 1950s Grene also began reading and publishing in what was, for her, a new area: the philosophy of biology. Grene's significance as an advocate and organizer for this field is widely acknowledged (for instance, Callebaut, 1993; Burian, 2009). These contributions included her own publications, as well as publication of several early edited volumes of essays (e.g. Grene and Mendelsohn, 1976c), and organization of meetings in the 1970s and 1980s that brought together biologists and philosophers and eventually led to the formation of ISHPSSB (the International Society for History, Philosophy, and Social Studies of Biology).[8]

Grene's output included book-length studies of major figures in the history of philosophy such as Heidegger, Aristotle, Sartre, and Descartes (Grene, 1957, 1963, 1973, and, 1985, respectively). In 1966, she published *Knower and the Known* [hereafter *KK*], which included chapters on the epistemologies of Plato, Aristotle, Descartes, Hume, and Kant (Chaps. 1, 2, 3, 4, and 5) as well as the fact-value distinction, Darwinism, ontological pluralism, and the ontological significance of temporality (Chaps. 6, 7, 8, and 9, respectively).[9] In *Approaches to a Philosophical Biology* (1968) [hereafter *Approaches*] she surveyed the work of five continental European writers that had engaged with contemporary biological science and drawn consequences for philosophy: Adolf Portmann, Helmuth Plessner, F. J. J. Buytendijk, Erwin W. Straus, and Kurt Goldstein. *The Understanding of Nature: Essays towards a Philosophy of Biology* (1974a) [hereafter *UN*] reprinted a chapter from *KK* and three from *Approaches* while adding papers on perception, causation, the structure of evolutionary theory, the reducibility of biology to chemistry and physics, the relevance of Aristotle to modern biology, the relevance of Darwinism to ethics and epistemology, and the continuities and differences between humans and non-human animals. And in 1995 she published a synoptic statement of her "mature" philosophy, presented in conversational style, entitled *A Philosophical Testament* (1995) [hereafter *PT*].

3 History of Philosophy

Perhaps the single most emphasized theme in Grene's writings on the history of philosophy is a phenomenon one might call "the forgetting of life." On Grene's reading, Aristotle represented a high point in appreciation of the complexity of organic life and its significance for metaphysics and human affairs. In the early modern period, however, physics became the discipline thought to have the deepest

---

[8] Recounted in Callebaut, 1993, 236–7, 467.

[9] Grene (2002, 25) later described this work as a recounting of the history of philosophy leading to a philosophy like that of Polanyi.

lessons for metaphysics and the greatest model for the human sciences. In particular, the interpretation of nature as Cartesian *res extensa*, or (as in Locke, Boyle, and Galileo) the domain of "primary qualities," in echo with ancient atomism, had rendered the conception of nature "lifeless." In the introduction to her first book-length statement of her own views, *KK*, she wrote:

> We have come, or are coming, at least to the end of this epoch, the epoch presided over by the concepts of Newtonian cosmology and Newtonian method. We are in the midst of a new philosophical revolution, a revolution in which, indeed, the new physics too has had due influence, but *a revolution founded squarely on the disciplines concerned with life*: on biology, psychology, sociology, history, even theology and art criticism. (1966a, 13, emphasis added).
>
> The revolution before us is a revolution of life against dead nature, and of understanding as against the calculi of logical machines (1966a, 13).
>
> [T]he Cartesian-Newtonian world was, in the last analysis, a world without life. That simple fact had, and still has, disastrous consequences for the conception both of the object of knowledge and of the subject who knows it (1966a, 14).

On the other side, this stripping of life from nature had left matters of the mind and human affairs as either "reducible" to physical processes or characterized by a "mentality" or "subjectivity" that floats free from nature itself. Such was the root of various idealisms – from Berkeley, to Hegel and Husserl, to contemporary postmodernism – which Grene rejected as thoroughly as reductive materialisms. Grene was fond of quoting Whitehead's description of the condition of modern philosophy as a "mystic chant over an unintelligible universe" (1966a, 14, 2002, 26). She also frequently echoed Helmuth Plessner's description of this inheritance as the "Cartesian alternative" between the mental and the physical – a duality that fails to include the "intermediate" or "mediating" category of *organic life* (as well as history and society) in the metaphysical picture (Grene, 1966b; Plessner, 1928/2019).

In *A Portrait of Aristotle* (1963), Grene took up d'Arcy Thompson's suggestion that Aristotle's biological research shaped his approach to metaphysics and epistemology. There Grene presented key concepts of Aristotle's *Posterior Analytics*, *Physics*, and *Metaphysics* such as nature (*physis*), substance (*ousia*), change, intuition (*nous*), induction, and others as shaped by a kind of field naturalist's experience with living things. In a late chapter of the 1963 work and an essay published shortly thereafter (reprinted as 1974a, Chap. 5), Grene also considered the extent to which Aristotelian metaphysics and epistemology required adjustment in the wake of modern Darwinian biology. Her conclusions were mixed. On the one hand, she argued that Aristotelian "formal causes" and some kinds of "final causes" were indispensable components of any adequate biological science, despite their frequent denigration by reductively-minded scientists and philosophers in favor of material and efficient causes (see Sect. 4 for discussion). On the other hand, however, Aristotle was of course unaware of much that contemporary biology had revealed – most notably, the historical and contingent dimension of evolution, and thus the historical and contingent character of our own human lives, situations, and epistemic access to reality (1963, Chap. 7, 1974a, Chap. 5).

Among the targets of Grene's criticism, Descartes stands out above all.[10] Grene criticized Descartes for ignoring the phenomena of living nature and of personal embodiment, and postulating a dualism of disembodied mind and de-vivified nature:

> The world of a human being is infinitely richer in directions and levels of reality than that of an amoeba, but in the last analysis the exchange of action and reaction, of adaptation, has essentially the same foundation in the situation of the living thing as such: in a centre of appetites, curiosities, gropings, satisfactions in which inside and outside, subjective and objective, mental and physical are inextricably intertwined. There is no such thing as a mind by itself; there is no such thing in the living world as a body by itself. It is from this cardinal metaphysical error of Descartes that his epistemological errors, with all their misleading consequences, flow (Grene, 1966a, 88).

On Grene's view, Descartes thus stood at the head of a catalog of common errors of modern philosophy. Grene's early rejection of both phenomenology and analytic philosophy could even be read as motivated by an objection to Cartesian errors still lingering in each (the epistemic privilege of the first-person standpoint, and the unsustainable attachment to rational certitude and a purely physicalistically or phenomenally construed "nature," respectively). Grene charged Sartrean existentialism, for instance, with a basically Cartesian emphasis on the self-enclosed and inviolable authority of first-personal experience and a denial of the reality of embodied, living nature, thus tracing many of its errors to the influence of Cartesianism (1973, 2005).

Empiricisms from Locke and Hume to the modern "logical empiricists" and Van Fraasian "constructive empiricists" typically remained hostage, on Grene's view, to one or another Cartesian error. These included a misguided view of *perception* as based on immediately perceived and perfectly known sensory atoms, followed by only mediately and imperfectly known inferences; and a refusal to grant legitimacy or reality to anything whose full reconstruction on the basis of "experience" so-construed was impossible. A problem with the latter feature was its denial of legitimacy to claims to *realism* of the objects of perception, as well as causality, time, the self, and much else besides.[11] (Grene's own account of perception – discussed further in Sect. 5 below – sought to avoid these errors.)

---

[10] In a late autobiographical reflection, she summarized the entire thrust of her philosophy almost from the beginning as an anti-Cartesian one: "The refusal to accept the *cogito*, and all it implies, as the unique starting point of philosophy "has been … a persistent theme in much of my work … [W]hat really put me off philosophy when I [first] tried it [as an undergraduate] was the instructor's insistence that I accept the *cogito*: that is, accept the notion that, setting aside all my everyday beliefs, I could have some special awareness of myself as something purely subjective, apart from my bodily existence. Again, in the fall of 1931 in Freiburg im Breisgau, in Werner Brock's prose-minar on Descartes, I had the same problem. I remember coming out after the seminar remarking: '*Was wär' ich ohne meine Umwelt?*'" (2002, 4).

[11] For Grene on empiricism, see her (somewhat conventionally Kantian) rejection of Hume in Grene 1966a, the synoptic bashing of empiricisms in Grene, 1983, and the chapter on empiricism in *PT*. See also Longino's (Longino, 2002) argument for a pluralist empiricism against Grene's pluralist realism.

Grene held that Kant had definitively answered Hume regarding the role of the *knower* in the phenomenon of knowing: "Kant has shown irrevocably that mind as agent shapes experiences, that the whole empiricist image of experience as purely passive was mistaken"; and, "[t]he deepest root of Hume's skepticism … lies in his want of a theory of the person, and similarly, the heart of Kant's answer to him lies, not so much in the specific arguments on cause, as in his demonstration that the existence of mind as agent is presupposed in the very analysis of experience itself" (1966a, 140–1). Yet Grene criticized Kant for the emptiness of his conception of this knowing subject, as well as the abstractness and incompleteness of his notion of "nature." In particular, Kant failed to recognize the full reality of living things, as evidenced both by his limitation of nature to Newtonian space-time-matter in the first *Critique*, and his account of the structure of living things as a merely "regulative" idea in the third. "[I]n the opposition of mind to merely inorganic nature," she wrote, "Kant was still a Cartesian thinker"; and, "[t]he encounter with plants and animals which is the first foundation of the biologist's knowledge is missing altogether from the primary range of experience Kant treats" (1966a, 147, 149). "How can we know what man is," she asked rhetorically, "if we restrict knowledge to non-living nature and so eliminate man himself, knowledge and all, from the object we are concerned to know?" (1966a, 152).

Furthermore, Grene argued, the enrichment of the "I" in the Kantian "I-think" ought to include not only the status of the I as *a living organism in an environment*, but also the peculiar *historicality, sociality, and contingency* of human "environments" and thus of the "I" that is involved with them. Grene described her proposed enrichment as follows:

> Man as historical person, rooted in man as living organism in a world of living organisms: only this double paradigm can give us a conceptual frame within which the activity of the knowing mind can be adequately understood. The knower is not simply the Transcendental Unity of Apperception, but myself, with my endowments, limitations, hopes, disappointments. It is a full, historical, not a mere logical 'I' (1966a, 152).

This adjustment is coupled, in Grene's revision, with an altered view of Kant's "transcendental object = X" or "object of possible experience" (from the A-edition and B-edition of the *Critique of Pure Reason*, respectively):

> And this transformation given, the object of possible experience becomes itself clearly the real reality. It is things in themselves I aim at knowing, even though I can never know for sure that it is things in themselves I have, in any given solution of any given problem, come to know. I can never know reality except through my own categorization, my own interpretation. And I can never know it as a whole (1966a, 152).

Kant was in error, therefore, in his attachment to the certainty and universality of knowledge and of the nature of human cognition. This point connects with Grene's *pluralism* in epistemology and philosophy of science (discussed further in Sects. 4 and 5), as well as with recognizing the role of historical situatedness in what we see or fail to see in nature itself. Thus, the required revision of Kant involves recasting his "synthetic a priori" not as timeless conditions of knowledge or experience known with "apodictic" certainty, but rather (as had Grene's teacher at Harvard,

C.I. Lewis) as historically variable assumptions, as "seeming" necessities. Yet Grene also found a support point for this pluralism within Kant's own system in his emphasis on the partiality and finitude of any human knowledge[12]:

> Kant himself has argued in the Antinomies that we can never know the totality even of the world as phenomenal, and if all our knowledge is partial it is always selective and always subject to revision and correction. What Kant has shown is that there are always some *a prioris*, some categorization by mind, at the basis of any experience of or within an organized world. But to have refuted Hume in the way he wished ... he would have had to demonstrate that the system of categories and principles is *the* system characteristic of all human minds everywhere and forever. In this proof we must admit he was unsuccessful, and if it is a question simply of necessity versus contingency or certainty versus conjecture, then Hume was right and Kant was mistaken (Grene, 1966a, 141).

Grene specifies, however, that far from entailing a limitless relativism about knowledge, the perspective she recommends rather allows for both the historicality *and* rationality of knowledge, with a relation between the two described as an "ineradicable ambiguity" (possibly borrowing the last term from Merleau-Ponty):

> Granted, the primacy of history puts reason in jeopardy; that is what it is to be 'in situation', and that is where we are. But the primacy of history does not contradict, or abolish, reason. It exhibits the risk of reason, not its non-existence. Logic and situation, the ideal and the factual, reason and history, live as aspects of our lives in tension with one another, in ineradicable ambiguity (1966a, 146, cf. also 1978a).

Together these three proposed substitutions – *situated organic human life* for the T.U.A.; *real objects in one's environment* for the "transcendental object = X"; and a *historically contingent set of basic assumptions, commitments and interpretive frameworks* (a "*historical a priori*") for the synthetic a priori – produce the revised semi-Kantian epistemology and metaphysics that Grene favored.

Grene frequently denigrated *logical positivism* and *analytic philosophy* in the forms given it by, for instance, Carnap, Quine, Sellars, and contemporary philosophers working in that genre. The main causes for her protest appeared to be its tendencies towards empiricism, physicalism, and reductionism; its tendency to float free of reference to empirical sciences or history[13]; and its rewarding of clever defenses of absurd positions (1976a, Chap. 2; *PT*, Chap. 3). Speaking of David Lewis she quipped that "Now I always enjoyed listening to David Lewis, because he was immensely clever. But it was quite clear that this was clever magic, in which what was pulled out of the hat, in fact, was only virtual rabbits" (2002, 222). Late in her career, however, Grene expressed some affinity with the later Wittgenstein, mostly in his arguments designed to shift philosophers' focus from semantic puzzles to their sources and resolutions in basic features of our (human) "form of life." But Grene mourned that Wittgenstein had not carried out the latter project in more detail and that his followers (such as the ordinary language school) often became

---

[12] This move was inspired in part by Heidegger's (Heidegger, 1929/1997) reading of Kant (Grene, 1957, Ch.4, 1966a, b, 143).

[13] Notice the difference between "reference to the empirical" and "empiric*ism*".

captive of practices antithetic to the master's own lessons (1976a, Chap. 2; *PT*, Chap. 3).

Though always critical of Heidegger for both philosophical and political reasons (e.g. 1938, 1948, 1957), Grene also acknowledged some debts to him, most notably Heidegger's emphasis on finitude and the schematism in his interpretation of Kant (Heidegger, 1929/1997), and the notion of *being-in-the-world* as a contrasting philosophical starting-point to the abstract, all-embracing consciousness treasured by Descartes and Husserl. However, she argued that being-in-the-world had ultimately been more adequately and defensibly developed by Merleau-Ponty (*PT*, Chap. 4). Her later writings on Heidegger (Grene, 1976b, 1995), though not the earlier (1948, 1957), further criticized Heidegger for failing to include features of Dasein's status as *living being* in his "existential analytic" (Heidegger, 1927/1962).

Grene periodically expressed agreement with figures whose views she did not discuss in detail, such as Whitehead, Peirce, Collingwood, and Cassirer. She characterized the texts of Dewey and Mead, on the other hand, as "dim and dated."[14] Though a close personal friend of Richard Rorty, she was severely critical of his position and style of argument. Her strongest ire was reserved for anything "postmodern": Haydn White and Stanley Fish, for instance, she referred to as "morbid self-reflection" and "wretched."[15]

4 Philosophy of Biology

Though Grene earned a BA in Zoology before pursuing graduate studies in philosophy, it wasn't until 1958 that she published on biological topics (though see Grene 1947 for a near exception). By Grene's account, the new theme emerged from two events of the 1950s: her study of Aristotle in order to effectively teach on his philosophy (1995, 91, 2005); and an assignment from Polanyi to look into "heresies in evolutionary theory, specifically critics of the evolutionary synthesis" (1995, 91, 2002, 2005). "And once I started reading that literature," she confessed, "I was unable to stop" (1995, 91).

An important context for these developments was the budding professional field of *philosophy of biology* at the time. While this emerging field was framed, by many if not most of the "philosophers" who participated, as an off-shoot of philosophy of science – a domain defined by logical positivism or, at the very least, by an analytic philosophical methodology – Grene's framing of questions in the field was

[14] "I tried reading Dewey and even Mead [in the 1940s], attempting to be a good American I suppose, but I soon found them as dim and dated as I do nowadays" (1995, 54).

[15] Smocovitis, 2009.

somewhat different.[16] The differences might be attributed to her scholarly background in the history of philosophy, her immersion and training in continental European philosophical traditions other than positivism (despite her own primarily critical approach to figures in those traditions), and her arrival at "philosophy of science" not by the usual positivist (or, increasingly, "post-positivist") route, but rather through contact with the disciplinary outsider Polanyi.[17] This raises the question of whether there was (and is) a "place" in professional philosophy of biology for influences of the Grenian sort.

Grene's engagements in the philosophy of biology in this period include (a) a critique of "Darwinism" for its over-reachings and unclarities, (b) arguments against the reducibility of biology to chemistry or physics, (c) defenses of the significance and indispensability to biology of "formal" as well as (d) "teleological" concepts, and (e) criticism of overly narrow or simplistic models of human behavior (such as E.O. Wilson's sociobiology or "evolutionary epistemology"). These engagements were, of course, theoretically interconnected. In the Preface to *Approaches*, she wrote

> What we think, basically, about the nature and functions of living things in general makes an important differences to what we think about ourselves. … But thinking about the subject matter of biology, and hence about ourselves as living things, is dominated, in the English-speaking world, by two dogmatisms, one of which confuses, while the other prevents, any attempt to achieve a new perspective. The first is the Darwinian tradition, with its accordionlike ambivalence; the other, especially in the last few years, is the outspoken physicalism … of all those who proclaims with gusto the imminent reduction of the biological sciences to physics and chemistry. … The European tradition, however, is broader, and I think we have something to learn from it. (1968, v-vi)

As in this passage, Grene's views were frequently presented as drawing from continental European traditions of philosophical reflection on biology. To an Anglophone audience, mention of such European traditions may call to mind Hans Driesch or Henri Bergson. Yet Grene only rarely appealed to Driesch or Bergson as positive sources. Her appeals to "life" were more likely to cite Whitehead's "philosophy of organism," Helmuth Plessner's "theory of organic modals" and "philosophy of the excentric position," or Merleau-Ponty on embodiment and lived experience, than Driesch or Bergson. Regarding Driesch, Grene sympathetically recounted Plessner's argument, *contra* Driesch on the one side and Wolfgang Köhler on the other, for the irreducibility of life as an "organic modal" – that is, an object

[16] By "philosophers" here I mean those affiliated with philosophy departments. The philosophical influences on participating biologists such as Levins, Lewontin, Gould, and Mayr was somewhat different – in some ways wider in the first three cases (e.g. Marxism) and in some ways narrower in the last case insofar as Mayr tended to be critical of all prior philosophy and sought to treat Darwin and Darwinism as a unique and original philosophical position. Honenberger, 2018 discusses the views of Mayr as well as the philosophers David Hull and Michael Ruse in their relation to positivism. For more on Grene's unusualness within the context of contemporary philosophy of biology, see Mèthot (this volume).

[17] Sloan, 2002 emphasizes the distinctiveness of the "continental" and "historical" influence that Grene brought to her work in philosophy of biology.

of perception with its own distinct character, comparable to qualitative states like colors or tones.[18] It was Driesch's reifying vitalism on the one side and Köhler's failure to recognize even a *Gestalt* distinction between "living" and "non-living" things, on the other, to which Plessner objected. Yet Grene did in passing affirm the validity of Driesch's "harmonious equipotentiality," provocatively equating this to what Plessner called "positionality" (discussed in Sect. 5 below).[19]

(a) *Darwinism*

Grene's first publications in philosophy of biology were presented as a critique of what she then referred to simply as "Darwinism" (e.g. 1958, 1959, 1961, 1966a). Grene's first publication in this area, "Two Evolutionary Theories" (1958), compared G.G. Simpson and Otto Schindewolf and argued (among other things) that Simpson had found it impossible to avoid employing "type" concepts despite his explicit refusal of them. Grene's apparent favoring of Schindewolf, with his very un-Darwinian "orthogenetic" evolution, over the canonical Darwinian Simpson, earned her the ire of many. The paper was soon criticized by several professional biologists (e.g. Bock & von Wahlert, 1963; Van Valen, 1963) and at least one philosopher (Hull, 1969). Subsequent discussion, however, has been more favorable to Grene's argument. Eldredge (1992) recalled frustration that he was unable to refute her position when he first read the paper in graduate school, but later perceived it as anticipatory to his own "punctuated equilibrium." Ruse (2007) suggested that the paper's earlier condemnation was perhaps unjust. Even Schindewolf subsequently received greater esteem: for instance, an English translation with University of Chicago Press boasting a sympathetic forward by Stephen Jay Gould (Schindewolf, 1993).

Grene's "Statistics and Selection" (1961) discussed R.A. Fischer's use of terms such as "fitness," "progress," and "adaptation." She argued that these terms were initially defined by Fischer by reference only to mathematical formulas with no substantive implications about patterns in nature, yet were elsewhere employed in ways that had such substantive implications. She thus distinguished, for instance, between "statistical selection" involving fitness and adaptation as defined in Fischer's formalisms, and "Darwinian selection" as describing fitness and adaptation in the more substantive sense.[20] The 1959 centennial of the *Origin of Species*

[18] Plessner, 1928/2019, Chap. 3; Grene, 1968, 66–70; this passage is not contained in Grene 1966b or 1974a.

[19] Polanyi had also favorably cited Driesch's "harmonious equipotentiality," and even sought to generalize its application to non-living complex systems, in Polanyi, 1958, Part IV. Other than the passage referenced above, I find no positive references (and quite a few negative or self-distancing references) to Driesch in *KK*, *Approaches*, or *UN*. For Bergson I find no positive references in *Approaches* or *UN*, but some sympathy expressed for Bergson's emphasis on the metaphysical significance of "time" in Grene 1966a, Chap. 9.

[20] Grene's criticism of Fischer has been cited as an early expression of the position now known as "statisticalism," e.g. Walsh et al., 2017.

found her commenting on "The Faith of Darwinism" (1961; reprinted as *KK*, Chap. 7).

Grene's critical stance towards Darwinism made her something of a heretic within the budding interdisciplinary research community in philosophy of biology. While she was at Davis in the 1960s, for instance, Ernst Mayr once came to her office and told her she knew nothing about biology and should "stay away from tinkering with evolutionary questions" (Grene, 1995, 1).

Between the publication of "Two Evolutionary Theories" in 1958 and the publication of *UN* in 1974, Grene's sophistication and competence in modern biology grew rapidly, in large part through her interactions and involvements with discussions that included professional biological researchers. In the introduction to UN (1974), she noted this change of view:

> since my return in 1965 from the isolation first of an Irish farmstead and then of British provincial universities to the more communicative milieu of a University of California campus, I have learned a good deal from talks with colleagues and students so that, I believe, I now understand better where the philosophical problems in evolutionary theory lie (vii).

It is not too much to say that she learned, through these dialogues, to see her earlier criticisms of "Darwinism" as unfair. Reflecting on this period later, she wrote,

> I would still defend both [Grene, 1958 and 1961] as reasonable pieces of conceptual analysis, … though some of what I published somewhat later on evolution now seems to me at least in larger part indefensible, especially a really bad chapter on "The Faith of Darwinism" in [*KK*]. … I realized later that what I had been trying to criticize was what … Gould calls "the hardened synthesis" … [A]fter I moved to Davis in 1965, I did learn better, thanks to colleagues like G.L. Stebbins and Theodosius Dobzhansky … Not that the hardened synthesis was not in fact as narrow and self-congratulatory as it seemed, but that there is very much more to the Darwinian tradition in evolutionary theory than that particular view allowed (2002, 17).

Likewise, she later described *Approaches* as "a book I should not have written" (2002, 80): "Not only did no one want to hear about my authors, but by now I wonder what, with one exception [namely Plessner, discussed further below], I thought I had learned from them" (2002, 18).[21]

Another reason for Grene's earlier errors in philosophy of biology, at least according to her 1974 account, was a mistake about the relation between biology and philosophy:

> [I]t is wiser, or so I have come to believe, to restrict evolutionary theory to what it can cleanly and clearly handle: changes in relative gene frequencies, and to abandon hope for a 'scientific' theory of emergence. That living systems can and should be studied at a number of levels is an important methodological and epistemological thesis. But systems analysis

[21] Nonetheless, "if one is looking for clues on what it is to be a person – clues that take due account of our situatedness, both in nature and in culture (itself within nature) – Portmann and Straus, as well as Plessner, do provide some evidence" (2002 19). Hence, the essays on Portmann, Plessner and Straus, though not the essays on Goldstein and Butendjik, were reprinted in *UN*. The role of *Approaches* as a transitional text between Grene's earliest papers in philosophy of biology and *KK*, on the one hand, and the more philosophy-of-science oriented papers of *UN*, calls for closer attention than I can give it here.

> and evolutionary theory are forced to coalesce, if at all, either through *ad hoc* principles or through the kinds of conceptual confusions I had earlier – and mistakenly – considered to be characteristic of evolutionary theory as such (1974a, vii).

And:

> I am in disagreement with some eminent evolutionists, even with some of those from whom I have learned most. But … this is a disagreement in philosophical interpretation, not in science, but *about* it. For though, admittedly, conceptual analysis and empirical investigation are not totally dissociable from one another, … their interrelations are both subtle and limited. On the one hand, philosophy reflects on the conceptual structure of scientific statements or theories; on the other hand, philosophical reflection, since it is situated, like every human activity, within the human world, is also influenced by the outcome of scientific research. The latter influence is a difficult one to specify. Scientists can't tell us how to solve our problems … But on the other hand we certainly cannot tell them (except when they try to philosophize) how to solve theirs (1974a, viii).

(b) *Reductionism*

In *KK* and then several chapters of *UN*, Grene argued against reductionism on the basis of a single argument drawn from Polanyi (Polanyi, 1958, Part IV; 1968). According to this argument, many systems – including machines, organisms, linguistic utterances, and human persons – follow laws that are unpredictable or unspecifiable in terms of "lower-level" sciences that apply to their parts (such as chemistry or physics); their behavior therefore cannot be fully known or explained on the basis of these lower-level sciences alone. The operation of such systems involves "boundary conditions," which can be conceived as defining the laws of a "higher-level" science (or set of descriptions or perspectives) on the system. Examples by Polanyi and Grene include the chemical bonds of the DNA molecule versus its significance as a "code" in cellular operations; and the letters making up a sentence versus the sentence's significance in the context of a communicative act.

Both Polanyi and Grene granted that all systems obey the laws of physics. They emphasized, however, that these laws are insufficient to predict or to meaningfully describe what happens with every kind of system: "All living systems do indeed obey the laws of physics, but without countervening the laws of physics they may well obey other laws as well" (1974b, 58). "[T]hat all living systems are *exhaustively explained by*" these physical laws "does not follow unless we know in addition that the laws of physics are the only laws there are" (1974b, 58, emphasis added).

Grene sometimes strengthened this argument to the point of claiming that reductionism is incoherent since it prevents meaningful recognition of the norms at work in inquiry, belief, and communication itself: "In short, were reductivism true, knowledge would be impossible, including the knowledge that reductivism is true. And were reductivism true, language would be impossible, including the formulation of the reductivist's thesis" (1974b, 54–55).[22]

[22] See also the similar argument in Grene 1966a, Chap. 7, about the epistemically self-undermining implications of Darwinism, which presages recent arguments by Street, 2011 and Nagel, 2012.

In "The Multiplicity of Forms" (*KK*, Chap. 8), Grene argued further that living things exhibit an *ontologically* distinctive status for this reason, placing them in a hierarchy of systems defined by increasingly specific boundary conditions: *non-living things*, *living things*, *persons* (in which she interestingly included animals), and *responsible persons* (of which human beings are paradigmatic). And in later works on hierarchies in biology, she distinguished the sort of "control" hierarchy involved in this argument from merely "classificatory" hierarchies such as those of Linnaean taxonomy (1987, 1988). Yet her work from 1974 onwards appears to drop the emphasis on a special ontological discontinuity between the non-living and the living suggested in *KK*, Chap. 8.

Grene described her ontology not only as *anti-reductionist* but also as *pluralist*. This theme appears as early as "The Multiplicity of Forms":

> What is important is not that there are precisely such and such kinds or grades of being in the world, so many and no more, but *that there are beings of more kinds than one*. There are many ways of being, as many, Aristotle said, as there are kinds of unity (1966a, 218).

In "Merleau-Ponty and the Renewal of Ontology" (1976b, 609), she reiterated this theme, distinguishing between a "centripetal" pluralism that posits "a universe that rises level upon level in a hierarchy of emergent realities" and a "centrifugal" pluralism that recognizes "not so much orderly and encompassing ascent as proliferation of structures, a multiplicity of forms" (609). In "Reductionism: Another Side Issue?" (1974b) she construed science as pluralistic due to its use of models of many different kinds, including both "reductive" and "non-reductive" ones.

Finally, this ontology was also *realist*. In the same essay (1974b), for instance, she described science as involving a search for mechanisms, which themselves should be understood as real causal powers involved in producing the phenomena of interest (rather than merely empirical correlations, as held by empiricism).[23] And, in *PT*, Chap. 6 as well as her previously mentioned discussion of Kant's "transcendental object = X" in *KK*, Chap. 6, she unabashedly defended a realism about the objects of perception and knowledge.[24]

(c) *Types and Forms*

Grene repeatedly argued that recognition of *form* was indispensable to biology, including evolutionary theories and models (1958, 1966a, Ch.8, 1990). Prior recognition of form in this sense included Aristotle's "formal causes" as well as the *Gestalt*s of Gestalt psychology, for which Merleau-Ponty could be read as providing an epistemological and ontological generalization (Grene, 1976b). Relatedly, she defended the legitimacy and even indispensability of "type"-concepts, including the notion of species' "natures," despite these natures being subject to change according to contemporary evolutionary theory (Grene, 1990). In defending that position, Grene's position ran against the grain of then-influential criticisms of "typological thinking" and "essentialism" by Ernst Mayr (1959) and David Hull (1965a, b),

---

[23] For discussion, see Brandon, 1984.

[24] For critical discussion, see Longino, 2002.

subsequently problematized as "the essentialism story" (Honenberger, 2015, 2018; Winsor, 2005). That commitment was further expressed in Grene's defense of real distinctions between *kinds* of things (Grene, 1976b), a topic often addressed under the heading of "natural kinds" or "anti-nominalism."

(d) *Teleology*

Grene defended the indispensability not only of Aristotelian *formal* causes, but also of *final* causes, to any adequate understanding of living things, including within biological science. In a late chapter of *KK* (Chap. 9: "Time and Teleology"), she argued for a distinction between teleology and purpose, as well as between teleology and any particular temporal order. She then noted three roles for teleology within biology: (i) part-whole relations (x is part of y), (ii) functional relations (x is for y), and (iii) "directed" developmental processes (x leads to y). In "Biology and Teleology" (1964), reprinted in *KK*, Chap. 9 and again in *UN*, Grene argued for the indispensability of teleology (in one or more of these three senses) to biology at six increasingly substantive "levels": reflective, regulative, descriptive, operational, explanatory, and ontological.

(e) *Human nature*

Grene repeatedly took up the topic of the relevance of modern biology to human nature and human behavior, arguing against views such as sociobiology and evolutionary epistemology for their reductive oversimplification of human behavior (e.g. Grene, 1978b). Nonetheless she emphasized the status of human being as living things, themselves part of a common nature and evolutionary history with other living things. Her view thus emphasized the status of humans as living beings, but also as beings with a pattern of life that was thoroughly and "naturally" (that is, as a matter of their own species' nature) mediated by contingent cultural environments (as in Grene's well-known "People and Other Animals" [1974c]). This part of Grene's view was informed primarily by the work of a Hans Driesch student, the German zoologist-turned-philosopher Helmuth Plessner (1892–1985) (Grene, 1966b, 1968, 1974c; discussed further below).

5 Epistemology, Metaphysics, and Philosophical Anthropology

As noted, Grene defended both sides of several seemingly opposed positions: *realism* about the objects of perception and knowledge, yet also a "centrifugal" *pluralism* about the nature of those objects themselves; and *naturalism* about persons and culture, yet also theoretical and explanatory *anti-reductionism* regarding relations between "lower-level" and "higher-level" sciences or models. Grene's work on epistemology and metaphysics, including her work on perception and human nature, appears to recognize these tensions insofar as Grene is often at pains to articulate how both sides of the apparent oppositions can be true.

Near the end of her career, Grene listed four figures as "guiding lights" for the development of her own views: Polanyi, Merleau-Ponty, Helmuth Plessner, and J.J. Gibson (Grene, 2002, 25), the last of whom appears to have provided the inspiration for the expression "ecological epistemology" itself. The question of the coherence of Grene's view can thus be addressed through consideration of the elements Grene borrowed from each of these sources and how she brought them together within the metaphysics and epistemology she defended.

(a) *Polanyi*

In addition to the anti-reductionist argument noted above, Grene frequently appealed to what Polanyi called the "from-to" structure of knowledge (Polanyi, 1958, Chap. 4; Polanyi, 1966/2009). This structure involves relying on one component, element, or aspect of a system or situation to illuminate another: for instance, a rabbit orienting its leap away from a predator on the basis of a loud sound in the brush. According to Polanyi's account, the "from" element typically recedes into the cognitive or experiential background, and what the perceiver experiences is rather the object "to" which one is attending (by the "from" element as means). For instance, readers tend to attend to the content of a text rather than to the marks (or even, in most cases, the precise words) by which the content is conveyed. Physicists looking at bubble-chamber photographs "see" sub-atomic particles rather than marks on paper. The object attended to is the "to" part of the from-to structure, while the means by which it is attended to is the "from" part.

The Polanyian from-to structure further allows for integrations between *pluralism* and *realism*. In our perception, experience, and knowledge, we are capable of and typically attuned to features of the environment that we really do perceive, experience, or know as "immediately" accessed objects, yet we are able to access such features only "mediately," despite the phenomenological receding of the mediating factors into our background (as the "from" part of the Polanyian "from-to" structure). This combination of commitments to realism and immediacy, yet also aspectivity and mediacy, further aligns with Grene's borrowings from Merleau-Ponty, Plessner, and Gibson.

(b) *Merleau-Ponty*

From Merleau-Ponty Grene drew the following lessons (Grene, 2002, 1976b, 1995):

1. perception exhibits features quite different from its characterization in classical empiricism, such as.
    - (i) *aspectivity* and hence *finitude*,
    - (ii) a *Gestalt* character (in contrast to the atomism of the Humean empiricist tradition, e.g. sense-datum theories), and
    - (iii) *ambiguity* of the contents and commitments that become available to us through it;

2. perception is of central significance for epistemology and metaphysics insofar as our experience and our access to the world are first and foremost components and operations of perception, and other cognitive acts may themselves best be understood as instances of or modulations of perception itself (what Merleau-Ponty called "the *primacy of perception*"); and
3. objects of perception corresponding to (1) ought to be recognized as fully (onto-logically) *real*, a move itself strengthened by the combination of (1) and (2).

Grene (1976b) argued, on the basis of these commitments, for an ontology that recognizes the reality of forms and distinctions among forms, the reality of things other than the objects of basic physics, and the legitimacy of multiple potentially cross-classifying categorizations, models, or schematizations of reality (see also the discussion of "multiplicity of forms" in *KK*, Chaps. 8 and 9 [1966a]). Merleau-Ponty thus serves as a corrective to both empiricist and rationalist misconstruals of the structure of perception itself, as well as a means of reconciling Grene's simultaneous commitments to anti-reductionism, naturalism, realism, and pluralism.

One might object that realism must entail an ontological monism in some sense insofar as a complete and accurate description of reality would have to involve recognition of the possibility of monistic reconciliation of the various "plural" perspectives on it. But Merleau-Ponty's arguments about the aspectivity and hence finitude of perception, coupled with their proposed implications for ontology, suggest an analogue of a thesis Grene had already drawn from Kant: the thesis that a "complete and accurate description of reality" is merely an idealistic fiction, what, in Kantian language, we might call an "idea of reason" pushed beyond the bounds of all possible experience. Rather than guide ourselves by such unreliable fictions, we ought rather to accept and seek to orient ourselves within the dynamic ambiguities of our real, embodied and embedded lived situations. Like Kant, Grene proposes that recognizing such finitude and partiality is not incompatible with a commitment to realism.

(c) *Plessner*

In an early paper (1966b), reprinted as a chapter of *Approaches* (1968) and in UN (1974a), Grene sympathetically recounted Plessner's "theory of organic modals," a sort of phenomenological analysis[25] of organic life that emphasized the role of something he called "positionality" and that amounted to an analysis of variations in the dynamic relationship between living things and their environments. Starting in 1974a, Grene further drew upon a series of linked concepts that Plessner (1928/2019), Chap. 7) had proposed as descriptors of distinctive features of human interactions with environments: *excentric positonality*, *natural artificiality*, and *mediated immediacy* (Grene, 1974c, 1992, 1995). "Excentric positionality" described human reflexivity, including self-reflection and self-criticism, as

[25] It should be noted, however, that neither Plessner nor Grene would have allowed a description of this theory as "phenomenology." Plessner compared but distinguished his approach from Husserlian phenomenology (e.g. Plessner, 1928/2019, 25–27, 107ff). And Grene frequently distanced herself from "phenomenology": regarding Merleau-Ponty's phenomenology of perception, for instance, she insisted it was not actually phenomenology but only titled as such to appease Merleau-Ponty's academic advisors (2005).

involving a passage through the perspectives of others and a shared world of culture. It also signified the way that humans live in a cultural and artifactual world, and thus in a sense carry out their lives "out beyond" their own bodily boundaries. The associated concept of "natural artificiality" highlighted the naturalness, for humans, of involvement with and mediation by artificial environments (which were themselves, on Grene's view, a "part of nature" – today we might see a parallel with the notion of "constructed niches" in the sense of Odling-Smee et al., 2003). "Mediated immediacy" signified the way that humans have access to features of their environments "immediately," yet the range of objects and features over which this success and immediacy of access is possible is expanded through complex, contingent, and artifactual mediators such as language, interpretive frameworks, and technologies.[26]

These Plessnerian concepts were significant for Grene's thinking as a means to effectively clarify humans' status as both living and encultured, and, relatedly, our knowledge (scientific and otherwise) as both adequate to reality and relative to historically contingent frameworks. Plessner's "organic modals" may also have inspired her position on the non-reducibility of life, a non-reducibility she eventually came to present, like Plessner, in epistemological and experiential terms more than ontological ones (comparing *KK* and *UN*). Yet the precise ontological status of "life" and "living things" remained something of an underspecified feature of Grene's mature view, despite all her early emphasis on its significance.

(d) *Gibson*

Grene's explicit references to J.J. Gibson's "ecological" theory of perception appear as early as the essays on perception and on reduction in UN (1974), but, as she later noted (2002, 22), her reliance on Gibson's theories in articulating her own views only began later (e.g. Grene, 1985, 1995, Chap. 7). The main attraction of Gibson's theory, for Grene, was its "ecological" and anti-internalist thrust, thereby eliding problems associated with traditional philosophical theories of perception (such as those of Descartes and Locke) and their psychological fellow-travelers. Unlike those theorists, Gibson treated perception not as primarily a phenomenon in the minds or brains of organisms and *representing* an external world, but rather as organisms' differential responsiveness to properties in environments (including "gradients," "differentials," and "invariants" in these properties) (Gibson, 1966, 1979). The "internal" mechanisms supporting this responsiveness are often much simpler, more direct, and more reliant on the organisms' action, than traditional theories of perception had supposed.[27] Other attractive features of Gibson's theory,

---

[26] Incidentally, this recognition and emphasis on "mediation" in integration with "immediacy" distinguishes Grene's (and Plessner's) views from some other efforts to deploy a Gibson-style "ecological realism" in epistemology, e.g. Dreyfus & Taylor, 2012.

[27] Historically speaking, Gibson's view might be read as a further development of earlier psychological approaches that emphasized the importance of environmental factors for the concept of behavior, such as the traditions of Darwinian animal psychology and of ethology; and as a precursor of the more recent turn toward "extended" and "embedded" theories of cognition (for examples of the latter that make explicit use of Gibson, see Chemero, 2009 and Cisek, 2019).

for Grene, may have included its non-atomistic characterization of the "contents" of perception itself (similar to Merleau-Ponty); its fully realist ontology and methodological standpoint in regard to perceptual objects and perceptual environments; its status as an empirically-engaged theory, subject to defense on experimental as well as theoretical grounds; and its applicability to both human and non-human perceivers.

In the context of her borrowings from Polanyi, Plessner, and Merleau-Ponty, Grene's appeal to Gibson promised to preserve realism despite recognition of the Gestalt-like, pluralist, and ambiguous character of perception as in Merleau-Ponty; and to update Plessner's and Merleau-Ponty's non-reductive "ontological" projects by enabling a link between (on the one hand) a philosophical account of perception and its philosophical implications, and (on the other) the facts of biology, ecology, and otherwise "empirical" scientific fields that must variously be informed by, or supply evidence for, philosophy.

Acknowledgments For comments on previous versions of this essay I am indebted to Chris Donohue, Phillip Sloan, Charles Wolfe, an anonymous reviewer for the press, and participants of the 2021 workshop on *Canguilhem and Continental Philosophy of Biology* organized by Charles Wolfe, Giuseppe Bianco, and Gertrudis Van der Vijver and hosted by Ghent University and Université Toulouse Jean Jaurès. All errors are my own.

Bibliography

Abbreviations for Works by Grene

KK = *Knower and the Known (1966).*
Approaches = *Approaches to a Philosophical Biology (1968).*
UN = *The Understanding of Nature (1974).*
PT = *A Philosophical Testament (1995).*

Works by Grene

Grene, M. (1938). (as Marjorie Glicksman). A Note on the Philosophy of Heidegger. *Journal of Philosophy, 35*(4), 93–104. Reprinted in 1976a.
Grene, M. (1947). On Some Distinctions Between Men and Brutes. *Ethics, 57*, 121–127. Reprinted in 1974a.
Grene, M. (1948). *Dreadful Freedom: A Critique of Existentialism*. University of Chicago Press.
Grene, M. (1957). *Heidegger*. Bowes & Bowes.
Grene, M. (1958). Two Evolutionary Theories. *British Journal for the Philosophy of Science, 9*, 110–127. Reprinted in 1974a.
Grene, M. (1959). The Faith of Darwinism. *Encounter*. November. Reprinted in 1966a.
Grene, M. (1961). Statistics and Selection. *British Journal for the Philosophy of Science, 12*, 25–42. Reprinted in 1966a, 1974a.

Grene, M. (1963). *A Portrait of Aristotle*. Thoemmes Press.
Grene, M. (1964). Biology and Teleology. *Cambridge Review*, 269–273. Reprinted in 1966a, 1974a.
Grene, M. (1966a). *The Knower and the Known*. Faber & Faber.
Grene, M. (1966b). Positionality in the Philosophy of Helmuth Plessner. *Review of Metaphysics, 20*, 250–277. Reprinted in 1968, 1974a.
Grene, M. (1968). *Approaches to a Philosophical Biology*. Basic Books, Inc.
Grene, M. (1969). Introduction. In M. Polanyi, *Knowing and Being: Essays by Michael Polanyi*, ed. M. Grene. University of Chicago Press.
Grene, M. (1973). *Sartre*. New Viewpoints.
Grene, M. (1974a). *The Understanding of Nature: Essays in the Philosophy of Biology*. D. Reidel Publishing Co.
Grene, M. (1974b). Reducibility: Another Side Issue? In Grene 1974a.
Grene, M. (1974c). People and Other Animals. In Grene 1974a.
Grene, M. (1976a). *Philosophy In and Out of Europe*. University of California Press.
Grene, M. (1976b). Merleau-Ponty and the Renewal of Ontology. *Review of Metaphysics, 29*(4), 605–625.
Grene, M., & Mendelsohn, E. (Eds.). (1976c). *Topics in the Philosophy of Biology*. D. Reidel Publishing Co.
Grene, M. (1978a). The Paradoxes of Historicity. *Review of Metaphysics, 32*(1), 15–36.
Grene, M. (1978b). Sociobiology and the Human Mind. In *Sociobiology and Human nature: An interdisciplinary Critique and Defense, eds.* Michael S. Gregory, Anita Silvers, and Diane Sutch. Jossey-Bass, Inc.
Grene, M. (1983). N-Dogmas of Empiricism. In R. S. Cohen & M. W. Wartofsky (Eds.), *Epistemology, Methodology, and the Social Sciences. Boston Studies in the Philosophy of Science* (Vol. 71, pp. 89–106). D. Reidel Publishing Co.
Grene, M. (1985). *Descartes*. Harvester Press.
Grene, M. (1987). Hierarchies in Biology. *American Scientist, 75*(5), 504–510.
Grene, M. (1988). Hierarchies and Behavior. In G. Greenberg & E. Tobach (Eds.), *Evolution of Social Behavior and Integrative Levels. The T.C. Schneirla Conference Series* (Vol. 3, pp. 3–17). Lawrence Erlbaum Associates, Publishers.
Grene, M. (1990). Evolution, "Typology" and "Population Thinking". *American Philosophical Quarterly, 27*(3), 237–244.
Grene, M., & Eldredge, N. (1992). *Interactions: The Biological Context of Social Systems*. Columbia University Press.
Grene, M. (1995). *A Philosophical Testament*. Open Court Press.
Grene, M. (2002). Intellectual Biography. In Auxier and Hahn 2002, 4–28.
Grene, M. (2005). Interview with *The Believer* magazine. (http://www.believermag.com/issues/200503/?read=interview_grene) Accessed February 6, 2011.

Works by Others

Auxier, R. E., & Hahn, L. E. (Eds.). (2002). *The Library of Living Philosophers: The Philosophy of Marjorie Grene* (Vol. XXIX). Open Court.
Brandon, R. (1984). Grene on Mechanism and Reductionism: More Than Just a Side Issue. *PSA: Proceedings of the Biennial Meeting of the Philosophy of Science Association, 1984*, 345–353.
Bock, W. J., & von Wahlert, G. (1963). Two Evolutionary Theories – A Discussion. *British Journal for the Philosophy of Science, 14*(54), 140–146.
Burian, R. (2009). Marjorie Grene: A Remembrance With Special Attention to Her Importance for ISHPSSB. *Biological Theory, 4*(2), 183–187.

Callebaut, W. (1993). *The Naturalistic Turn: How Real Philosophy of Science is Done*. University of Chicago Press.
Chemero, A. (2009). *Radical Embodied Cognitive Science*. MIT Press.
Cisek, P. (2019). Resyntheizing Behavior Through Phylogenetic Refinement. *Attention, Perception & Psychophysics, 81*, 2265–2287.
Dreyfus, H., & Taylor, C. (2012). *Retrieving Realism*. Harvard University Press.
Dupré, J. (1993). *The Disorder of Things: Metaphysical Foundations of the Disunity of Science*. Harvard University Press.
Eldredge, N. (1992). Marjorie Grene, 'Two Evolutionary Theories' and Modern Evolutionary Theory. *Synthese, 92*(1), 135–149.
Gibson, J. J. (1966). *The Senses Considered as Perceptual Systems*. George Allen & Umwin, Limited.
Gibson, J. J. (1979). *The Ecological Approach to Visual Perception*. Houghton Mifflin.
Heidegger, M. (1927/1962). *Being and Time* (J. Macquarrie & E. Robinson, Trans.). Harper and Row.
Heidegger, M. (1929/1997). In R. Taft (Ed.), *Kant and the Problem of Metaphysics* (Churchill J. S. Trans.) (5th ed.). Indiana University Press.
Honenberger, P. (2015). Grene and Hull on Types and Typological Thinking in Biology. *Studies in History and Philosophy of Biological and Biomedical Sciences, 50*, 13–25.
Honenberger, P. (2018). Darwin Among the Philosophers: Hull and Ruse on Darwin, Herschel, and Whewell. *HOPOS: The Journal of the International Society for the History of the Philosophy of Science, 8*, 278–309.
Hull, D. (1965a). The Effect of Essentialism on Taxonomy – Two Thousand Years of Stasis: Part 1. *The British Journal for the Philosophy of Science, 15*(60), 314–326.
Hull, D. (1965b). The Effect of Essentialism on Taxonomy – Two Thousand Years of Stasis: Part 2. *The British Journal for the Philosophy of Science, 16*(61), 1–18.
Hull, D. (1969). What the Philosophy of Biology is Not. *Journal of the History of Biology, 2*, 241–268.
Longino, H. (2002). Marjorie Grene's Philosophical Naturalism. In: Auxier & Hahn.
Mayr, E. (1959). Typology versus Population Thinking. In *Evolution and Anthropology: A Centennial Appraisal* (pp. 409–412). The Anthropological Society of Washington.
Mitchell, S. (2002). Integrative Pluralism. *Biology & Philosophy, 17*, 55–70.
Mullins, P. (2002). On Persons and Knowledge: Marjorie Grene and Michael Polanyi. In: Auxier & Hahn.
Nagel, T. (2012). *Mind & Cosmos: Why the Materialist Neo-Darwinian Conception of Nature is Almost Certainly False*. Oxford University Press.
Nye, M. J. (2011). *Michael Polanyi and his Generation: Origins of the Social Construction of Science*. University of Chicago Press.
Odling-Smee, F. J., Laland, K. N., & Feldman, M. W. (2003). *Niche Construction: The Neglected Process in Evolution*. Princeton University Press.
Plessner, H. (1928/2019). *Die Stufen des Organischen und der Mensch: Einleitung in die philosophische Anthropologie*. Walter de Gruyter & Co. [*The Levels of Organic Being and of Man: Introduction to a Philosophical Anthropology*. Trans. Millay Hyatt. Fordham University Press].
Polanyi, M. (1958). *Personal Knowledge: Towards a Post-Critical Philosophy*. University of Chicago Press.
Polanyi, M. (1966/2009). *The Tacit Dimension*. University of Chicago Press.
Polanyi, M. (1968). Life's Irreducible Structure. *Science, 160*(3834), 1308–1312.
Ruse, M. (2007). Relecture de "Deux theories évolutionnistes" de Marjorie Grene. In J. Gayon & R. Burian (Eds.), *Conceptions de la science: hier, aujourd'hui, demain: hommage à Marjorie Grene* (pp. 368–386). Brussels.
Schindewolf, O. (1993). *Basic Questions in Paleontology: Geologic Time, Organic Evolution, and Biological Systematics*. Translated by J. Schaefer. University of Chicago Press.

Sloan, P. (2002). Reflections on the Species Problem: What Marjorie Grene Can Teach Us About a Perennial Issue. In: Auxier & Hahn.

Smocovitis, V. B. (2009). Marjorie, Matriarchy, and 'Wretched Reflection': A Personal Remembrance of Marjorie Grene. *Biological Theory, 4*(2), 191–195.

Street, S. (2011). Evolution and the Normativity of Epistemic Reasons. *Canadian Journal of Philosophy, 35*, 213–248.

Van Valen, L. (1963). On Evolutionary Theories. *British Journal for the Philosophy of Science, 14*(54), 146–152.

Walsh, D., Ariew, A., & Matthen, M. (2017). Four Pillars of Statisticalism. *Philosophy, Theory, and Practice in Biology, 9*(1).

Winsor, M. (2005). The Creation of the Essentialism Story: An Exercise in Meta-History. *History and Philosophy of the Life Sciences, 28*(2), 149–174.

Phillip Honenberger is a professional philosopher with specializations in philosophical anthropology, the history and philosophy of biological science, and the history of European and American philosophy from Kant to present. His work has appeared in *Biology & Philosophy*, *Studies C*, *HOPOS*, *Synthese*, *International Journal of Philosophical Studies*, and other forums.

"Dilettantes of Life." Franco-German Refractions of Anthropogenesis in Twentieth Century Philosophy

Thomas Ebke

Abstract It is intriguing to observe that the massive rise of a contemporary set of so-called *life sciences* at the turn of the twenty-first century has not only spurred philosophers of vastly different backgrounds worldwide to rethink the very concepts of "life," "the living" or of "lived experience," to name but a handful. What is more, the literal revival of this terminology has allowed historians of modern philosophy to shed new and unwont light on one genealogical compound in particular: that of the quite often charged and complex relations between French and German positions in twentieth-century philosophical thought. My paper aims at a specification of this new map of Franco-German philosophy that has secretly and insistently been centered upon the concept of life. On *one* level, it discusses an antagonism: whereas the paradigmatic tradition of modern German philosophical anthropology (Scheler, Plessner, Gehlen) has coined the notion of a philosophical biology (*philosophische Biologie*), French thought, in reverse, witnessed the project of a biological philosophy (*philosophie biologique*), most pronouncedly in the historical epistemology of Georges Canguilhem. The paper will reflect on the diametrical antagonism between these two formulae. In its extended first part, however, it also tries to unearth the shared roots and sources of these two genealogies, that is of the French and the German constellation respectively. Attention will be drawn to the empirical biologies that resonated stronly both with thinkers such as Scheler, Plessner and Gehlen in Germany, and with figures such as Jacques Lacan or Raymond Ruyer, if it comes to the "French connection." Thus, the overall discursive refraction between the conceptions of philosophical biology and biological philosophy does not rule out, but rather imply a corporate genealogy that traverses the

This text is a strongly revised and abdridged variation on Ebke (2017).

T. Ebke (✉)
Universität Potsdam, Potsdam, Germany

G. Bianco et al. (eds.), *Canguilhem and Continental Philosophy of Biology*, History, Philosophy and Theory of the Life Sciences 31,
https://doi.org/10.1007/978-3-031-20529-3_4

works of Hermann Klaatsch, Paul Alsberg and Louis Bolk, and the history of the radical divergence between modern French and German philosophy on the subject of the philosophy and the science of life is finally attenuated by a space of unsuspected encounters.

Keywords Philosophical biology · Biological philosophy · Philosophical anthropology · Historical epistemology · Anthropogenesis

In his essay "Note sur la situation faite en France à la philosophie biologique" (1947; here: Canguilhem, 2015), Georges Canguilhem blames modern French philosophy for its inveterate scientism and academic rationalism (see *ibid.,* 313) that continue to stymie any attempt to take the very idea of life into a serious account and to appraise the "speculative" quality of this concept. In fact, any assessment of the German-speaking tradition of philosophical anthropology, encompassing the works of Max Scheler, Helmuth Plessner or Arnold Gehlen, that might be expressed from the specific angle of Canguilhem's "attitudinal vitalism,"[1] amounts to its portrayal as a cathartic force for an epochal self-reflection of the rationalist bias in the modern constellation of French philosophy: after all, any comprehension of man as, first and foremost, a *living being* carries with it an insistence on the position that man's "spiritual" dimension, too, needs to be accounted for in terms of a "logic of life" (although decidedly not by way of a naturalistic causal reduction of mind to matter). As Canguilhem duly remarks (*ibid.,* 314), the singular author in modern French philosophy that ventured into this direction was Henri Bergson (see Zanfi, 2013). Notwithstanding Canguilhem's bleak inventory of the "lifeless" scenery of modern French philosophy, it strikes us as an interesting enterprise to fathom the ways in which later stages of the French discussion have indeed incorporated the ever-rising pertinence of biological frameworks into the social and human sciences, in political philosophy, in technology etc. If, from Canguilhem's point of view, the intellectual panel of the 1940s may have revolved around the blatant absence of any philosophy of life or philosophy of biology, the 1950s and 1960s may justly be perceived as an age that witnessed the assimilation of biological models and principles into social theory (Erdur, 2018).[2] The pertinence of the biological model, in turn, became fully transparent in the arena of modern French philosophy under the influence of the research in epigenetics conducted by François Jacob, Jacques Monod and André Lwoff, who were even awarded the Nobel Prize in physiology in 1965. Against this backdrop, it is exciting to see the rise of a systematic interest in and a paradigmatic significance of biological rationality within the genesis of

[1] See the definition and the polemical implications of this term in the editors' introduction to this volume.

[2] Derrida (2016), for that matter, reconstructed the linguistic idea of the text (or *écriture*) as the common paradigm of the biological and the human sciences at a point when "structuralism" prevailed on both fields from the 1950s on and way into the 1960s.

philosophical thought in France in the 1950s and 1960s (apart from Canguilhem's notable role in this context, one might evoke the writings of Maurice Merleau-Ponty, Jacques Lacan, Raymond Ruyer, or Gilles Deleuze for ample substantiations of this "vital turn"[3]).

In any attempt to map the concrete philosophical issues that might lend a structure to such a Franco-German debate, it may come as a surprise to find that, whereas German philosophical anthropology accommodates within its boundaries the project of a *philosophical biology*, it is, on the contrary, the formula of a *biological philosophy* that prevails in the French landscape, spearheaded in this regard to a certain degree by Canguilhem himself. The considerations that I wish to unfold in the following article tackle this terminology in an effort to question whether these different wordings – "philosophical biology" as opposed to "biological philosophy" – are, in effect, interchangeable or rather indications of a truly conceptual divide. In that case, the "close encounter" between the paradigm of German philosophical anthropology and the turn(s) towards a philosophy of life in contemporary French thought is likely to reveal itself as a complex history both of affinities and schisms, neither of which are anything but superficial.

I will argue that a conceptual split between the model of a "philosophical biology" (whose most concise configuration can be derived from Helmuth Plessner's version of philosophical anthropology), as opposed to the framework of a "biological philosophy" (espoused, most tangibly, by Georges Canguilhem), indeed exists. This distinction captures two different visions of the ways in which the empirical sciences are thought to tie in with (a) philosophy (of the sciences): Whereas the conception of a "philosophical biology" posits a *correlation* between the phenomenologically accessible *positional form* of living phenomena and their material structure of *organization* – mirroring both the compatibility between biology and philosophy and the latter's transcendental supremacy – the idea of a "biological philosophy" seems to be driven by the attempt to reinscribe the specificity of the living (its productive normativity) both into the life sciences *and* into the very fabric of a philosophy which thus, according to Canguilhem, would be transposed into a peculiarly "vital rationalism" (see Canguilhem, 2000).

Spread out between, on the one hand, the historical protagonists of philosophical anthropology in Germany, and, on the other hand, the relief of classical French authors such as Foucault, Lacan, Canguilhem, Deleuze or Raymond Ruyer, whose positions might all be cautiously subsumed under the banner of a modern "critique of humanism," a wide-ranging network of mediators unfolds: an intriguing

[3] Claire Colebrook's monograph on *Deleuze and the Meaning of Life* (Colebrook, 2010) is a fine example of the prolific focus of recent contemporary research projects geared to the dimension of a "biological rationality" in the works of the above-mentioned authors (whom I have evoked here with a view to their international aura, and as the phalanx of a broader current in modern French thought). Colebrook succeeds in arguing that certain major figures of thought deployed by Deleuze entail a vitalist implicature: to the extent that, in modern French thought, "life" has generally not been appreciated as a relevant category for the discouse of philosophy, it is all the more important to unearth the implicit repercussions of this concept for the sake of a better understanding of the philosophical approaches pursued by Deleuze and others.

historical mesh that comprised biologists, physicians, neurologists, psychologists, anatomists, paleoanthropologists and representatives of a large variety of related disciplines. Out of this vast and sprawling prism of "life scientists" interlinking modern German and modern French philosophy, my text will zero in on three exemplary figures. It will tackle – in this order – the theorem of "ancestrality" expounded by the paleoanthropologist Hermann Klaatsch (1863–1916), Paul Alsberg's (1883–1965) hypothesis of "body disconnection" as the driving force of human evolution, and the concept of "neoteny," coined by the Dutch comparative anatomist Louis Bolk (1866–1930) as a token of the physiological "premature natality" of the human foetus. Traces of a significant reception of precisely these three researchers and their major hypotheses can be discovered both in the writings of Scheler, Plessner and Gehlen, and in the works of several prominent French twentieth-century philosophers, who placed their ideas (as did the philosophical anthropologists) at the intersection of philosophy, biology, and the human sciences. For the sake of brevity and a more concentrated type of argument, I will not examine the impact of all these three authors on the two fields of German philosophical anthropology and on modern French (biological) philosophy. Although I am indeed going to sketch a brief "map" of the influences of this entire group of biologists – Klaatsch, Alsberg and Bolk – on the constellation of philosophical anthropology, I will restrict myself to the exemplary role of Louis Bolk when it comes to the tableau of twentieth-century French philosophy – which, what is more, I will only tackle in the representative case of Jacques Lacan.

1 Philosophical Biology Vs. Biological Philosophy

It is historically interesting to witness the American philosopher of biology Marjorie Grene inscribe the conception of Helmuth Plessner into her vision of a "philosophical biology," systematically expounded in her book *Approaches to a Philosophical Biology* (1968).[4] In 1966, Grene had even devoted a meticulous paper to the category of "positionality" that is, indeed, a gist within Plessner's entire philosophy of nature (Grene, 1966), thereby setting the stage for the intense reception of Plessner's ideas that is (noticeably) beginning to take hold only today, after the English translation of his major monograph, *Die Stufen des Organischen und der Mensch*. In fact, we can immerse ourselves into an original passage from Plessner's book to unearth both the term and the effective function of a "philosophical biology." The passage finds its context in a subchapter entitled "The need for a revision of the Cartesian dichotomy." Drawing on the contemporary biological debate about the "life plan" (Jakob von Uexküll, Erich Becher), Plessner anticipates the agenda of furnishing the "vital categories" (*ibid.*, 60 f.) that turn out to be irreducible in the interrelation

[4] See the precious investigations of Grene's thought by P. Honenberger and P.-O. Méthot in this volume. For a concise account of Grene's relationship with Plessner, see Honenberger (2022) (forthcoming). See also Krüger (2019, 28–31) for an appraisal of Grene's reception of Plessner.

between the organism and its environment, without being materially rooted in either end of this dynamic dyad. In terms of Plessner's philosophical argument, this move is delicate because what he envisages here is a claim to a "quasi-transcendental" (cf. Krüger, 2019, 35) stance which, however, bypasses the classical Kantian option to ground the categorical forms in the ultimate unity of transcendental apperception. Hence, Plessner distinguishes the *function* of the (vital) categories from their aprioric *form* that seems to be inextricably intertwined with transcendental consciousness.[5] As far as I can see, one would not be misguided in extracting the following core from Plessner's strategy: according to Plessner, the relation between the individual organism and its correlative environment hinges upon *a priori categories*. These categories, however, are not grounded in the synthetic unity of transcendental apperception (more simply put, in the *conscioussness* of the subject that experiences this relation). Instead, they are relational functions that are wound up in the concrete interaction of an organism with its environment. Interestingly, however, Plessner will go on to demonstrate that the facilitation of the relation between organism and environment is in itself a peculiar activity performed by living entities – who, thus, turn out to be indeed the "subjects" whose spontaneity provides the source from which the "vital categories" arise. Yet, *as* subjects the living organisms present themselves as being inextricably decentered and immanently bound by their environments, which is why Plessner emerges with a conception of transcendental subjectivity that is decidedly transformed in comparison with Kant's idealism. Plessner achieves an emphatic "vital turn" (Ebke, 2014; Fischer, 2018) in the light of which the "subject" experiences its own position as one of radical immanence, namely within the life process as a whole.

This preliminary clarification finally enables us to do justice to the proper status of what Plessner designates as a "philosophical biology:"

> The task of a philosophical biology as the science of the essential laws of life, as well as of the foundational discipline of a possible animal "psychology" lies in the systematic grounding of such vital categories. Closer observation will show that the life plan – if it is actually worked out to mean the spheric unity of living subject and counterworld and not only the individual stimuli and reactions by which this unity takes on tangible contours for the sense perception of the experimenter – also represents the foundation of those relationships between subject and world that determine the former's consciousness (Plessner, 2019, 61).

With a view to this vital correlationalism – that defines categories as "bridges" between the individual organism and its environment but also, as it were, as expressions of their reciprocal entanglement (cf. Schürmann, 2017, 66 f.) – Plessner goes on to propose "process" and "type," "individuality," development in terms of "aging

[5] "Ought we then to dismiss entirely the notion that there are preconscious a priori forms, categories of existence, vital categories belonging to to the deeper layers of existence of the bearers of life – that is, organisms (understood not as existing objects but as living subjects), upon which the togetherness and cooperation of the organism and its environment rest? They would in any case have the value of categorical functions, as they, while neither being taken from the counterworld [*Gegenwelt*] nor applied to the counterworld by the living subject, determine the structure of this counterworld along with the structure of the living subject that fits into it"(Plessner, 2019, 61).

and death," systematicity and "self-regulation" as a concrete ensemble of the "vital categories." For the sake of this topic, it is important to underscore that Plessner does not, however, associate the reality of these categories with biological facticity as such. His project of a "philosophical biology" does *not* coincide with a biological empiricism in the sense that the identification and description of the categories could be left to the science of biology. This represents, in fact, a significant observation because, as I will larify at a later stage of this essay, Plessner introduces a clear separation of philosophy from biological discourse: instead of transforming philosophical biology into an empiricism, the empirical science of biology is "framed" and even grounded in a generically philosophical description that, in Plessner's methodology, takes the shape of a specific constellation of phenomenology, hermeneutics and even a dialectical approach (see Krüger, 2019).

Regarding this profile of a "philosophical biology" in Plessner's specific modulation, we are provided with ample clues as to why this concept is anything but a synonym for the idea of a "biological philosophy" (*philosophie biologique*). Ever since Camille Limoges introduced this wording in the prominent context of his preface to the sixth volume of Georges Canguilhem's *Œuvres complètes*,[6] it has inspired contemporary scholars, mostly with a firm interest in Canguilhem (Bognon-Küss & Wolfe, 2020), to explore the modern constellation of philosophy and biology with the notion of a "biological philosophy" as a heuristic thread. It is true that Canguilhem himself embosses this terminology at various instances in *The Knowledge of Life* (English translation: Canguilhem, 2009) and also, prominently, in his article about the situation of a "biological philosophy" in France which he contrasts rather delicately to the philosophical scenery in twentieth-century philosophical thought in Germany (Canguilhem, 2015, 311). In fact, Limoges contends that the vision of a "biological philosophy" is the centre of gravity within Canguilhem's entire trajectory, outshining and incorporating both his turn towards his concrete studies in medicine (including its history) and his affiliation with Bachelard's model of historical epistemology (Limoges, 2015, 31).

What, then, is the precise function and status of "biological philosophy" along the lines of Canguilhem? If we dwell for a moment upon a pivotal observation formulated by Limoges, biological philosophy presents itself, in Canguilhem, as an immanently normative project to the extent that it insists on the elementary significance of *valorisation* for the life process of the individual (Limoges, 2015, 35). Plausibly, Limoges ties in with Canguilhem's insistence on *negative* values and *negative* evaluation as the key to an understanding of the generic phenomenon of life. According to Canguilhem, every desire [*besoin*], already on the most primitive levels of animal life, manifests the tendency of the living individual to select and, simultaneously, to exclude certain elements in its environment; desire itself is the expression of a preference which, as such, is intrinsically a matter of (establishing

[6] Limoges (2015), 25 f. Limoges paraphrases Canguilhems "biological philosophy" as a the position that "tout vivant se voit reconnu comme centre d'activités polarisées, acteur de valorisations, sujet de sa propre normativité" (*ibid.*, 30).

and pursuing) *values*.[7] In other words: every living entity positions itself in its own milieu in operations of preference and avoidance, of affirmation and rejection. The living perceives and pursues values, positively appreciating situations in which its own generative spontaneity is, under the given circumstances, only minimally (if at all) impaired (which amounts to an understanding of what being "healthy" means for the living individual). In commenting on passages from Canguilhem's writings that clearly substantiate his perspective on valorisation, Limoges pinpoints the relevant moment where the explicitly "biological" dimension of Canguilhem's argument spills over into the larger (or, for that matter, narrower) problem of "judgment" (Limoges, 2015, 35 f.).

The implications of Canguilhem's association with the semantic field of judgment as the generic structure of "thought" itself to the activity of biological normativity are indeed both far-reaching and delicate. And it is here that we find the genuine *enjeu*, the critical stake of Canguilhem's philosophical outlook. As it turns out, his primary concern is to reinscribe the originary quality of life (as valorisation) into a philosophical rationalism that bases itself upon a radical neutralisation and, as it were, a deflation of the living. Canguilhem's vitalism goes out to an academic positivism (in early and mid-twentieth century France, that is) that holds sway both in philosophy and in the natural and human sciences which, in turn, furnish the problematic paradigm of philosophical reflection itself. This is why Canguilhem's focus on biological normativity is potentially explosive for the excessively positivistic mindset that had been dominating the philosophical landscape in France, especially since the rise of a sociological scientism in the nineteenth century.

At this juncture, we are prepared to grasp the notion that a "biological philosophy" is, indeed, an intrinsically critical, that is *normative* enterprise which, far from construing a specific framework for the description and conceptualisation of living phenomena, tackles the scientistic bias of those "rationalities" that lay claim to an objectivation of life. A "philosophical biology," on the other hand, reveals itself as a genuinely foundational approach towards living phenomena: in Plessner's case, it assumes the shape of a "quasi-biology," that is an open-minded phenomenology of the irreducible constituents of life that is not translatable into a natural science. Instead of exploiting this polarity any further, it shall be my strategy to appeal to it in the background of the pithier distinction between an *anthropology of desire* and an *anthropology of need* that I wish to distill from the historical constellation of biological debates in the twentieth century which supplies the theme of the following chapter.

[7] In making this point, Limoges draws on an unpublished lecture under the title "La Biologie" from 1942–43 that finds itself in the Canguilhem archives, located at the CAPHÉS in Paris. See (Limoges, 2015, 35).

2 The Sources. Klaatsch, Alsberg, Bolk

It is a first milestone of this text to interlink the philosophical anthropologists from Germany with a few of the prominent twentieth-century French philosophers who referred to the same "biological" authors as they did. Let us begin this analysis with Hermann Klaatsch, whose *aperçu* that, among the variety of biological forms, man stands out as a "dilettante of life" was famously quoted by Max Scheler in his essay *Zur Idee des Menschen*.[8] Having been a professor of human anatomy in Heidelberg since 1895, Klaatsch, encouraged by his Heidelberg colleague, the prehistorian Otto Schoetensack, embarked on an expedition to Australia in 1904 which was to last for three years, having been expected to yield clues to an issue that had been controversially discussed in the aftermath of Darwin's *Origin of Species*. Klaatsch's hope was to shed light on the question of the geographical continent that might most likely have been the area of origin of *homo primigenius*, that hypothetical prehistoric man whose emergence was speculated to have been the starting point of anthropogenesis in the sense of the evolution of the recent species of *homo sapiens*, as opposed to the anthropoids.[9] Klaatsch's expedition to the Aborigines was supposed to furnish proof of Schoetensack's central assumption, who had identified Australia as the initial territory of anthropogenesis.[10] In anthropological terms, the summation of the anatomical and morphological comparisons conducted by Klaatsch among several populations of Aborigines is contained in his book *Der Werdegang der Menschheit und die Entstehung der Kultur (The Evolution of Humanity and the Emergence of Culture)*, published posthumously in 1920, but finalized already in 1914. The book's gist consists in the argument that certain morphological traits of *homo sapiens* are not explicable on the basis of his presumed direct kindredship with higher primates, but rather, on the contrary, the late indications of a primitive (and hitherto undiscovered) prototype from whom the line of the anthropoids, that is to say the higher primates, and the line of *homo sapiens* had evolved independently of one another. "In *one* point of his organisation, man tends to tie in with one particular form of apes, while in *another* point he resembles another form. We must not consider any species of primates now living categorically as being man's closest next of kin" (Klaatsch, 1899, 337, my translation). Klaatsch construes a polygenesis of *homo sapiens*, supposing him to have descended in part from the higher primates and in part from an archaic form, from which the evolution of the higher primates, too, was thought to have sprung, even though higher primates also featured characteristics

[8] In English: Max Scheler, "On the idea of man" (1915), translated by Clyde Nabe, in: *Journal of the British Society for Phenomenology*, 9/3, October, 1978, pp. 184–198.

[9] While Darwin had supposed this territory to be located in Africa, Haeckel, in his *History of Creation* (1868, English translation 1880), placed his bet on Southeast Asia as the more likely spot – a supposition that was proven true by means of the findings of fossils at the banks of the Solo River in the East of the island of Java in 1891.

[10] Due to health issues, Schoetensack himself had to resign from the expedition, which had been explicitly initiated by him. Klaatsch, as his friend and colleague, stepped in for Schoetensack on the journey. See (Erckenbrecht, 2010).

that were entirely different from those of *homo sapiens*. Therefore, Klaatsch underscores the atavistic traits in the morphology of *homo sapiens*.[11] For instance, he considers the pentadactyly of *homo sapiens* to be a feature that would connect him more closely with Eocene mammals (such as bats) than with primates, and in the realm of the latter, in turn, at best with prosimians. Another example pointed out by Klaatsch are the molars of the human maxilla, which cannot be found among anthropoids, but, on the contrary, among Eocene carnivores such as sabre-toothed cats. In short: *Homo sapiens* is described by Klaatsch as a species in whose form atavisms have persevered which do not show any signs of the elaborate functional specialization that is typical of anthropoids. In Plessner's *The Levels of Organic Life and the Human,* we can find a comment on the philosophical dimension of this thesis of Klaatsch's, where Plessner assesses it as one variant of the "positive modification" of a naturalistic theory of human culturality (Plessner, 2019, 290). And it is indeed the case that, in *The Evolution of Humanity and the Emergence of Culture*, Klaatsch puts forward the argument that man's culturality has been rendered possible precisely through a certain internal dephasing in the genesis of *homo sapiens*: namely the non-simultaneity of the specializations that he shares with higher primates, and the atavisms which have come down on him in the prehistory of the most ancient tetrapod vertebrates (Klaatsch, 1920, 17).

The bacteriologist Paul Alsberg, who had emigrated from Nazi Germany to Great Britain in 1934 and later settled down as a physician in Berlin, finds himself confronted with quite a similar research question to the one Klaatsch had inquired into. While Klaatsch traces back the specific condition of the genesis of the cultural expressivism of *homo sapiens* to the latter's relative resistance to organic specialization and differentiation, Alsberg, in reverse, raises the opposite question, namely how the survival of a living being that seems so biologically puny in terms of its organic equipment had been possible in the first place. Alsberg's answer attaches itself to the poignant hypothesis of an inverted nexus between man's organismic structure and the technical mode of his form of life. In contrast to those biologists who conceived the emergence of technical and artificial procedures within human action on the dynamic basis of an "organ projection" (Alfred Espinas), Alsberg specifies the role of technicity not as an extension of the organic body. On the contrary, he interprets the morphological constitution of the human body as the product of a technical intervention. In his book *Das Menschheitsrätsel* (English translation 1970) from 1922, Alsberg distinguishes the somatic adaptation of the animal organism to its environment from the external origin of the very means that man has to operationalize in order to obtain an adaptation to his environment in the first place.

---

[11] There is, by the way, a remarkable reference to Klaatsch in Deleuze's and Guattari's *Thousand Plateaus* (Deleuze & Guattari, 2005, 172) in the context of the authors' attempt to conceptualize the morphology of the human organism as an effect of "deterritorialization"of animal forms. Apparently, Deleuze and Guattari came across excerpts of Klaatsch's book on the basis of the French translation of Hans Kraemer's edition of canonical texts from the tradition of natural history "Weltall und Menschheit" (1900).

This circumvention of the body is presented by Alsberg is being at the same time the cause of its "atrophy:"

> Primeval man inherited from his predecessor, and hence originally possessed, a perfectly protective bodily outfit such as would have guaranteed his survival. However, when he took to using artificial tools for defence and other purposes instead of resorting to his natural protective means, he was no longer in urgent need of, and therefore neglected, his original adaptive equipment, which thus became liable to subsequent deterioration. Our theory provides a simple and yet adequate explanation of the strange contrast in the appearance of the well-equipped animal body and that of the non-equipped human body, in that their divergent development is here being taken as the direct result of two contrasting evolutionary schemes – that of the Animal growing its adaptive means within the body, and Man taking them from without. While the animal scheme makes the body extremely fit to meet the exigencies of life, the human scheme leads the body into a state of extreme unfitness (Alsberg, 1922, 34 f.).

Indeed, in Alsberg's account, the dynamic principle of anthropogenesis and the evolutionary principle of the animal form drift apart: the adaptation of the body via successively refined specialization on the one hand, the body's suspension by means of technical artifacts on the other hand. It is here that we can catch a first glimpse at what moved Arnold Gehlen to portray Alsberg as a an "ingenious outsider"[12] in his lexicon article entitled *Philosophische Anthropologie*: "No matter whether I suspend the body by means of the fire in the tile stove, a fur coat or an electric heat pack, it is all essentially the same thing: in all of these cases, I am using an extra-bodily means for the sake of the suspension of the body, in other words: a tool" (Gehlen, 1983c, 238; my translation).[13] These passages testify to an upheaval that Gehlen will rearticulate in his own conception by way of uncoupling the "drive" from its "fulfilment." Artificial substitutes bypass the biological deficiencies of the human species; yet they operate in such a way that what fills the void will inevitably transform, too, into the object of biological desires.

We can conclude this brief parcours of distinctly *biological* conceptualizations of anthropogenetic difference with a sideglance at Louis Bolk's category of neoteny.[14] The Greek term *νέος* means young, the verb *τείνειν* signifies "to stretch" or "to expand." This word formation contains what the Dutch anatomist Louis Bolk identifies as the decisive trait of ontogenesis (Bolk, 1926): ontogenetically and anatomically speaking, man is not only a retarded ape, but, on close inspection, even a premature apish foetus whose embryonic genesis has been precociously externalized in its final stage (*ibid.*, 471). In contrast to Klaatsch and Alsberg, Bolk, however, does not aim at a phylogenetic argument: countermanoeuvring any Darwinian explanation, his considerations concentrate not on mechanisms of variation and selection pertaining to the level of entire populations and their environmental niches, but rather on intrinsic modifications in the morphology of individual organisms.

---

[12] Arnold Gehlen, „Philosophische Anthropologie", in: Id., *Gesamtausgabe,* Bd. 4 (*Philosophische Anthropologie und Handlungslehre*), p. 238.

[13] In the absence of an official English translation, I am working with my own translation here as well as in all the other cases where no official English translation is available.

[14] For an account of the impact of Bolk's biological theories on modern philosophy cf. Verhulst, 1993.

Based on facts such as the overall scarce mass of body hair in *homo sapiens* (the absence of pelt layers), the vertical position of teeth rows and jaw arches, the centrally subcranial localization of the foramen magnum or the weak extent of skin pigmentation (*ibid.,* 468), Bolk illustrates his point of view according to which morphological particularities that tend to manifest themselves only temporarily with other primates persevere, ontogenetically, in man. In a similar vein to Klaatsch, albeit not under phylogenetic aspects, man, in Bolk's conception, comes into view as an atavistic figure (Verhulst, 1996), in which that seems to endure which, in the play of evolution, ceases in other, seemingly closely related species. However, in Bolk's conception, as opposed to Klaatsch's approach, the idea of a certain coagulation of specific groups of features – which therefore seem strangely divested from the evolutionary cycles that bind other primates – crystallizes in such a way that it transforms into its own opposite. For it is true that the organic form of that living being whose morphological evolution proceeds in a protracted way cannot be but a precocious and thereby fragile, or, as Bolk has it, a "fetal" (Bolk, 1926, 468) form.

3 Philosophical Anthropology between an Anthropology of need and an Anthropology of Desire

3.1 Gehlen

In his text *Ein Bild vom Menschen* (1941; here cited as Gehlen, 1983a; English title: *An Image of Man*) Gehlen programmatically defines the agenda of his very own type of philosophical anthropology. The presuppositions made by Gehlen in this article are important to register if one wishes to understand his reading of the three authors I just surveyed:

> It is true that anthropology must not dismiss the individual sciences, such as morphology, psychology, linguistics etc., which content themselves with narrow questions about man. Instead, anthropology has to presuppose these sciences including their results. If, however, anthropology deems it reasonable to make "man" the object of its own research, this signifies a twofold claim whose importance is easy to register: on the one hand, the claim to prefer a single science (with verifiable, not with poetic statements) to those other partial sciences, or, if you will, to constitute it across them; a science about whose method, technique of questioning and selection of objects nothing is certain yet, because the tradition of philosophical anthropology is one of tendency rather than results. And secondly: the claim to seize, with one science, both "sides" of man, the psychic and the corporeal [*leiblich*] one or whatever we call them, whose habitual separation is, after all, not only a popular prejudice, but has become institutionalized in view of the coexistence of psychology and physical research on man [*Menschenkunde*] (*ibid.,* 44, my translation).[15]

[15] Again, I am employing my own translations from the German original in this context, in the absence of an official English version.

Clearly, the salient point of this quotation is Gehlen's definition of philosophical anthropology as a meta-science of the human sciences and the bio sciences; in other words, his conceptualization of philosophical anthropology, as he puts it elsewhere, as a "science of facts" (Gehlen, 1983c, 236), based on empirical procedures. According to Gehlen, there are two alternatives to this project: on the one hand, a reflective anthropology containing "metaphysical bullet-holes" (*ibid.*) or, on the other hand, an approach which, although "in its explicit intention it wishes to move within the boundaries of the emprical realm alone" (*ibid.*), in the end fails to do so in a consistent manner. It doesn't take much imagination to make sense of these allusions: Gehlen's critique aims, in the first case, at Scheler and, secondly, at Plessner. From Gehlen's point of view, philosophical anthroplogy needs to remain constantly in tune with the discourses of contemporary empirical science for two reasons: on the one hand for the sake of synthesising the outcomes of research processes as they occur in the particular sciences, and on the other hand in the name of generating a guideline for any categorical determination of man – a guideline in the sense of a substrate of all the empirical discourses synchronically at hand, which would be in a position to *systematically* raise the question of anthropological difference. As the quotations suggest, this question, in turn, would have to be theorised in a *unified* way, bypassing the Cartesian dualism and its modern legacy as represented in the dichotomy between psychology and physiology that has become so familiar to the human sciences.

Against this methodological backdrop, the mainly approving tone of Gehlen's references to Alsberg, Klaatsch and Bolk becomes entirely plausible:

> The specifically human physical organization had already been placed into a new view by an ingenious outsider, P. Alsberg, in 1922. Man alone, according to Alsberg, had suspended his organs from the struggle for existence [*Daseinskampf*] by means of tools, language and his ability to use concepts [*Begriffsfähigkeit*]. Language, too, is considered by Alsberg as an immaterial tool for the suspension of the sense organs. As far as the causes of this development in the direction of the suspension of the organs are concerned, Alsberg identified the deficient organic equipment of man – with regard to the anthropoid apes that are his next of kin – in terms of the absence of pelage, of the prehensile foot and the canines. Hence, Alsberg has been the first to capture the compensatory relationship between a biologically deficient organic equipment in man and his linguistic and actional intelligence, and he has drawn the methodical comparison between animal and man even before Scheler (*ibid.*, 238 f.).

In this quote, Gehlen's point that the dynamics between the biological deficiency of man and his "linguistic and actional intelligence" are of a *compensatory* character doubtlessly plays the crucial role. This is why the following features are the inaugural motives sui generis of a philosophical anthropology à la Gehlen: the relegation of the ontological clarification of anthropological differences to the means of empirical discourses (which a philosophical anthropology does not aim to thwart, but rather to synthesise); and a sharp sense of the fact that the core of the structure of human behaviour is formed by a compensation which "produces the incentive towards its own continuation" (ibid., 242), thus recursively deferring itself precisely to the extent that its own source is a "too little:" radical biological underdetermination. This is how the framework for Gehlen's fructification of Bolk's neoteny thesis constitutes itself. According to Gehlen, Bolk basically achieved nothing less than the empirical verification of Herder's much quoted formula picturing man as a "being of lacks" (*Mängelwesen*). In *Der gegenwärtige Stand der anthropologischen Forschung* (English title: *The Current State of Anthropological Research*), Gehlen prizes Bolk's hypothesis about factors of hormonal inhibition as being causally responsible for the fixation of the quasi-embryonic level of the development of the human organism as an – again – "ingenious theory" because, in Gehlen's view, it had been able to "link human particularities that, hitherto, had always remained unexplained with the anatomically exceptional position [*Sonderstellung*] of man" (Gehlen, 1983b, 119; my translation). Because of its accomplishment in creating a link of this type, Bolk's conception, in Gehlen's evaluation, "after it had seemed to fall into oblivion for a while, is today again acknowledged as one of the most significant anthropological hypotheses" (*ibid.,* 120).

It is a striking feature of Gehlen's reflections that he renders the groundworks conducted by Alsberg and Bolk – there is no comment on Klaatsch in his writings, though – more precisely in terms of an anthropology which, in turn, conduces to the elucidation of these empirical findings. This is an anthropology that Gehlen himself clearly identifies as one of lack and desire. In his encyclopaedia entry entitled *Philosophische Anthropologie* (1971; *Philosophical Anthropology*), Gehlen introduces the stances of Alsberg and Bolk successively in order to radicalise them, substantialised by some considerations of Adolf Portmann on man's "extra-uterine year," in the direction of his own basic anthropological thesis according to which man is an *animal corrompu*:

> The book *Man. His Nature and Place in the World*, published in 1940, seems to owe its publicity to the yield of the view to interpret man on the basis of action. In so doing, the theme of the dualism (body-mind) had, firstly, been bracketed with the help of a psychophysically neutral (an expression of Scheler's) fact; secondly, man could now be comprehended against the background of his real doings, namely the intelligent transformation of randomly discovered circumstances of nature into that which serves the purposes of life. Hence an elementary concept of culture arises. And thirdly, this approach made it possible to incorporate the existing findings about the organic deficiency and the 'embryonic' habitus (Bolk, Alsberg etc.) into a coherent view. For it turned out to be perspicuous that a 'being of lacks' (regarded in comparison with higher, specialized animals) should be able to subsist, already on the biological level, only through intelligent transformations of its environment. This is how man appears to be a being of culture already by nature and his exceptional position rooted in this situation. [...] In describing action as an activity that ameliorates on the factual level based on perceived and re-felt [*rückempfundenen*] successes or failures, thereby producing the incentive to its own continuation, philosophical anthropology was placed into a relationship with a theme that later became known as "cybernetics" (Gehlen, 1983b, 239 f.; my translation, *TE*).

At this point, it is plain to see why the centre of Gehlen's anthropology can be addressed as a theory of desire. If the core of his conception is the idea of *action*, we have to specify this category as implying *both* an "activist" excess *and* a need – a gap which forces us to act *here and now*. Man is not only a cultural being by nature: he also naturalises his own artefacts and transforms them into objectivities which, on their part, then become the targets of his drive to further cultural production. The barb of this productivity, then, is *lack*: biological lack in the sense of a blank space which drives the play of culture into a situation of permanent hysteria, constantly transposing compensation into over-compensation. Gehlen distinctly speaks of an "*surplus of drives* yielding enormous amounts of energy that cannot be accommodated in immediate species conservation" (*ibid.*, 241).

If at this point my contribution moves on from Gehlen's considerations to those developed by Scheler, this step does not so much amount to a mere shift of emphasis within a register of philosophical anthroplogy that would comprise both their positions, but rather a leap from a philosophy of *desire* to a philosophy of *need*.

3.2 Scheler

In contrast to Gehlen, Max Scheler did not deem Paul Alsberg an ingenious anthropologist, but rather, as his comments on Alsberg's hypothesis of organ suspension suggest, as a rather brute "disciple of Schopenhauer's" (Scheler, 1962, 98). In *Man's Place in Nature* (published posthumously in German in 1928), Scheler classifies Alsberg's approach as a "negative theory of spirit:"

> [Alsberg] has expanded Schopenhauer's doctrine by asserting that "the principle of humanity" consists exclusively in man's being capable of releasing his organs from the struggle for survival as an individual or as a species in favor of developing tools, language and concepts. The latter are explained in terms of the principle of canceling sensory organs and functions [...] Alsberg refuses explicitly to define man in terms of spirit and reason. Reason, which he identifies erroneously – as did his teacher Schopenhauer – with discursive thought, especially with the formation of concepts, is for him the consequence of language, not its source. Language itself is conceived as a nonmaterial tool whose purpose is to eliminate the work of the sensory organs (*ibid.*, 98 f.)

Where Gehlen finds a reason to interpret Alsberg's work as a genuine contribution to a modern empirical anthropology, Scheler sees in it a circular argument: namely an instrumentalist conception of language as a "tool" that defines noumenal syntheses as epiphenomena and not, on the contrary, as conditions of possibility of linguistic representation. While Gehlen pays tribute to Alsberg's naturalistic strategy as an exemplary footstep in the direction of a unified, non-dualistic account of man – which, in Gehlen's understanding, had historically never been fully achieved – Scheler criticizes Alsberg for a *petitio principii*: his shift towards a cause for the phenomenon of language that would be entirely inherent to nature already benefits logically from what it aims to explain in the first place. For Scheler, it is not an option to expound the "suspension" of the natural body itself as an internal operation within nature. Rather, this suspension presupposes something other than organic forces in and by themselves – something that Scheler apostrophises as the spirit's "own nature and autonomy" (*ibid.*, 97):

> Alsberg, too, must answer the questions: What is it that produces the atrophy of the organic functions? What is it that invents tools both material and symbolic? Are organs really discarded, and this only for the sake of the values and goals characteristic of animals, for the sake of survival? The concept of need[16] alone will need not do as an explanation. (It was already vastly overrated by Lamarck to explain the formation of organs) (*ibid.,* 101).

Indeed, the antithesis favored by Scheler in his discussion of Alsberg boils down to the outdoing of a dialectics of lack by way of a philosophy of need. For Scheler, bypassing the body *via* the body implies a moment of numinosity. Yet, it is important to see that he does not think spirit in the tradition of occasionalism as a nous ex machina, but rather in the structure of a *folding*:

> It is precisely the *spirit* that initiates the repression of instincts. It does so in the following manner: Subject to its own ideas and values, the spiritual "will" withdraws from the opposing vital impulses the images necessary for action. At the same time, it lures the drives with a bait of appropriate images in order to coordinate the vital impulses so that they will execute the project set by the spirit. The process that consists in the inhibition and release of vital impulses through the spirit will I call "direction "(*Lenkung*). The process of presenting ideas and values, which are then realized through the impulses, I call "guidance "(*Leitung*). There is one thing the spirit cannot do: it cannot generate or cancel the instinctual energy; it cannot enlarge or diminish it. It can only call upon energy complexes which will then act through the organism in order to accomplish what the spirit "wills" (ibid., 101 f.).

Thus, it is an *indirect* virulence of spirit in life that is indicated here. If the spiritual moment is able to "switch off," as it were, the vital impulses, this happens precisely through solliciting configurations of those impulses which guide and conduct their very dynamics. In principle, life is a drive (*Drang*). However, strivings emerge within this drive that open life up from within itself onto its other – the dimension of spirituality – while they also form the only possible, namely the indirect mode through which spirit may debouch in the midst life: such a mode can be described as one of "folding" or, more technically, as implication. Thus, the biological constitution of man is no longer determined by the play of lack and its displacement into a type of compensatory action which desires its own artefacts. One might perhaps speak of a *striving* within living nature, a focus within being onto non-being that comes to bear *only* within this very focus. This striving may well be characteristic of what one may sketch very broadly (and, by the way, in an exact reversal of the Hegelian categories, at least if one follows the way in which Kojève reads them) as a philosophy of "need" in opposition to a philosophy of "desire."

[16] At this point, it needs to be emphasized that, in the terminological framework of this essay, the concept of "need"should precisely be substituted with the concept of "desire". It is my claim that Gehlen's anthropology, which turns out to be closely and implictly affiliated with the field of modern French philosophy that we are goind to investigate, is a philosophy of "desire," whereas the more plausible term to portray the approaches of Scheler and Plessner is the concept of "need."

3.3 Plessner

It is interesting to see Gehlen unfold one point of the critique that he levels against Plessner's variant of philosophical anthropology in direct recourse to the ideas of Louis Bolk and Adolf Portmann. As Gehlen argues, Plessner's theorem of "excentric positionality" does not sufficiently account for the peculiar *form* of the human body. In Gehlen's view, it is just not the case that, as far as man is concerned, we are faced with "the closed form of organization being taken to the extreme" (Plessner, 2019, 270). For Gehlen, Bolk's concept of neoteny demonstrates the "exceptional status" of the human organism already on the level of its morphological constitution. Because Plessner does not locate the momentum of anthropological difference on the level of the form of the factual organization, instead linking it with the level of the phenomenologically pertinent form of *positionality* (with which the "organization" is correlated), his terminology fails to register the sharp problem of the way in which human nature spontaneously disrupts into artificiality (see Gehlen, 1987). It is here that Gehlen indeed hits on a phenomenological distinction which profoundly sets Plessner's approach apart from Gehlen's own position: that is, precisely, the difference between the forms of the *organization* and of the *positionality* of living entities. In fact, the description envisaged by Plessner by means of this separation draws attention to the problem that a living entity never coincides materially with its own functional organization. On the contrary, the living always needs to equipoise with regard to the structure of its organization – due to the circumstance that every living being is structurally placed or rather *displaced* within itself, which means that it is detached from its own boundaries that mark it off against its environment.

As far as the excentric positionality of man is concerned, Plessner exposes the fact that the chasm between the "organizational" and the "positional" form, which is quasi-spontaneously mediated within centric positionality, is itself experienced, thereby transforming into the very "problem" that confronts the living being (man, in this case). In this light, Plessner employs a twofold terminology of "centering" and of "excentricity" in order to do justice to the singularity of human phenomena. On the one hand, man participates in the "centric" form of animal organization: he does not possess a more complex organic "design" than what is represented in the centric situation of the animal organism. However, what ultimately sets man apart from the animal is precisely the ecxentric structure of his *positionality*. As opposed to the level of centric positionality, this ambivalence entails a fortiori a pressure towards mediation, of generating correlations and of counterbalancing sharply divergent directions in behaviour. Yet, in excentric positionality, these possible and at the same time necessary "centering" operations are acted out precisely to the effect that they are (and have to be) carried out by living beings that are "absorbed in nothingness" (*ibid.*, 292). These entities "centralize" their own behaviour only because they constitutively move within a sphere that itself know no centre, but rather marks a radical vacancy and exteriority. Plessner's term for this sphere that is determined by nothingness, but which out of nothingness renders any determination possible is the concept of the "world" as opposed to (biologically natural and/or artificial) "environment" (*Umwelt*) (*ibid.*, 294).

In this light we can comprehend Plessner's critical reception of the biological insights into human ancestrality, organ suspension and neoteny. Interestingly, it is in the preface to the second edition of the *Levels* that he discusses precisely these empirically validated and empirically relevant concepts en bloc, and he does so at a point that aims at the rejection of the main tenet of Gehlen's anthropology: the conceptualization of action as compensation.[17] In the logic of Plessner's argument, we would arrive, then, at a twofold critique of a reductive anthropology that narrows down the human position to the drama of biological lack and its own overflowing compensation qua action, but also of the scientistic naturalisation of the philosophical framework in which the question of anthropological difference can be posed.

Yet, what precisely is the alternative to this compensatory anthropology of desire that Plessner envisages? By and large, he pursues a path that is quite similar to Scheler's course: Plessner spells out an ontology of the intertwining of strata (in this case: strata of forms of organization and of positionality) that is largely modeled on Scheler's figure of the "folding" of numinous suspension within the drive of life. In Plessner's thought, we find a generic topography of the way in which the living is, as it were, "thrown back onto itself "on all of its "levels:" living things are thrown back upon the non-living, the organic upon the inorganic, the human upon the animal, spirit upon life, history upon nature ... Thus, the "lived body" (*Leib*) is precisely not a function of the body, but the very interlayer that links the organism with its milieu in such a way that the living being *has* its body *as* lived body (Plessner, 2019, 270–275). Analogically, excentric positionality does represent two moments simultaneously: both a rupture with the dimension of the "lived body" and the fact of its being folded into that very dimension. Excentric positionality earmarks a breakthrough into a beyond where the fundamental structure of life itself becomes reflexive and at the same time precarious. But still, the living (= human) beings that rise up into this beyond always already find themselves thrown back onto that which is given to them but in an unsurmountable exteriority – namely onto the registers of their body and their lived body, in the absence of which they would be unable to *live*.

Thus, contrary to Gehlen, there is, in Plessner's thought, no reversal of biological lack into artificial compensation, but rather, similar to Scheler, the folding of non-being into being (ultimately: of the environment into the world), with the crucial

---

[17] "The concept of action also avoids the fateful cleavage of human being into a bodily and a non-bodily region. Whether it is merely evaded and banished from view, as it were, is another question. If, like Gehlen, one wants to be an empiricist, one has the right to do just that. His theories are wellknown and can all be grouped around the notion of compensation, for which Herder provided the label "deficient being." Gehlen's skillful combination of Hermann Klaatsch's notions of the characteristic ancientness and relative lack of specialization of the build of the human body with [xvi] Bolk's ideas about retardation and fetalization, Portmann's about the extrauterine spring, and Scheler's about weak instincts, surplus drives, and world- openness add up to a creature to whom Herder's 'invalid of its higher powers' seems less fitting than my characterization of a combatant of his lower ones. Gehlen conceives of the *homo* species exclusively in terms of its potential to act" (Plessner, 2019, XXVII).

implication that being is insistent but at the same time constantly transcended. It is precisely this motion that I have been interpreting as the basso continuo of a philosophy of need. Whereas desire depends upon a structural lack which introduces into itself the generative, ecstatic and risky propensity towards lack's interminable banishing, need points not so much to the dualistic interruption as to the immanent neutralization of the dynamics of "life:" in Scheler's case as the ever heteronomous, spiritual "striving" of drives, in Plessner's case in terms of the argument that living beings that are excentrically positioned find themselves being thrown back onto the centricity of the vital, in which, nevertheless, they never vanish.

4 Fractionaries of an Anthropology of Desire from Lack. The French Constellation, with an Exemplary Focus on Jacques Lacan

The most comprehensive inventory of passages in which Jacques Lacan's œuvre signals subsurface or explicit recourses to Louis Bolk's theory of fetalisation has been supplied by Levivier (2011). According to Levivier, we are justified in saying that a direct appropriation of Bolk's guiding theorem is at work in the very epicenter of the formation of Lacan's philosophical thought, namely in his key text "The mirror stage as formative of the I function as revealed in psychoanalytic experience" (Lacan, 2005). As is well known, the initial step of the theory of narcissistic, but at the same time symbolically fractured genesis of subjectivity, which Lacan lays out in this text, consists in the hypothesis of a "jubilant assumption of his specular image by the kind of being – still trapped in his motor impotence and nursling dependence – the little man is at the *infans* stage" (*ibid.*, 76). Lacan aims at the "transformation that takes place in the subject when he assumes an image" (*ibid.*); the subject here being the infant identifiying "himself" in the mirror. His theme is the ecstatic relation that binds the still dependent and, as it were, dismembered body to the projection of its imaginary integrity and intactness. If, in this originary act of subjectivization, the "total form of his body, by which the subject anticipates the maturation of his power" (*ibid.*), presents itself to the subject purely as an image, as "*Gestalt*," the question arises how such a "destructive fascination with images" (Berz, 2016, 299) can possibly come to pass for specifically human beings in the first place, creating a situation in which the constitution of the self depends upon *imagines* which as such remain forever uncatchable. It is at this point that Lacan, surprisingly, comes up with the straightforwardly anthropological assumption that we must account for "the effect in man, even prior to this social dialectic, of an organic inadequacy of his natural reality" (Lacan, 2005, 77).

Indeed, his whole model is based upon the fundamental idea of a difference of hominisation as opposed to the ontogenesis of other higher mammals. The decisive moment is the time shift that comes to pass in the first "extra-uterine year" of the human toddler, as analyzed by Portmann: while in the first year of its life the human baby grows significantly faster than other new-born mammals, its development begins to falter at the close of its first year until it continues much more slowly in relation to the growth of other species of mammals. The consequences of this change of tempo, as it were, consist in the dilated duration of the toddler's organic dependence upon external providers and, closely related with this aspect, in a decoupling of the inner world from the outside. In this context, we can see quite well in what way Lacan resorts to Bolk's insight into the neoteny of the human organism: "In man, however, this relationship to nature is altered by a certain dehiscence at the very heart of he organism, a primordial Discord between betrayed by the signs of malaise and motor uncoordination of the neonatal months. The objective notions of the anatomical incompleteness of the pyramidal tracts and of certain humoral residues of the maternal organism in the newborn confirm my view that we find in man a veritable *specific prematurity of birth*" (*ibid.*, 78). Looking back on Lacan's text about "The family complexes in pathology" from 1938, we find a clarification of the circumstance that Lacan indeed signalises a direct connection to Bolk's theory of foetalisation where he evokes the term "dehiscence" [*déhiscence*] (Lacan, 1984, 34). Immediately after his mention of Bolk as the source of the motive of the precautiousness of human birth comes the following passage:

> Contingent on this developmental delay, the precocious maturation of visual perception takes on the value of a functional anticipation. The result of this is, on the hand, the distinct prevalence of the visual structure in terms of, as we have seen, such a precocious recognition of the human form. On the other hand, what receives, if I may say so, a decisive support from all of this, are the prospects of an identification with this form, which will constitute within man that absolutely essential imaginary knot which psychoanalysis, darkly and across inextricable contradictions, has nevertheless admiringly designated under the name of *narcissism* (ibid., 34 f.; my translation).

As highlighted by Peter Berz, what is surprising about Lacan's arguments (throughout diverse texts) is his tendency towards a biological grounding of psychoanalytically relevant processes: a remarkable rupture with the culturalistic and psychologistic axioms that used to prevail, and still do prevail today, in rivalling readings of Freud (Berz, 2016, 297).

We are almost compelled to pursue a parallel between the philosophies of Lacan and Gehlen at this juncture (Edinger, 2019). Because just like Gehlen, Lacan turns out to think in terms of a paradigm of lack and desire (Owens & Almqvist, 2018) which could hardly be explicated better than with the help of his concept of "dehiscence." Strictly speaking, Lacan's approach is aware of two points of divergence between *sutures* (see the technical meaning of the term "deshiscence"): Firstly the rise of the non-simultaneity between the temporal relations of the inner and the outer world of the organism, and secondly, what is more, the intrinsic disruption between the somatic "inadequacy" (Lacan, 2005, 77) of the organism and the psychic anticipation of its plenitude and potency that will forever remain anticipated and deferred. In this sense, a radical, basically twofold lack is at the origin of the constitution of subjectivity sui generis, which will primarily and in fact incessantly operate in terms of "phantasm" "(*ibid.*). The paradigmatic situation that renders the identification of the "I" possible is presented by Lacan as the establishment of a disruption that will never close upon itself: The "I" (in the sense of Freud's "ideal I," which will be terminologically stabilized by Lacan as "*le moi*" from 1953 on), the "I" is someone else, is always already an agent that differs from the one in which the I that speaks (as the "I" within language, "*le je*") attempts to coincide with itself.

This last point, by the way, the non-coincidence of "*moi*" and the "*je*" in Lacan's conception, already clears the view for the way that he will go on to deal with the anthropologically insatiable lack (and, at the same time, the barb) of the imaginary. For on the one hand, Lacan describes a pluralisation of the anthropological dimensions: Although the imaginary formation of the identity of the I is, as it were, primordial, the hypothesis of the mirror features knows another, entirely different implication, namely the fact that this imaginary structure becomes thwarted by the so-called symbolic order. While the imaginary relation, as it were, *singularizes* to the extent that it characterizes every individual on its own behalf, it is true that it also moves within a totalizing, linguistically mediated "shared world," a *Mitwelt*, to use Plessner's formula. Thus, the visual fixation of the mirror stage is refracted through the order of language, which in itself is constituted in terms of symbolic structures and anonymously functional regularities; inversely, though, the imaginary is constantly kept alive within the linguistic and socio-cultural practices of the community. While the result of this always already performed entry into the space of the "great Other" is a pluralisation – to the extent that the subject is never perfectly incorporated into the imaginary – this mediation of the imaginary with the symbolic, does precisely not signify, on the other hand, a de-radicalization of anthropological lack. On the contrary, we could rather speak of a shift of the modalities of anthropological lack and the site of its experience.

5 Conclusion

"Philosophical biology" and "biological philosophy" are by no means synonymous terms. The former defines itself as the systematic construal of a categorical framework that expresses the irreducible qualities of living beings, allowing to ground and to incorporate the descriptions deployed within biology itself. The latter, on the other hand, is a generically normative enterprise that aims at "vitalizing" the philosophical and scientific discourses about life to the extent that it re-introduces and re-inscribes into these rationalities the irreducible problem of valorisation (which, as such, is permanently obviated within science in the name of positivistic claims).

I have illustrated (rather than fully exploited) this antagonism by scrutinizing the ways in which a certain ensemble of biological anthropologies between, roughly, 1900 to 1920 (Klaatsch, Alsberg, Bolk), had been received by the German tradition of philosophical anthropology and by specific strands within twentieth century French philosophy, respectively. In narrowing down the focus on the exemplary case of Bolk, we have witnessed a certain alliance between the positions of Arnold Gehlen and Jacques Lacan, as opposed to the proximity that broadly unites the perspectives of Max Scheler and Helmuth Plessner. It would be adequate to discuss

Gehlen and Lacan as paradigms of a *biological philosophy* to the extent that they both reach out to a concrete biological grounding of their stances on anthropology. Interestingly, the views of Lacan and Gehlen, that are so intimately replete with direct adoptions from the biological anthropologies of their time, converge upon the idea of man as a being of lack, desire and symbolic compensation. On the other wing of this heuristic division, we find Max Scheler and Helmuth Plessner as advocates of a *philosophical biology* which, for various reasons, preserves its distance from the discourse of biology per se. In Scheler's case, I have observed an insistence on the autonomy of the qualities of the *spirit* which, according to Scheler, cannot plausibly be phrased in terms of an evolutionary naturalism (Alsberg). Against this background, philosophy is confronted with the task of supplying categorical descriptions of life that create bridges into the concrete discourses of biology, without ever being replaceable by them. This critical gesture is sustained in the work of Helmuth Plessner, who, on the basis of a demarcation between the positionality and the organization of the living organism, draws a permanent boundary between philosophy and the natural sciences. After all, it is a striking fact that we find no trace of a compensatory anthropology of vital desire in the philosophies of Scheler and Plessner. Instead, they seem to deploy a philosophy of need which, in Scheler's case, addresses the quality of the spirit to suspend the vital structures of desire, while, in Plessner's case, it panders to the idea that the human form is “thrown back upon” the subhuman levels of organic life.

We would not be justified in aligning the opposition “philosophical biology” versus “biological philosophy” too closely with the juxtaposition between the two types of anthropology that we encountered: on the one hand in Gehlen and Lacan and, on the other hand in Scheler and Plessner. But in the end, these attributions might conduce to a sharper orientation on a nascent research field (Charpentier et al., 2022) where the traditions of German philosophical anthropology and of modern French thought finally coalesce into a new, truly European panorama of the philosophies of life and man that kept the intellectual history of the twentieth century in suspense.

References

Alsberg, P. (1922/1970). *In Quest of Man. A Biological Approach to the Problem of Man's Place in Nature*. Pergamon Press.

Berz, P. (2016). L'imaginaire animal. In T. Ebke, S. Edinger, F. Müller, & R. Yos (Eds.), *Mensch und Gesellschaft zwischen Natur und Geschichte: Zum Verhältnis von Philosophischer Anthropologie und Kritischer Theorie* (pp. 275–301). de Gruyter.

Bognon-Küss, C., & Wolfe, C. T. (2020). The Idea of Philosophy of Biology Before Biology: A Methodological Provocation. In C. Bognon-Küss & C. T. Wolfe (Eds.), *Philosophy of Biology Before Biology* (pp. 4–22). Routledge.

Bolk, L. (1926). On the Problem of Anthropogenesis. *Proceedings of the Royal Academy (Amsterdam), 29*, 465–474.

Canguilhem, G. (2000). *A Vital Rationalist. Selected Writings from Georges Canguilhem*, Zone Books.

Canguilhem, G. (2009). *The Knowledge of Life* (S. Geroulanos & D. Ginsburg, Trans.), Fordham University Press.
Canguilhem, G. (2015). Note sur la situation faite en France à la philosophie biologique (1947). In *Œuvres complètes,* tome IV: *Résistance, philosophie biologique et histoire des sciences, 1940–1965* (pp. 307–320). Vrin.
Charpentier, A., Andrea, D. P., & Pagan, M. (2022). *Alternaturalismes. Dewey, Canguilhem, Plessner*. Éditions Rue d'Ulm.
Colebrook, C. (2010). *Deleuze and the Meaning of Life*. Continuum Press.
Deleuze, G., & Félix, G. (1980/2005). In B. Massumi (Ed.), *A Thousand Plateaus. Capitalism and Schizophrenia*. University of Minnesota Press.
Derrida, J. (1967/2016). *Of Grammatology*. Trans. Gayatri Chakravorty Spivak. Baltimore: Johns Hopkins University Press.
Ebke, T. (2014). Life, Concept and Subject. Plessner's Vital Turn in the Light of Kant and Bergson. In J. de Mul (Ed.), *Plessner's Philosophical Anthropology. Perspectives and Prospects* (pp. 99–110). Amsterdam University Press.
Ebke, T. (2017). Dilettanten des Lebens. Deutsch-Französische Brechungen der Anthropogenese in der Philosophie des 20. Jahrhunderts. In T. Ebke & C. Zanfi (Eds.), *Das Leben im Menschen oder der Mensch im Leben? Deutsch-Französische Genealogien zwischen Anthropologie und Anti-Humanismus* (pp. 265–338). Universitätsverlag Potsdam.
Edinger, S. (2019). Symbolische Individuation zwischen Philosophischer Anthropologie und Psychoanalyse. Die konstitutive Funktion der Symbolik bei Arnold Gehlen und Jacques Lacan. In T. Ebke & S. Hoth (Eds.), *Die Philosophische Anthropologie und ihr Verhältnis zu den Wissenschaften der psyche* (pp. 53–85). de Gruyter.
Erckenbrecht, C. (2010). *Auf der Suche nach den Ursprüngen. Die Australienreise des Anthropologen und Sammlers Hermann Klaatsch 1904–1907*. Wienand.
Erdur, O. (2018). *Die epistemologischen Jahre. Philosophie und Biologie in Frankreich, 1960–1980*. Chronos.
Fischer, J. (2018). Plessner's *Vital Turn*. Ekstatik der Exzentrischen Positionalität. In H. Delitz, F. Nungesser, & R. Seyfert (Eds.), *Soziologien des Lebens. Überschreitung – Differenzierung – Kritik* (pp. 167–198). Transcript.
Gehlen, A. (1940/1987). *Man: His Nature and Place in the World* (C. McMillan & K. Pillemer, Eds.). Columbia University Press.
Gehlen, A. (1941/1983a). Ein Bild vom Menschen. In *Philosophische Anthropologie und Handlungslehre. Gesamtausgabe,* Band 4 (pp. 50–62). Vittorio Klostermann.
Gehlen, A. (1951/1983b). Der Gegenwärtige Stand der Anthropologischen Forschung. In *Philosophische Anthropologie und Handlungslehre. Gesamtausgabe,* Band 4 (pp. 113–126). Vittorio Klostermann.
Gehlen, A. (1971/1983c). Philosophische Anthropologie. In *Philosophische Anthropologie und Handlungslehre. Gesamtausgabe,* Band 4 (pp. 236–246). Vittorio Klostermann.
Grene, M. (1966). Positionality in the Philosophy of Helmuth Plessner. *The Review of Metaphysics, 20*(2), 250–277.
Grene, M. (1968). *Approaches to a Philosophical Biology*. Basic Books.
Haeckel, E. (1868/1880). In D. L. Schmitz (Ed.), *The History of Creation: Or, the Development of the Earth and its Inhabitants by the Action of Natural Causes* (Vol. 2). D. Appleton.
Honenberger, P. (2022). Marjorie Grene (1910-2009). In J. Fischer (Ed.), *Plessner-Handbuch. Leben – Werk – Wirkung*. Metzler (forthcoming).
Klaatsch, H. (1899). Die Stellung des Menschen in der Reihe der Säugetiere, speziell der Primaten und der Modus seiner Herausbildung aus einer niederen Form. *Globus, 96*(22), 329–341.
Klaatsch, H. (1920). *Der Werdegang der Menschheit und die Entstehung der Kultur*. Bong & Co.
Krüger, H.-P. (2019). *Homo absconditus. Helmuth Plessners Philosophische Anthropologie im Vergleich*. de Gruyter.

Lacan, J. (1949/2005). The Mirror Stage as Formative of the I Function as Revealed in Psychoanalytic Experience. In B. Fink (Ed.), *Écrits. The First Complete Edition in English* (B. Fink, H. Fink, R. Grigg, Trans.) (pp. 75–81). W.W. Norton and Company.

Lacan, J. (1984). *Les complexes familiaux dans la formation de l'individu.* Bibliotheque des Analytica.

Levivier, M. (2011). La foetilisation de Louis Bolk. *Essaim, 26*, 153–168.

Limoges, C. (2015). Philosophie biologique, histoire des sciences et interventions philosophiques. Georges Canguilhem, 1940–1965. In G. Canguilhem, *Œuvres complètes,* Tome IV: *Résistance, philosophie biologique et histoire des sciences, 1940–1965* (pp. 7–52). Vrin.

Owens, C., & Almqvist, N. (Eds.). (2018). *Studying Lacan's Seminars IV and V: From Lack to Desire*. Routledge.

Plessner, H. (1928/2019). *The Levels of Organic Life and Man. an Introduction to Philosophical Anthropology* (M. Hyatt, Trans.). Fordham University Press.

Scheler, M. (1915/1978). On the Idea of Man. Transl. Clyde Nabe. *Journal of the British Society for Phenomenology, 9/3*, 184–198.

Scheler, M. (1928/1962). *Man's Place in Nature* (H. Meyerhoff, Trans.). Farrar, Straus & Giroux.

Schürmann, V. (2017). Der Cartesianische Einwand und die Problemstellung. In H.-P. Krüger (Ed.), *Helmuth Plessner: Die Stufen des Organischen und der Mensch* (Book Series "Klassiker Auslegen") (Vol. 65, pp. 55–70). de Gruyter.

Verhulst, J. (1993). Louis Bolk Revisited II: Retardation, Hypermorphosis and Body Proportions of Humans. *Medical Hypotheses, 41*(2), 100–114.

Verhulst, J. (1996). Atavisms in Homo Sapiens: A Bolkian Heterodoxy Revisited. *Acta Biotheoretica, 44*, 59–73.

Zanfi, C. (2013). *Bergson et la philosophie allemande 1907–1932*. Armand Colin.

Thomas Ebke focuses in his research on the histories of German and French contemporary philosophy, with a particular emphasis on their intersections. His major areas of research include German-speaking philosophical anthropology (notably Plessner) and French historical epistemology (Canguilhem). His dissertation *Lebendiges Wissen des Lebens*, presenting a systematic dialogue between Plessner and Canguilhem, appeared in 2012 (Berlin, Akademie Verlag). He is currently writing a book about the idea of a "metaphysics of difference" with a particular interest in the thought of Jean Hyppolite.

Part II
Canguilhem's Philosophy of Biology

"Unknown Material"? Georges Canguilhem, French Philosophy and Medicine

Giuseppe Bianco

Abstract In the introduction to the *Normal and the Pathological*, Canguilhem's doctoral dissertation in medicine, defended in 1943, he claimed, "philosophy is a reflection for which all unknown material [*matière étrangère*] is good." In this case the "unknown material" was precisely medicine; "a technique or art at the crossroads of several sciences" which was supposed to provide "an introduction to concrete human problems." Canguilhem had started studying medicine six years before, while he was a high-school professor in Toulouse. At the time he was distancing himself from the philosophical framework that had marked his studies and writings during the previous decade. This framework implied an anti-vitalist, Kantian and Cartesian approach to man, strongly influenced by his mentor Emile Chartier, also known as Alain. In this chapter, I try to provide concrete explanations concerning his decision to study medicine. I will not rely on those proposed by the existent scholarship, which frequently relate his decision to his interest in technology and technique. On the contrary, by examining unpublished material, such as a series of lectures given between 1933 and 1935, I claim that the motivation of his turn has to be related to the readings of works in psychology and ethology undertaken during this period.

Keywords Medicine · Psychology · Canguilhem · Ethology · Concrete

1 Introduction

At the very beginning of the *Normal and the Pathological*, originally intended as a doctoral dissertation in medicine, defended in 1943 in Strasbourg's Faculty of Medicine, Georges Canguilhem (1904–1995) famously claimed that "philosophy is

G. Bianco (✉)
Department of Philosophy and Cultural Heritage (DFBC),
Cà Foscari Università, Venice, Italy
e-mail: giuseppe.bianco@unive.it

G. Bianco et al. (eds.), *Canguilhem and Continental Philosophy of Biology*, History, Philosophy and Theory of the Life Sciences 31,
https://doi.org/10.1007/978-3-031-20529-3_5

a reflection for which all unknown material [*matière étrangère*] is good" (Canguilhem, 1991, 33). During the last two decades, this sentence has become a catchphrase, and it has even been used to name a French book series from the Vrin publishing house. This definition of philosophy as a peculiar discipline able to treat all kinds of "unknown materials," much like a gigantic stone-crusher, is often presented as particularly original and radical. However, I will try to show that this is a very classical definition of philosophy, proper to Kant's heritage in France. In the case of Canguilhem, the "unknown material" belonged to medicine, that he considered neither an academic discipline, nor an independent science – as Claude Bernard (1813–1878) and, before him, François Magendie (1783–1855) or François Broussais (1772–1838) believed – but rather as "a technique or an art at the crossroads of several sciences." This "unknown material" was thus supposed to provide "an introduction to concrete human problems" (Canguilhem, 1991, 33).

Canguilhem started studying medicine six years earlier, during the autumn of 1937, at Toulouse's medical school, after having obtained, in Toulouse University, the previous year, his "certificat d'études physiques, chimiques et biologiques" (*certificate of physical, chemical and biological studies*), mandatory to study medicine [*Faculté de médecine*]. At this time, he was teaching at the Fermat high-school. He then continued studying at the more important Strasbourg medical school (from 1941 until 1943), where, under the invitation of his friend Jean Cavaillès (1903–1944), he taught philosophy and logic at the local university. As for all philosophers studying medicine, teaching philosophy was, for Canguilhem, a way to finance his studies. The development of his career as a philosopher – from high school to preparatory undergraduate school, or *khâgne*, up to university – was accompanied by the progression of his medical training. Once he had gained a basic knowledge of medicine during the period spent in Toulouse's medical school, he encountered the determinant "unknown material," through the intercession of different figures at Strasbourg University: his friend, the psychopathologist, Daniel Lagache (1903–1972), who played a major role in his discovery of Kurt Goldstein's (1878–1975) work; the histologist Marc Klein (1905–1975), and the physiologist Charles Kayser (1899–1981).

Nonetheless Canguilhem never practiced the "art" of medicine – but he was registered in the Order ot the Physicians. One exception was during the Resistance in the French region of Auvergne, where he treated the wounds of his comrades.[1] He never took an active part in the discipline of medicine, since he never published anything concerning practical contemporary medical problems. From the *Normal and the Pathological* onwards, medicine, and especially its history, were a simple source of "unknown material" for him, which had to be treated philosophically.

My central question is: why did Canguilhem choose medicine as a source of "unknown material", instead of another practice or science? In this chapter I will try to provide some answers concerning this decision and attempt to correct other hypotheses I formulated in an essay published almost a decade ago (Bianco, 2013).

[1] See Limoges, "Introduction," in Canguilhem (2015), 15.

The treatment of this question is not a simple matter of detail – a question reserved for "canguilhemologues." On the contrary, it provides an opportunity to investigate the history of the relation between philosophy and medicine in France, and to understand how this relation changed during the peculiar period of the interbellum.

This paper is structured in three parts. In the first, I try to elucidate the meanings of "*matière*" [matter or material] and "concrete." I insert these terms into a semantic field which emerged during the French interbellum that was trying to render what the role of philosophy was at that moment. In the second part, I consider the "unknown material" that Canguilhem's generation chose to treat during the 1930s; paying particular attention to sociology and psychology, as well as the peculiar relationship that medicine had with philosophy. In the third and final part, I try to provide some answers concerning his choice for medicine.

2 Matter/Material, Concrete

The term "matter" [*matière*] was progressively becoming a buzzword during the interwar period, since it was at the centre of the writings of an author who was slowly gaining success, namely Karl Marx (1818–1883), who, after half a century of stigmatization, started to be read by philosophers. In fact, one can find expressions like the ones used by Canguilhem in the writings of many authors belonging to the same generational cohort, many of whom were influenced by Marx, Engels (1820–1875) and Lenin (1874–1925). Two examples are striking, specifically, the philosopher Georges Politzer (1903–1942), as well as Canguilhem's schoolfellow at the *École normale supérieure,* Paul Nizan (1903–1942). Politzer and Nizan authored two of the most devastating – and influential – philosophical pamphlets of the interbellum: *La fin d'une parade philosophique: le Bergsonisme* (1929) [*The End of a Philosophical Parade: Bergsonism*], that Canguilhem reviewed and praised,[2] and *Les chiens de garde* (1932) [*The Watchdogs*].

In a short essay from 1925, which anticipated *La fin d'une parade philosophique*, simply entitled "Introduction," Politzer (2013, 59) wrote that "spiritual life needs matter to nourish itself"; that "philosophy needs a real matter to be valid". At the same time, French philosophy had been denounced and considered "lacking matter." In *Les chiens de garde*, Nizan (1932, 13–14) claimed something similar: "Philosophy is a type of exercise of synthesis which consists in bringing together and ordering elements of any kind: there is no such thing as a proper "philosophical matter". However, he also stated that "philosophy in general is what remains of the various philosophies when they have been emptied of all matter". Nizan stigmatized contemporary French philosophy, epitomized by figures such as Henri Bergson (1858–1940), Dominique Parodi (1870–1955), André Lalande (1867–1963) and

[2] See Canguilhem, 2011, 221–226.

Léon Brunschvicg (1869–1944), equally, since he considered it a philosophy "emptied of all matter."

Both Canguilhem, Nizan and Politzer, conceived of philosophy as a way of problematizing and synthesizing the "materials" coming from the different specific sciences. As anticipated, this is an extremely traditional definition of philosophy, which I would name "idealist," since it states that all problems can be treated in a "philosophical way." This is in contrast to the regional sciences, which limit themselves to their own objects. In an essay from 1938, "Activité technique et creation" ["Technical Activity and Creation"], Canguilhem (2011, 501), quoting the Catholic spiritualist philosopher René Le Senne (1882–1954), an important reference for him at that time, claimed that the philosopher has to be a "professor of unity," someone who looks for the unity of human experience in its scientific, moral, and aesthetical aspects.[3] During the 1930s, Canguilhem thought that this unity had to be found in a transcendental consciousness; a source of three different values – aesthetic, scientific and moral. This is particularly clear in the preliminary notes of a series of lectures that he gave at Valenciennes' high-school between 1933 and 1935. Here, Canguilhem (1933–1935, 2) writes:

> Philosophy does not have its own object, if by object we mean a specialisation of judgement or thought. It [philosophy] was originally the ambition of a total explanation [...]. Philosophy would thus be defined as an inventory and critique of possible values or standards of affirmation.

Canguilhem inherited this idealist or transcendentalist conception of philosophy from a series of French authors, such as René Le Senne, Jules Lagneau (1851–1897), Emile Boutroux (1845–1921), Emile Chartier *aka* Alain (1868–1951), and finally Léon Brunschvicg. In the essay "L'éducation et la liberté," originally published in 1902 and then republished in the book *Nature et liberté* [*Nature and Freedom*], Brunschvicg (1921, 123) wrote that "philosophy has no material of its own; for its material is the mind as it has been formed by the study of history, the discipline of science, and aesthetic culture; it is on this mind that it [philosophy] exercises its reflection in order to show its unity". It is probably this essay that Canguilhem, Politzer and Nizan had semi-consciously in their minds when they insisted on the importance of "matter" in philosophy. Let us not forget that at the end of the "Introduction" to *The Normal and the Pathological*, Canguilhem obliquely quotes an excerpt taken out of Brunschvicg's essay "La méthode dans la philosophie de l'Esprit," republished in the book *L'idéalisme contemporain* [*Contemporary Idealism*] (1921), which would be a frequent reference in the preparatory notes for his high-school classes from the 1930s. Here Brunschvicg (1921, 179) stated that "philosophy is the science of solved problems", namely, a way of unfolding the conditions that allowed mind (*Esprit*) to overcome these problems. Brunschvicg also stated that, according to idealism "all problems remain open, because Mind [*Esprit*] does not cease to live and work in all of us."

[3] See Roth, 2013.

Nonetheless, despite the respect that Canguilhem felt for Brunschvicg,[4] there was a big difference between the two thinkers and between their ways of treating the "unknown material" coming from the past. According to Brunschvicg, the history of the sciences is nothing but the history of the conquering march of human Mind, or Spirit [*Esprit*]; the progressive, though not teleological, unfolding of truth. Therefore, once one theory wins against the others, these other theories end up being relegated to the past, like archaeological remains. The role of philosophy is to explain the transcendental conditions according to which one theory won over another. This vision is clearly depicted in one of his most important books, *Le progrès de la conscience dans la philosophie occidentale* (1927) [*The Progress of Consciousness in Occidental Philosophy*], a book which was mocked by Nizan (1932, 56) and described as a "philosophical breviary of the Universe where all is well that ends well." Similarly, starting from the 1940s, Canguilhem realized that the march of human knowledge was discontinuous, and philosophy's role was to unearth old debates and re-interrogate the winning theories. In *The Normal and the Pathological* (Canguilhem, 1991, 35), one of these winning theories was the one "according to which pathological phenomena are identical to corresponding normal phenomena save for quantitative variations."

More generally, a big gap separated the generational cohort of Lalande, Brunschvicg and Bergson from Canguilhem, Nizan and Politzer. This latter group of young men, born during the first decade of the twentieth century, wanted to oppose the "philosophy without matter" of their mentors, in the name of the "unknown material" and the "concrete." The term "concrete" was, in fact, tied to the term "matter." Starting from the interwar period, it turned into a buzzword; constantly opposed to the "abstract."[5] In France, the oppositional couple, concrete/abstract, only started to be used obsessively starting from the 1920s, because of the importation into France of texts coming from Germany, especially Hegelian, and then Marxian texts. Hegel (1770–1831) had reactivated the old semantic layers present in the verb *concrescere*, at the root of "concrete." According to his absolute idealism, abstract thought consists in separating one term from others, thus ignoring the totality of the dialectical relations between them. Conversely, the concrete is reality itself, or what has grown together from several determinations. Reason is the faculty of the "concrete universal," namely the one able to grasp this organized unity, the "concrete."

The peculiar social and ideological context of the post-war period was at the root of the interest of a new cohort of intellectuals studying "unknown material". The trauma of the conflict created a neat divide, separating the dark post-war years from the Belle Époque's splendours, and the new cohort born at the dawn of the twentieth

[4] Brunschvicg was always praised by Canguilhem. For instance, in the 1988 conference "La problématique de la philosophie de l'histoire au début des années 30" ["The problem of philosophy of history at the beginning of the 1930s"] Canguilhem (2018, 1123–1141) considers Brunschvicg as the academic philosopher who, during the 1920s, was the most respected by him and his school fellows.

[5] For the history of the oppositional couple concrete/abstract, see Bianco, 2023a.

century from their mentors. This cohort was aware of the horrors experienced by their older peers during the conflict. Moreover, the long months soldiers spent in the trenches had put young intellectuals in contact with men from the working class. On top of this, the Bolshevik Revolution started to be considered by many of these men as the proof that a radical change, led by the principles of Marxism, was possible. This transformation provoked a sudden increase in the use of the term "concrete", accompanied by critiques addressed to the academic "abstractions" produced during the Belle Époque.

The new cohort of scholars, all born during the first decade of the century, attacked both the idealism of Kant-influenced authors such as Brunschvicg and Lalande – accused of confining themselves to epistemological problems considered from the "idealist" standpoint of an anonymous transcendental consciousness – and the "intuitive" and "irrational" philosophy of Bergson and his followers – attacked for having practiced a useless introspective psychology that invited inaction. The accusations were theoretical, but also moral and political. In fact, most of the authors targeted by the younger scholars took active part in war propaganda and were accused of being the voice of the bourgeoisie or at least of being no more than disengaged cowards who justified the war's massacres. Addressing themselves to a wider readership, these young men took advantage of new, and more generalist publishing houses and journals. They created new and, in many cases, short-lived periodicals. They sometimes resembled the surrealist avant-garde who inaugurated the roaring twenties, publishing caustic tracts and manifestos where they opened an invitation to violent action and messianically invoked a forthcoming revolution. In some cases, they made use of a violent language and became more and more politicized. Being "concrete" meant being actively involved in politics, or being "engaged" [*engagés*], another buzzword of the period. This new cohort displayed some of the same features as the previous critics of "abstraction", including the hatred of "idealism", now used as synonym of "spiritualism."

Finally, the gap between the growing number of students and the stable number of academic positions produced a perceptible decrease in the possibility of pursuing academic careers and resulted in the phenomena of de-professionalization.[6] While continuing to teach philosophy in secondary education, many graduates tried to find alternatives outside of the philosophical field. Some examples are: Jean-Paul Sartre (1905–1980), Emmanuel Mounier (1905–1950), Pierre Morhange (1901–1972), Norbert Guterman (1900–1984), Georges Friedmann (1902–1977), Paul Nizan (1905–1940), Henri Lefebvre (1901–1991) and Georges Politzer. Each engaged in the creation and editing of journals addressed to a wider public, the latter six having become communist militants. Most of them, along with Simone de Beauvoir (1908–1986), also wrote novels and journalistic articles. Friedmann and Lefebvre, along with Raymond Aron (1905–1983), Claude Lévi-Strauss (1909–2009) and, briefly, Raymond Polin (1910–2001) played an important role in the consolidation of the social sciences.

[6] See Fabiani, 2010.

3 Different Material: Sociology or Psychology

The young philosophers were looking for "unknown material" in political action, or in new topics of inquiry. It was especially the two new twin disciplines of sociology and experimental psychology, which attracted many young graduates. The first would not be institutionalised as a separate curriculum from philosophy until 1957. Until then, it had simply been a sub-discipline of philosophy since it belonged to the "moral sciences" [*sciences morales*].[7] Canguilhem had an ambivalent attitude towards the social sciences,[8] an attitude he inherited from his mentor Alain, who was a role model of philosophical and political rigour during the 1920s. I wrote about Canguilhem's relation to Alain elsewhere (Bianco, 2013), but let's briefly revisit some facts.[9] During the period spanning from 1924 to the mid-1930s, Canguilhem embraced Alain's Cartesian and Kantian philosophy, as well as his militant pacifism and his political radicalism. Canguilhem contributed to his journal *Libres Propos*, and even played the role of chief editor from 1930 to 1932. Alain provided him with a philosophy – specifically, an anthropology – and a political ethics. He also provided him with political contacts in the pacifist movement, figures such as: George Demartial (1861–1945), Romain Rolland (1866–1944), Félicien Challaye (1875–1967) and Jean-Michel Bloch (1913–1987). Alain even helped him publish articles in other pacifist journals, such as *Europe*. In 1934, Canguilhem started disagreeing with many pacifists concerning the attitude one had to adopt towards the rise of fascism. Alain himself was against any violent uprising. In 1935, Canguilhem joined the Comité de Vigilance des Intellectuels Antifascistes [Vigilance Committee of Anti-Fascist Intellectuals] and published a, booklet *Le fascisme et les paysans* [*Fascism and Peasants*] (Canguilhem, 2011, 535–593). This happened just one year before Canguilhem's decision to start his medical training. Elsewhere (Bianco, 2013), I claimed that his decision must be directly connected to a political change which made Canguilhem question Alain's philosophy. Without simply discarding this bold hypothesis I formulated ten years ago, I'll try to correct it on some points.

According to Alain, philosophy is an ethics. It aims at the realization of wisdom, achieved through a purification of mind from the passions which affect it and divert rational judgment. Alain's philosophy depended on an idea of subjectivity as pure agency. Without will, perception and knowledge are impossible. Will must be educated and purified from the passions which affect it. Alain condemned all the sciences of man that produced laws explaining human behaviour through simple causation. These sciences negated the existence of the willing subjectivity, reducing the subject to an object. Therefore, Alain, and Canguilhem after him, were extremely

[7] The term "human sciences" [*sciences de l'homme*] become common only at the end of the 1940s, as a translation of Wilhelm Dilthey's (1833–1911) *Geisteswissenschaften*.

[8] See Bianco, 2024.

[9] For these aspects, see Braunstein, 2000, Roth, 2013.

sceptical towards certain trends within sociology, and especially Durkheim's, which considered that "social facts must be treated as things."

Nonetheless, Alain had a lot of respect for Auguste Comte's (1798–1857) sociology[10] which relied on the positivistic appropriation of François Broussais's "principle." According to this principle, there is merely a difference of degree, which can be quantified, between the healthy organism – called "normal" – and the one affected by a pathology. According to Broussais, the structure of an organism cannot be changed, and the transition from pathology to normality is gradual. Following Comte's application of the principle to sociology, Alain considered that society cannot be changed abruptly, and that brutal changes would lead to social pathology. Finally, Alain inscribed Comte's sociology into his philosophy of freedom. This is the reason why Canguilhem chose Comte as the topic of his master's degree dissertation (*Diplome d'études supérieures*) and chose, as a supervisor, Céléstin Bouglé (1870–1940), a sociologist sympathetic to Durkheim, but not as dogmatic as other Durkhemians. Just like his friend Alain, Bouglé was close to the centre-left radical party. As I have shown elsewhere (Bianco, 2023a), his dissertation *La doctrine de l'ordre ou du progrès chez Auguste Comte* [*The Doctrine of Order and Progress in Auguste Comte*] (1926) depicts a fully Alainian Comte and turns the supposed fatalism of the doctrine of order and progress into a voluntarist theory proving human freedom and the necessity of *desiring* progress. Between 1927 and 1934, the references to "social theory" become frequent in Canguilhem's work, but he never praises Durkheim's sociology, privileging other authors, such as Paul Vidal de La Blanche (1845–1918), the godfather of the French school of geography; Max Weber (1864–1920) and Maurice Halbwachs (1877–1945). In 1931, Canguilhem (2011, 375–382) reviewed Halbwachs' *Les causes du suicide* (1930) [*The Causes of Suicide*] and extensively used this book and his *La théorie de l'homme moyen, essai sur Quételet* (1913) [*The Theory of the Average, Essay about Quételet*] in the *Normal and the Pathological*.

Let us now turn to psychology, which, since the 1830s, was a mandatory discipline for all philosophy students. Psychology only became independent from philosophy in 1947, when the first independent curriculum in the discipline was instituted by Daniel Lagache. Alain, just like Canguilhem, appreciated Comte's and Kant's condemnation of psychology,[11] both of whom saw it as a contradictory pseudo-science. According to Alain, in the study of man, there are two options: physiology – including brain physiology – and philosophy, understood as a reflexive analysis of the conditions of possibility of knowledge and action. Therefore, there was no space for introspective psychology, which had been condemned by Broussais and, after him, by Comte. Professor Georges Dumas' (1866–1946) famous "presentations" of clinical cases at Sainte Anne hospital, enthused some of Canguilhem's school fellows like Daniel Lagache, Maurice Merleau-Ponty and Jean-Paul Sartre. Canguilhem attended to them just once, accompained by Lagache, and never came

---

[10] See Bianco, 2024.

[11] For this, Braunstein, 2012 and Sturm, 2001.

back. According to Canguilhem, psychology, even more than sociology, reduced man to a fact. By mobilizing determinants to explain human behaviour and knowledge, psychology reduces the human ability to act. It constitutes an invitation to inaction and becomes a means of controlling men. This thought would be a constant throughout Canguilhem's career, which involved the denunciation of behaviourism and cognitivism.[12]

Nonetheless, during the ten years in which he taught in secondary education, Canguilhem *had* to teach psychology, given that this "sub-discipline" was part of the high school and college curricula. The number of high school lessons he dedicated to psychology, and more particularly to the difference between physiology, psychology and philosophy, is impressive. A large part of the high school lectures given by Canguilhem until 1935–6, as well as a thick yet unpublished textbook he finished writing in 1932, start with a definition of psychology. Canguilhem's strategy consisted in separating introspective (or eclectic) psychology – considered a pathetic pseudoscience – from physiology, which was considered useful. This included making physiological psychology depend on an epistemological framework provided by philosophy, namely reflexive or transcendental analysis. On the third page of the textbook, Canguilhem (1929–1932, 3) sarcastically writes:

> Psychology [...] can be considered a science if [...] one makes the soul a subtle object, but nevertheless an object. If, on the contrary, one understands that the object of psychology is, without any pun, the subject, one understands, at the same time, that psychology is necessarily a reflexive inquiry. It then becomes difficult to distinguish between psychology and philosophy. Psychology would be the study of the soul considered in union with the human body, in other words, the knowledge of the subject insofar as it is linked to conditions of a lower order, from which it is the task of reflection to progressively free itself, but which constitutes a starting point that none is allowed to neglect.

Concerning the physiological bases of psychology, Canguilhem praised Descartes' mechanistic idea of the reflex movements – considered a "reasonable theory" at the time – and its development in authors such as Broussais, Claude Bernard, Jacques Loeb (1859–1924), Ivan Pavlov (1849–1936), Vladimir Bekhterev (1857–1927) and Henri Piéron (1881–1964).[13] Nonetheless he stigmatized these approaches when they tried to explain human behaviour without considering will, which was exclusive to human beings. The "error of psychological method," he wrote in his manual, "is undoubtedly only the consequence of a more serious moral error. To treat the study of man objectively is very explicitly to hold and treat man as a means, asking him to renounce his essential quality of subject". Finally, he added that that this type of approach was acceptable only if interpreted in the framework of transcendental philosophy, what Canguilhem (1929–1932, 15–16) calls here "reflection."

---

[12] For this, see Braunstein, 1999.

[13] Canguilhem would go on to criticize this approach in his Ph.D. dissertation on the notion of reflex (Canguilhem, 1994).

Canguilhem neatly separated human beings from other animals which, according to him:

1. Do not use instruments, but act using their organs.
2. Rely on instinct and not on intelligence.
3. Do not possess anything comparable to human language.
4. Mechanically react to the environment, without being able to shape it, since they are deprived of teleological behaviour.

As I have explained elsewhere (Bianco, 2013), during the 1930s, Canguilhem rejected all vitalist or proto-vitalist doctrines claiming the originality of life.[14] He thus concluded that "in short, nothing obliges us to attribute to the animals, perception, memory and reasoning, and this because all these apparently separate functions are in fact united by their relation to the same principle, reflection, implied in even the lowest forms of human knowledge of the world" (Canguilhem, 1929–1932, 28), and therefore, not in animals.

The mechanistic view of the organism, and more particularly of the human body, was enough for Canguilhem. In the manual he wrote, laconically, that "it is useless to insist on notions which any physiology manual can explain" (1929–1932, 11).

4 Medicine

What about medicine? What was its relationship with philosophy? Medicine was considered a peculiar body of knowledge taught in one of the four faculties of the French University; the other three being: the Faculty of Law, the Faculty of Science and the "Faculty of Letters," (*Faculté des Lettres*) where philosophy, along with literature, geography, and history was taught). It could be stated that medicine was, since the Napoleonic reform of 1808, philosophy's big Other. The transformation of French philosophy into a separate discipline from the beginning of the nineteenth century onward,[15] was mainly triggered by the debates philosophers had with physicians, mostly physiologists, pathologists and alienists. At the turn of the nineteenth century, the Ideologists, who played a major role in the Revolution, had combined the analysis of ideas with physiology and pathology. To stop the possible reconstitution of the legacy of Ideology, Napoleon created two neatly separated faculties: the Faculty of Sciences and the Faculty of Letters. Starting from the 1820s, the philosopher Victor Cousin (1792–1867) and his men played a major role in avoiding all the

[14] See Canguilhem (1929–1932, 25): "Generally speaking, any vitalist doctrine that maintains the originality of life and instinct does so only by negations, by exposing the difficulties and limits that any positive method of explanation encounters. But, in addition to the fact that faculties and limits are necessarily relative facts which must not be transformed into principles, one can consider unacceptable an attitude which amounts to attributing as a proper character to the object of one's research the very fact that nothing can be said about it".

[15] For this see Bianco and Wolfe (2023b).

possible influences that some physicians – the first figure in this tradition being Victor Broussais – were having on the development of philosophical doctrines. These doctrines denied human agency and the existence of a unitary mind. For half a century, starting from 1820 until 1870 at the earliest, *philosophical psychology* opposed the fragmentation of the mind proposed by brain neurology and alienism, in order to defend the unity and agency of the human mind. I cannot go into the main episodes of the long series of controversies here,[16] but we should note that during the nineteenth century, the philosophers' attitude towards the medical "matter" was not objective. Rather, it was militant and aimed at defending the epistemic pillars of the philosophical field; namely the unity of mind, the absence of determinism and the existence of free will, if not of divine providence. It is from this perspective that the academic philosophers reacted to the physiological theory of the localization of faculties in the brain, the development of alienism, the theory of evolution and even to Claude Bernard's ground-breaking *Introduction to Experimental Medicine* (1865).

During the nineteenth century, the idea of an "historical epistemology" of medicine was simply unthinkable, and even the histories of medicine, which started appearing at the beginning of the century, were produced by physicians. To better respond to the potential threat coming from the physicians - who viewed the philosophers trained inside the Faculty of Letters as scientifically ignorant - philosophers had to learn some basic medical notions. This ended up in the emergence of a new discipline: psychology. This is the reason why, during the 1880s, the reforms of the curricula in philosophy, embodied in manuals, aimed at including basic notions of physiology, brain-anatomy, and psychopathology. Théodule Ribot (1839–1816), the godfather of French experimental psychology, who had occupied a chair in "experimental psychology" since 1885, possessed only a second-hand knowledge of medicine. However, he invited the younger philosophers to engage in medical training. These men started their curriculum with an *agrégation* in philosophy,[17] then a training in medicine, ending in a Ph.D. dissertation in medicine, and then, eventually, a second Ph.D. dissertation in philosophy.[18]

In 1893, Pierre Janet (1859–1947) was the first *agrégé* in philosophy to earn a doctorate in medicine. Just one year later, in 1894, his friend Georges Dumas (1866–1946) became a philosopher-physician as well. During the following 40 years, only seven other scholars were able to complete their medical training after a philosophical training. They were Charles Blondel (1876–1939), who became a doctor in medicine in 1906, Henri Wallon (1879–1962) in 1908, Henri Piéron (1881–1964) in 1912, André Ombredane (1898–1958) in 1924 and, finally, Daniel Lagache (1903–1973) in 1934. Lagache, one of Canguilhem's school fellows at the *Ecole Normale*, created the first independent curriculum in psychology, in 1947.

[16] for this, see Bianco and Wolfe (2023b).

[17] The *agrégation* is the selective test a graduate in philosophy had to pass if she wanted to teach this discipline in secondary and higher education.

[18] For this, see Bianco, 2019.

Therefore, Canguilhem was not an exception in his interest in medicine, given that he was one of the ten French "philosophers" who earned a doctorate in Medicine before WW2. Nonetheless, two main differences neatly separate the case of Canguilhem from the other philosophers with a doctorate in medicine. The first difference had to do with the great gap separating Canguilhem's philosophical training (suspended after 1927, when he received his *agrégation* in philosophy) and his medical training, which started almost a decade later. The second difference, which must be explored here, deals with the reasons behind the decision to undertake medical training.

Now, another detail must be stressed. Because of the effort that a long training in medicine required, the graduates in philosophy who were embarking on it had to first possess the material and economic means to do so. They then had to understand that this endeavour would lead to certain results in terms of their careers. For example, Théodule Ribot and Henri Bergson, who both wanted to pursue said training (Bianco, 2019), were either too economically unstable or were teaching in towns in which there was no medical school. On the other hand, Dumas came from a family of physicians, and Janet had a physician brother and was already well inserted into the medical field thanks to his powerful uncle Paul. Even Lagache came from an extremely wealthy family. Other philosophers who started by studying psychology at university, such as Sartre and Merleau-Ponty were, in a way, either too ambitious, too quickly inserted into the philosophical field, or simply too involved in other projects to undergo seven years of medical training.

Canguilhem was from a middle-class *milieu*, but he could not pursue medical training while teaching so many hours in high schools, located in small towns such as Albi or Valenciennes. It was only once he got appointed to his job teaching preparatory classes in Toulouse, a town where there was a Medical School, that he was able to start his training. But had he already planned to do this over the past decade? We cannot be certain. Nonetheless, in a review of *Orientation des idées medicales* [*The Orientation of Medical Ideas*], a work by the physician and psychoanalyst René Allendy (1889–1942) from 1929, Canguilhem (2011, 248–51) declared preferring this work to Bernard's *Introduction*. He praised Allendy's ideas on synthetic medicine, addressing the concreteness of the "individual". Canguilhem's claimed resonate with parts of Alain's work. For instance, in *Elements of Philosophy* (Alain, 1941, 113), a work originally published in 1916 and that had a tremendous influence on Canguilhem, Alain claims that "the great problem for a doctor is to discover the concrete, namely the singular patient he has in front his eyes". In his books Alain frequently considers the philosopher to be a physician, who must be aware of the basic notions of pathology and physiology. However, he must remain prudent, since he could influence the patients (i.e., the students and the citizens) by announcing a pathology and provoking a reaction that could worsen it. I have cited most of these passages – all relying on a mechanistic physiology inspired by Broussais – elsewhere (Bianco, 2013), so it is not worth revisiting it here.

Before concluding, I would like to mention two elements which may have motivated the choice of medicine. The first element is regarding WW1. Because of his ties with Alain – who fought against the war, became a militant pacifist, and authored the important book *Mars ou la guerre jugée* (1921) [*Mars or War Judged*] – and with other pacifist figures, Canguilhem was extremely sensible to the absurdities of the war and its effects on the human body and psyche. Two physicians who operated during WW1 turned into novelists and their descriptions of shattered bodies and of the heroic mission of the military doctors reached the wider public, Georges Duhamel (1888–1966), author of the excruciating *Civilisation* (1918), who won the Goncourt prize, and Louis-Ferdinand Céline (1894–1961), author of *Journey to the End of the Night* (1932).

We now come to the second motivating factor in pursuing a medical career, which is theoretical. In 1933, Canguilhem was neatly separating intelligent, willing human beings from other non-sentient living beings, while asserting that, concerning the functioning of the body, the law of reflex proposed by Descartes and the few explanations contained in any manual of physiology addressed to medical students would be enough. In a series of lectures he gave in Valenciennes in 1934–35, in the part concerning the relation between, on the one hand, philosophy and psychology, and on the other, sociology and biology, entitled "Dépendance et indépendance de la conscience" ["Consciousness' dependency and independency"] Canguilhem (1934–38) began to hesitate. The old references were substituted by new ones. The new authors invoked in the lectures questioned the difference between intelligence and instinct, teleological behaviour and simple reaction, man and animal. For example, he quoted *Nature* (1934) by the physician Charles Nicolle (1866–1936), *Problems of Instinct and Intelligence in Insects* (1931) by the physician and ethologist Richard W. Hingston (1887–1966), and *The Genesis of Instincts* (1917) by the founder of animal psychology Pierre Hachet-Souplet (1869–1947). These works each criticized the conception of instinct as a simple mechanical adaptation and introduced the hypothesis of the existence of animal intelligence. The research of Herbert Spencer Jennings (*Life and Death: Heredity and Evolution in Unicellular Organisms*, 1930) on the behaviour of protozoa and on the use of the "trial and error" method, were also invoked to prove the existence of a rudimentary intelligence in even the simplest organisms. Canguilhem also mentioned the research of entomologists like Charles Ferton (1856–1921), Jean-Henri Fabre (1823–1915) and Morton William Wheeler (1865–1937), especially his book *Ants: Their Structure, Development and Behavior* (1910). He also referred to zoologists such as Jacques Delamain (1874–1956), the author of *The Days and Nights of Birds* (1932) and Louis Roule (1861–1942), author of *The Life of Rivers* (1930) – who considered ant-hills, swarms, spider webs and bird and fish nests to be instruments constructed in order to satisfy needs and, therefore, achieve goals.

At that moment, the French psychologist Paul Guillaume (1878–1962) had introduced the work of the *gestaltist* Wolfgang Köhler (1887–1967) concerning intelligence in primates. He achieved this by translating Köhler's ground-breaking *The Mentality of Apes* (1917) in 1927. Between 1930 and 1937, in the *Journal de psychologie normale et pathologique*, Guillaume co-authored a series of essays concerning the usage of instruments by monkeys (1987) with Ignace Meyerson (1888–1983). These essays contradicted the idea that only humans can create and use tools.

Through these multiple sources, two certitudes – proper to Alain's Cartesian and Kantian philosophy – become inadmissible: that of the absolute exceptionality of man, and that of the reduction of animal life to tropisms and reflexes. The idea that, in order to understand the function of the human body "whatever manual of physiology" would suffice, was no longer acceptable. At this moment Canguilhem realized that medical training was thus necessary, and that this training would imply the encounter with new, unknown material; first in Clermont, then in Strasbourg. Canguilhem would go on to reject mechanistic physiology, the principle of Broussais and its usage in understanding society. This path would lead him to the formulation of a holistic theory of the organism influenced by medical vitalism and by *Gestalt* theory. In the years following *The Normal and the Pathological*, biological philosophy would provide the ground for the development of an historical "continental" philosophy of the life-sciences.

References

Alain. (1941). *Éléments de philosophie*. NRF.

Bianco, G. (2013). The Origins of Georges Canguilhem's 'Vitalism': Against the Anthropology of Irritation. In S. Normandin & C. Wolfe (Eds.), *Vitalism and the Scientific Image in Post-Enlightenment Life Science, 1800-2010* (pp. 243–270). Springer. https://doi.org/10.1007/978-94-007-2445-7_10

Bianco, G. (2019). What was 'Serious Philosophy' for the Young Bergson? In A. Lefebvre & N. Schott (Eds.), *Interpreting Bergson: Critical essays* (pp. 22–47). Cambridge University Press. https://doi.org/10.1017/9781108367455.003

Bianco, G. (2023a). Concrete/Abstract. In M. Sinclair & D. Whistler (Eds.), *The Oxford Handbook of Modern French Philosophy*. (forthcoming).

Bianco, G., (2023b). Life: Modern French Philosophy and the Life-Sciences. In M. Sinclair & D. Whistler (Eds.), *The Oxford Handbook of Modern French Philosophy*. (forthcoming).

Bianco, G. (2024). Georges Canguilhem's First Reading of Comte and Positivism's Fortune in the French Philosophical Field (1830–1930). *Revue internationale de philosophie,* (forthcoming).

Braunstein, J. F. (1999). La critique canguilhemienne de la psychologie. *Bulletin de psychologie, 52*, 181–190.

Braunstein, J. F. (2000). Canguilhem avant Canguilhem. *Revue d'histoire des sciences, 53*(1), 9–26.

Braunstein, J. F. (2012). L'invention française du 'psychologisme' en 1828. *Revue d'histoire des sciences, 65*(2), 197–212.

Canguilhem, G. (1929–1932). *Ouvrage de philosophie écrit par Georges Canguilhem à Charleville, Albi, Paris*. Georges Canguilhem Archive, CAPHES library, Paris, GC. 8 1. Typescript of the book, 271 handwritten sheets.

Canguilhem, G. (1933–1935). *Introduction à la philosophie*. Lectures given at Valenciennes high school in 1933–1934. Georges Canguilhem archive, CAPHES library, Paris, GC. 10. 2.

Canguilhem, G. (1934–38). *Psychologie*. Lectures given in Valenciennes and Toulouse high schools. Georges Canguilhem Archive, CAPHES library, Paris, GC. 10. 3.

Canguilhem, G. (1991). *The Normal and the Pathological* (C.R. Fawcett, Trans.). Zone Books.

Canguilhem, G. (1994). *La formation du concept de réflexe aux XVIIe et XVIIIe siècles*. Vrin.

Canguilhem, G. (2011). *Œuvres complètes*, vol. I.: *Ecrits philosophiques et politiques (1926–1939)*. Vrin.

Canguilhem, G. (2015). *Œuvres complètes,* vol. 4: *Résistance, philosophie biologique et histoire des sciences (1940–1965)*. Vrin.

Canguilhem, G. (2018). *Œuvres complètes,* vol. 5: *Histoire des sciences, épistémologie, commémorations (1966–1995).* Vrin.

Fabiani, J.-F. (2010). *Qu'est-ce qu'un philosophe français? La vie sociale des concepts (1880–1980).* EHESS.

Guillaume, P., & Meyerson, I. (1987). *Recherches sur l'usage de l'instrument chez les singes.* Vrin.

Halbwachs, M. (1913). *La théorie de l'homme moyen, essai sur Quételet.* Alcan.

Nizan, P. (1932). *Les chiens de garde.* Rieder.

Politzer, G. (2013). *Contre Bergson et quelques autres. Écrits philosophiques 1924–1939.* Flammarion.

Roth, X. (2013). *Georges Canguilhem et l'unité de l'expérience.* Vrin.

Sturm, T. (2001). Kant on Empirical Psychology: How Not to Investigate the Human Mind. In E. Watkins (Ed.), *Kant and the Sciences* (pp. 163–184). Oxford University Press.

Giuseppe Bianco is senior researcher at Cà Foscari University, Venice. He was awarded his PhD in philosophy from Lille3 University and worked in several European and American Universities. His area of interest is 19th and 20th century history of European philosophy and the history of the relation between philosophy, psychology, sociology and medicine. He has worked on the history of concepts, problems, authors, texts, intellectual clusters, chairs, educational systems and, in general, cultural objects that have to do with philosophy. He is the author of *Après Bergson* (Puf, 2015), he edited books on the history of 20th century French philosophy and a monographic issue of the Revue philosophique on Georges Canguilhem (Georges Canguilhem. Les traces du métier). He was part, along with Gertrudis Van de Vijver and Charles T. Wolfe of two research projects dealing with the relation between philosophy and the life-sciences (both funded by the Flanders Research Foundation and based at Ghent University): "Vitalism. A counter-history of biology" (2019-2022) and "Human life? From philosophy of life to philosophical anthropology" (2022-225). He was the recipient of the Marie Skłodowska-Curie Individual grant for his research project INTERPHIL, "The international congresses and the transnational shaping of philosophy (1900-1948)." The project will start in 2023 and will involve the collaboration between Cà Foscari University and the University of Québec in Montréal (UQAM). He is currently writing a book on Gilles Deleuze.

Life, Concept and Purpose: The Organism as a Connection in Kant's Critical Philosophy and Georges Canguilhem's Historical Epistemology

Giulia Gandolfi

Abstract The focus on the concept of organism, considered as a totality, is a common feature both in the work of Georges Canguilhem and Immanuel Kant. Given Canguilhem's strong Kantian training, in this text I try to locate the similarities and differences between the two authors. In the first part, I show the close connection between concept, life and knowledge in Canguilhem's philosophy. I then try to demonstrate that, for both authors, knowledge is formed from vital force, and consequently from the organism. In part two, I analyze Kant's idea of the organism from his pre-critical works up until the *Opus Postumum*. Here, Kant considers the fundamental connection between organism and conceptual knowledge. In the *Opus Postumum,* for the first time, the organism is presented as the condition of possibility of knowledge; as a material *a priori*. In the last section, once I have discussed the Kantian tradition Canguilhem was part of, I will explore his reception of the Kantian idea of the organism and the development that he proposes of the idea of organic totality after the discovery of DNA.

Keywords Organism · Purpose · Kant · Canguilhem · Biology · Life · Concept · Knowledge

Addressing the philosophical concept of the organism raises two issues: firstly, whether vital organization can be rigorously conceptualized and secondly, whether something escapes this conceptualization. For example, certain types of vitalism have attempted to investigate that which exceeds rationalization. How does one conceive of a science of the organism, and would that science be biology, physiology, medicine, or a philosophy of nature? These questions are common to Georges

G. Gandolfi (✉)
Ca'Foscari University, Venice, Italy
Paris 1 Panthéon Sorbonne University, Paris, France

G. Bianco et al. (eds.), *Canguilhem and Continental Philosophy of Biology*, History, Philosophy and Theory of the Life Sciences 31,
https://doi.org/10.1007/978-3-031-20529-3_6

Canguilhem and Immanuel Kant.[1] In the first part of this paper, I attempt to show the close connection between concept, life and knowledge in Canguilhem's philosophy. I demonstrate that, for Kant and Canguilhem, knowledge is based on the idea of a vital force and, consequently, on that of organism. In the second part, I analyze Kant's idea of the organism from the pre-critical works up until the *Opus Postumum*. Here, Kant considers the fundamental connection between the organism and conceptual knowledge. In the *Opus Postumum,* the organism is presented for the first time as the condition of possibility of knowledge. Finally, in the third part, after a discussion of the Kantian tradition to which Canguilhem belongs, I will explore his reading of the Kantian idea of the organism.

1 Life, Concept and Organic Knowledge in Kant and Canguilhem

In both Kant and Canguilhem, the role of the organism in knowledge is fundamental. For Kant, the organism serves as a filter for the formation of concepts. Alternatively, according to Canguilhem, the development of concepts is grounded in the capacity of the living[2] to choose and prefer. In other words, it is grounded in the organism's normative power (vital normativity). There are two ways in which concept and organism relate to one another: either concept and organism remain two separate entities, or they overlap, leading to the assimilation of knowledge by life, and vice versa.

According to Kant, the formation of concepts can be investigated on two different levels: that of the *CPR*, where concepts result from sensible intuitions elaborated by the intellect, or on the "systematic" level of the *CJ* and the *OP*. In this latter case,

---

[1] Regarding Canguilhem, I refer to his works in the English translation. If the English translation is unavailable, I will translate it myself (all translations are mine, unless otherwise indicated). The following abbreviations will be used in the text: *Knowledge of Life (CV) (Canguilhem,* 2008*), The Normal and the Pathological (NP) (Canguilhem,* 2011*), Études d'histoire et de philosophie des sciences (EHPS) (Canguilhem* 1994*), Fonds Canguilhem (GC).* All of Canguilhem's unpublished texts are gathered at Caphés, ENS Paris. Here, they are cataloged with the initials GC, along with the reference card number.

Concerning Kant, I refer to the English translation that follows the *Akademie-Ausgabe.* For the *Critique of Pure Reason,* I will cite the §, followed by an A or B indication. For the *Critique of the Power of Judgement,* I will cite the §. For the *Opus Postumum,* I will follow the pagination of the standard German edition of Kant's works, *Kant's Gesammelte Schriften,* edited by the Royal Prussian Academy of Sciences (Berlin: Georg Reimer, 1900). The following abbreviations will be used in the text: *Critique of Pure Reason (CPR) (Kant* 1781*), Critique of the Power of Judgment (CJ) (Kant,* 1790*), Opus Postumum (OP) (Kant* 1936–1938*).*

[2] Canguilhem does not clarify in his writings what he means by living. In some cases it seems that the living being he refers to is the human being (Canguilhem, 2008, 77–82; 100–106) in others to any living form starting with single-celled ones ("Even for an amoeba, living means preference and exclusion" Canguilhem, 2011, 136). I will consider that only human living beings are capable of producing concepts, while all living organisms have normative capacities.

the activity of understanding on sensible intuitions is mediated. Sensible data are sorted according to the purpose of the organism, which is both an organic body, and part of the unity of nature. As a result, not all the object's features become part of the sensible data submitted to understanding. Only the object's features that fulfill the organism's purpose are taken into account. As I will show, both Kant and Canguilhem conceive of knowledge as normative, and state that the productive center of this normativity is located in the organism. However, considering the status of the organism in Kant and Canguilhem, this leads to a supplementary issue: is it possible to speak of the organism as a concept?

Marina Brilman (Brilman, 2018, 25–46) has recently denied this possibility, claiming the impossibility of a shared definition of the concept of organism in Kant and Canguilhem. To Brilman and Rachel Zuckert (Zuckert, 2007, 45), Kant states that a concept is a way of unifying the manifold; a concept subsumes this manifold under a category of the understanding (Brilman, 2018, 29). In contrast, for Canguilhem, a concept does preserve the manifold. Brilman refers to Canguilhem's concept as a fossil: within it, it is possible to discover the original context from which it emerged. For Brilman, Canguilhem's concepts differ radically from Kant's. While, in Kant's work concepts are tools of judgment to build knowledge – and prior to the world they seek to make intelligible - in Canguilhem, they are the opposite. Concepts for Canguilhem, according to Brilman, (Brilman, 2018, 30–35), are produced by humans in response to challenges posed by their environment – analogous to instruments and tools produced by technical activity. The focus on the role of the concepts is not a constant topic for Canguilhem: it emerges only during the second half of the 1940s, with the development of his historical-epistemological approach. Only by considering a concept in the light of a historical-epistemological approach – the manner of grasping the existing relationship between human sciences, scientific and extra-scientific fields – can said concept eventually be defined as a "problem container," as Brilman states (Brilman, 2018). Only by considering the concept as a historical element can we grasp its containing function, as a guarantor of its background. Canguilhem considers the concept as antecedent to the scientific theory related to it, and treated, since the second half of the 1940s, the concept as autonomous from the scientific field. As the author illustrates in his analysis of the concept of reflection, a concept fundamental to a scientific theory can be derived from extra-scientific fields such as magic and mysticism (Canguilhem, 1977b, 22–26).

This does not mean that there is no Kantian influence in Canguilhem's idea of the concept. In 1945, Canguilhem held a series of lectures on "Nature et valeur du concept" ["Nature and value of the concept"] (Canguilhem, 1945) at the University of Clermont-Ferrand. During this lecture, he analyzes the "function of the concept," considering it as a tool to move from subjective sensible perception to objectivity. Canguilhem states that "the problem of the concept is the very problem of the rule and the meaning of value of the identity applied to the organization of perception"[3]

[3] "Le problème du concept c'est le problème même de la règle et de la signification de la valeur d'identité appliquée à l'organisation de la perception."

(Canguilhem, 1945, 84). By analyzing both perception and abstraction as knowledge structures, Canguilhem suggests that they can be considered equally valid counterparts. One does not have to choose between the truth proposed by perception and the one proposed by abstraction. Rather, one can prefer one to the other depending on the specific context. The object of knowledge depends on the *purpose* of its knowledge and there is no single and absolute way to know the object, but multiple ones according to the purpose of knowledge.

As I will discuss in Sect. 2, the idea of purposiveness underlies knowledge and is crucial in the *OP*.[4] Here, the organism becomes the condition of possibility of knowledge. According to Canguilhem, knowledge is the imposition of values and preferences. Knowing is the fulfillment of a choice. The same is true for life: life is the creation of values due to preferences proper to the living, confronted by the environment (Canguilhem, 1989, 122). The creation of a value, which in the case of knowledge can be defined as a truth value, does not negate what is left out: "The concept is not the destruction or the annulment of a certain order of the perceived qualities, it would be the subordination of this order of qualities to another order"[5] (Canguilhem, 1945, 84). Canguilhem adds that "the normativity of the living is translated by the organization of its *Umwelt*. A psychological order is a hierarchy of polarized reactions by tendencies (values). The specificity of a psychological order is to admit beside and concurrently other orders of preference and hierarchy"[6] (Canguilhem, 1945, 85).

Knowledge – the creation of concepts – stands in a continuum with life. Both life and knowledge unfold through the same pattern. There is, nevertheless, a difference between the animal and the human being. The animals create what Bergson defines as a "vital order" of preferences, while the human is capable of rationalization: "What is proper to man is to be capable of another order which remains a psychological order, that is to say, an order oriented by values, but which is precisely the inversion of sense of the previous orders. This order is that of Representation or Knowledge"[7] (Canguilhem, 1945, 85). The difference between the two "orders" lies in the differences in judgment, understood here as the possibility of subsuming the particular under the general (Kant, *CPR* A132/B171).

---

[4]Although Canguilhem does not refer to Kant's *OP* in his writings he is familiar with its content. Canguilhem was the supervisor of Jack Ligot's 1957 thesis, entitled *La téléologie comme fondement de la biologie selon Kant*. Canguilhem commented on and corrected this thesis, the last chapter of which focuses on *OP*. A copy of the thesis -with corrections and annotations by Canguilhem- is deposited at CAPHES (CAN 3937).

[5]"Le concept ne serait pas la destruction ou l'annulation d'un certain ordre des qualités perçues, il serait la subordination de cet ordre des qualités à une autre ordre."

[6]"La normativité du vivant se traduit par l'organisation de son *Umwelt*. Une ordre psychologiquement une hiérarchie de réactions polarisés par des tendances (valeurs). Le propre d'un ordre psychologique c'est d'admettre à coté et concurremment d'autres ordres de préférence et de hiérarchie."

[7]"Le propre de l'homme c'est d'être capable d'un autre ordre qui reste chez lui aussi un ordre psychologique c'est à dire orienté par des valeurs, mais qui est précisément l'inversion de sens des ordres précédents. Cet ordre c'est celui de la Représentation ou Connaissance."

Judgments map mental reality onto the supposedly 'independent' reality we receive through sensible data. They make up the cognitive order of humans. On the other hand, judgments used by living beings to form their *psychological* order are judgments of "convenience", between the living being and the subjective. Through the judgments of "adequation," humans "convert" an order of preference into an order of measurement. Therefore, according to Canguilhem, purpose determines the creation of a psychological order, while the capacity of rationalization provides abstraction from said order. In this sense, Canguilhem follows Kant's trajectory: the knowledge of an object starts from the experience gathered in a sensible intuition. Only some of the object's features are then submitted to the intellect to be subsumed under categories.[8]

Brilman correctly claims that Canguilhem's concept always maintains a connection to what it overcomes, but this also occurs in the *OP*, precisely thanks to the organism (as I shall discuss in Sect. 2). Like Kant, Canguilhem believes that abstraction does not cease to relate to the sensible; not because it results from it, but because it must always be applied to it. This application resembles a lever used to flip a stone (Canguilhem, 1945, 87). Canguilhem adopts the difference established by Kant between regulative and constitutive principles by stating that "the concept is not an essence, it is only a norm of identity which appears as an essence only by forgetting existence, not to destroy it but to organize it by stabilizing it"[9] (Canguilhem, 1945, 89). The concept is what allows knowledge to be organized according to a given norm. Important to note is that, for Canguilhem, a norm always has a teleological value.

According to Canguilhem, until the 1960s, knowledge and life were two normative activities. From the 1960s onwards, knowledge and life became part of the same domain, though they remained distinct. In 1966, in the essay "Le concept et la vie" (Canguilhem, 1994), Canguilhem changed his way of conceiving knowledge and life; concept and organism. He began to think of genetic information as the possibility of transmission, or communication of a code which is inscribed into life. Canguilhem writes: "There is then no difference between the error in life and the error in thought, between the errors of informing and informed information. The first provides the key to the second" (Canguilhem, 1989, 210). Since the concept of error is already embedded in information, the possibility of being ill – the inability to replace a disadvantageous norm with a favorable one – is also embedded within the organism, understood as the "vital space." Kant supports this idea when reflecting on disease and health in the *OP*. He describes disease and health as peculiarities of organisms – with organisms being defined as "beings in which there is life." He writes:

---

[8] I refer here to the process of comparison and abstraction as described in the *Jäsche Logik* (Kant, 1992e).

[9] "Le concept n'est pas une essence, il est seulement une norme d'identité qui n'apparaît essence que par oubli de l'existence, non pour l'anéantir mais pour l'organiser en la stabilisant."

> One feels the state of being sick, although the sickness could be quite hidden. Health itself is not felt, but only its hindrance – *agilitas*. Discomfort is itself not a sickness but often only the desire to increase one's well-being – not the negative but the *contrarie oppositum*. One can think of *health* and *sickness* with regard to organic bodies (not organic matter), since they possess a vital force, be it vegetative or animal, and for this reason also death or decay [...]. The principle of the possibility of such bodies must be immaterial since it is possible only through *purpose* (Kant, *OP* 22:100).

Guillaume le Blanc affirms that error reveals how life and knowledge contain their own limitations (le Blanc, 2008, 112). However, this does not involve a reduction to the biological-physiological level, against which Canguilhem argues vehemently in *NP* - as I will discuss in Sect. 3. Rather, we must understand knowledge production from within life (as an organic feature), as it is a result of *vital normativity* (i.e. the interaction between the living being and its environment).

2 Organism and Purpose from Kant's Pre-Critical Writings to the *Opus Postumum*

Studying a living being and trying to understand its functioning opens up a thorny question for Kant. The organism is not just inanimate matter, although it is composed of it. Kant questions the difference between the living and the non-living, and what this difference consists of. He wonders how to conceptualize materiality and the organizing force that makes this matter alive. In this section, I will examine Kant's idea of the organism, starting from his pre-critical writings: *Universal Natural History and Theory of the Heavens; The Only Possible Argument in Support of a Demonstration of the Existence of God;* and *Inquiry Concerning the Distinctness of the Principles of Natural Theology and Morality* (Kant, 1992a, b, c). From these writings, we can grasp the relevance of the concept of unity for the idea of purpose. Firstly, I will discuss the idea of a unity of nature as presented in the *CPR*. Then, I will explain the evolution of the definition of organism from the *CJ* to the *OP*.

2.1 *The Organism in the Pre-Critical Writings and in the* CPR

Between 1755 and 1764, in *Universal Natural History and Theory of the Heavens; The Only Possible Argument in Support of a Demonstration of the Existence of God;* and *Inquiry Concerning the Distinctness of the Principles of Natural Theology and Morality,* Kant stressed that the explanation of the organism is even more complex than the one of the solar system. The organism cannot be explained by the same mechanical theories that can be used to understand planets and the cosmos. As Kant argues in *CJ* §80, when we look at an organism, we perceive a feeling of wonder. We are puzzled by the complexity of its functioning and by the irregularities it creates. In the pre-critical writings, Kant already focused on the peculiarity of the organism in the context of a first sketches of a theory of the unity of nature. In the pre-critical writings, Kant argues for the necessity of unity as a regulative condition for the spontaneous unification of essences into one world. The idea of God and the unity of the world are taken up in the *Appendix to the Transcendental Dialectic* (Kant, *CPR* A642/B670). Here, Kant distinguishes between a constitutive and a regulative use of reason.[10] The constitutive use consists in the possibility of broadening knowledge through apprehension. The regulative use is the possibility of grasping the connections between objects within a whole, although this does not produce objective knowledge (Kant, *CPR* A 646/ B 674). For objective knowledge, there must be a condition of systematicity of phenomena. This condition is the transcendental apperception in the *CPR*. A transcendental unity and regularity, proper to apperception, does not have empirical regularity and unity as a correlate.[11] As L. W. Beck points out, a different condition is needed for empirical regularity (Beck, 1981, 452).

As Canguilhem noticed in a series of lectures on Kant (Canguilhem, 1941, 16), the regulative or hypothetical use of reason makes it possible to find a unity of particular knowledge.[12]

Kant defines the assumptions that are fundamental for the regulatory use of purposiveness. First, this regulatory use has no objective validity; though it does not conflict with its heuristic and positive value. Secondly, the idea of unity does not come from an external object. Following Silvano Marcucci, unity and purpose – for which Kant proposes a transcendental deduction in the *CJ* - have transcendental value. This guarantees a formal and objective validity (Marcucci, 1990, 123). Although many have objected[13] to the transcendental objective value of unity and purpose, one might nevertheless admit that neither can be derived from experience. On the contrary, they are conditions of possibility for experience.[14] The idea of unity and purpose are inseparable, as we can see in Kant's efforts to formulate three transcendental principles[15] for the unity of nature. On the one hand, the principle of purposiveness is fundamental for the regulation of the unity of nature. On the other, purpose is already contained within the idea of unity.

---

[10] It is worth noting that Kant speaks of constitutive and regulative use with respect to reason. Where he applies the concept of the regulative, as opposed to the constitutive use, as in the CJ, he will do so with respect to the faculty of judgment.

[11] See in particular Kant CPR, A 122.

[12] Reason is the faculty to derive the particular from the universal, but there are two manifestations of this: either the universal is already certain and given, and then a judgment under which the particular is subsumed is sufficient, or, the universal is assumed only problematically, as an "idea." In this case, what is certain is the particular but not the universal. One then tries to lead the particular cases back to a rule and, if "all cases are referable to the rule then the universality of the rule is inferred extending it to all cases even to those not given." See Kant, CPR A 646/B 674.

[13] In particular (Illetterati, 2014; Lotfi, 2010; Ginsborg, 2014).

[14] Kant, CPR A 648/ B 676.

[15] Kant, CPR A 658/ B 686.

In *Determination of the Concept of Human Race* (1785) and *On the Use of Teleological Principles in Philosophy* (Kant, 2007), Kant returns to the concepts of race and species by referring to Buffon's ideas. Kant supports the thesis that all races have a common origin. Interestingly, for Kant, differentiation occurs through a purposive principle of adaptation to the environment. In every organism, there are traits (e.g. skin color in humans) that are purposive in relation to the environment. When actualised, these traits can reorganize the organism. These traits are potentialities already present in the individual since they were transmitted by the parents. Some potentialities remain defunct, while others are activated (depending on the organism's relation to the environment). According to Kant, there is an inheritance of potential adaptive capacity. In this sense, Kant asserts that only pre-adaptive changes – already available as dispositions – can be transmitted. Their purpose is already inscribed within the organism.[16] In the pre-critical writings, there is a hint of the essential shift from a formal *a priori* purpose, to a material *a priori* purpose that would occur in the *OP*.

2.2 *The Organism in the* CJ

Kant presents a comprehensive formulation of organism theory in the *CJ*. He writes:

> Organisms are, therefore, the only beings in nature that, considered in their individual existence and apart from any relation to other things, cannot be thought possible except as ends of nature. It is they, then, that first afford objective reality to the concept of an end that is an end of nature and not a practical end. Thus they supply natural science with the basis for a teleology, or, in other words, a way of judging its objects on a special principle that it would otherwise be absolutely unjustifiable to introduce into that science—seeing that we are quite unable to perceive a priori the possibility of such a kind of causality.
>
> This principle, the statement of which serves to define what is meant by organisms, is as follows: an organized natural product is one in which every part is reciprocally both end and means. In such a product nothing is in vain, without an end, or to be ascribed to a blind mechanism of nature (Kant, CJ §65–66).

This passage is useful in understanding Kant's concept of the organism. Kant asserts the impossibility of scientifically understanding the organism according to purpose. As I have already stated, purpose does not have a constitutive value, merely a regulative one (Ginsborg, 2014; Illetterati, 2014). Although Kant speaks of a principle of purposiveness, this must be understood as a maxim.[17] Kant distinguishes between principle-laws and principle-maxims as solutions to the problem of the unity of nature.[18] The three principles concerning the unity of nature follow this distinction.

[16] See also McLaughlin, 1990.

[17] Following what Kant says in the *Critique of Practical Reason*, a maxim is a subjective principle of action that must be distinguished from an objective principle.

[18] It is important to note that a wide range of critical literature conceives regulative purpose *as if* it were a law (maxim-law). In this regard, the functionalist reading of Kant proposed by Hans Vaihinger is fundamental for Canguilhem (Vaihinger, 1935).

In the investigation of nature, if a scientist uses a different principle than another scientist, there can be no contradiction between the two. On the contrary, if one speaks of principles as laws, then they are constitutive and objective, and can therefore contradict one another. If, on the other hand, one considers principles such as purpose and unity as maxims, contradiction is avoided, since a regulative and heuristic use of such principles is then adopted. In this way, Kant manages to ensure two fundamental aspects, which Canguilhem will follow:

> Firstly, the autonomy of biology with respect to physics. The former has regulative maxims and a particular object of study, while the latter functions merely by laws. Secondly, Kant draws attention to the epistemological value of one's standpoint. A principle-maxim is both absolute and partial for the standpoint taken. A scientist who decides to employ a particular principle, such as homogeneity, will create an absolute and closed knowledge system; though only when concerning that particular principle. Kant stresses that absolute objectivity does not exist as a product of *choice*, which prefers one principle as opposed to another. Instead, objectivity is only constructed. Likewise, Canguilhem, taking up the Bachelardian concept of "epistemological regionalism" [*régionalisme épistémologique*], discusses the existence of different regions of knowledge. Within different disciplines, different research models can be applied. These models are partial and relative only when compared to other epistemological areas. In particular, according to Canguilhem, it is crucial to guarantee the independence of each discipline without causing a fracture between the various regions. Canguilhem analyzes the emergence and development of scientific concepts. In many cases, scientific concepts are initially used within a limited epistemological field, but later they are also used in other, different fields. Canguilhem is concerned with understanding continuities and heterogeneities in the development of concepts, and their application within a given discipline, as well as in their relationship to other disciplines (Canguilhem, 1977a, b, 25–27). Hence, in addressing principles as maxims, Kant discusses an indeterminate objectivity (Kant, CPR A 669/ B697) instead of a scientific objectivity.[19] An indeterminate objectivity[20] does not rely on the object of its application, but on the scheme used to apply the maxim.

The second aspect regards the organism as the product of a natural purpose, which differs from technical and artistic productions. It is necessary to distinguish the natural purpose of nature, not only from art, but from technique. In §63 and §64, Kant distinguishes between purpose as the anticipation of the product, concepts and representations that guide the production, and purpose as the anticipation of the effects of said production. Purpose-as-anticipation is peculiar to the technical act, since an agent that takes part in the production is guided by a concept. However, this is not the case with natural purpose (Kant, *CJ* §63). Canguilhem follows this distinction in his essay "Machine and organism." Canguilhem argues that a machine is always animated by an external purpose; it does not have an internal purpose and follows mechanistic laws. This is because every machine is produced for a specific purpose imposed by its creator. Canguilhem, against the Cartesian idea of the man-machine, states that a machine must be interpreted by the organism and not vice versa. In

[19] "The ideas of reason, of course, do not permit any deduction of the same kind as the categories; but if they are to have the least objective validity, even if it is only an indeterminate one, and are not to represent merely empty thought-entities (*entia rationis ratiocinantis*),a then a deduction of them must definitely be possible, granted that it must also diverge quite far from the deduction one can carry out in the case of the categories" Kant, CPR A 670/ B698.

[20] For an analysis of Kantian objectivity, see Montuschi, 2018.

other words, the author proposes we understand a machine from the perspective of a living being.

In "Machine and organism" Canguilhem attributes to Kant (Canguilhem, 2008, 92) the idea "of the irreducibility of the organism to the machine" and the overturning of "[t]he Cartesian relation between machine and organism;" the machine interpreted through the organism (Canguilhem, 2008, 88). Canguilhem remarks that "so long as the construction of the machine is not a function of the machine itself, so long the totality of an organism is not equivalent to the sum of its part, it seems legitimate to hold that biological organization must necessarily precede the existence and meaning of mechanical construction" (Canguilhem, 2008, 90). In § 64, Kant further specifies that the purpose of nature must not be understood in the light of rational utility and that it must be "both cause and effect of self." Although it can be considered as independent, each part of an organism has an action that modifies the rest of the whole and vice versa.[21] This internal autonomy is fundamental to understanding medicine according to Canguilhem.

Since the 1920s, Canguilhem considered medicine to be the discipline "of the real man." It should not be an abstraction that only considers physiological analysis: "all men became equal in front of the disease and equal also for the cure" (Canguilhem, 1929, 250). If, on the other hand, real human beings are considered in their singularity, the search for a cure would become much more complicated. As Pierre-Frédéric Daled (Daled, 2021, 89) notes, the clinician should consider the environment, the work activity and the diet preceding the disease, because human beings are "units-wholes'" and not "units-parts." For this reason, disease is not "somewhere in man," as a locatable part. On the contrary, "it is every-where in him; it is the whole man" (Canguilhem, 2011, 40). According to Canguilhem, even if physiological data and statistical determination play a crucial role in medical practices, especially in the clinic, they are not enough on their own. It is necessary to combine physiology with the idea of organic totality, both internally (between parts and their whole) and externally (between the organism and the environment). As Claude Bernard argued "statistics teach absolutely nothing about the mode of action of medicine" (Bernard, 1927, 138). Since a qualitative shift occurs between part and organism, physiological analysis can only guide knowledge of the part. A pathologist cannot determine in abstract the effects of a treatment and its result on a single patient. Canguilhem attributes this idea directly to Kant. In *EHPS*, Canguilhem considers a passage from *The Conflict of Faculties* (Kant, 1992d, 28), which argues that the clinician must use knowledge derived from medicine, as well as from the "Faculty of Philosophy,"[22] to deal with the living being's capacity to self-organize

[21] See also Kant, OP 22: 100.

[22] The Kantian division of faculties is the following: higher faculties (theology, law, medicine) and lower faculties (Philosophy, what today is referred to as letters and sciences, in French "sciences et lettres,"). Among the higher faculties, Canguilhem recalls, Kant considers the faculty of medicine to be the "freest" of the three and the one that most resembles that of philosophy. "The doctor," says Kant, is an 'artist' and as such he must use a skill that makes him dependent not only on his own Faculty, but also on the Faculty of Philosophy" (Canguilhem, EHPS, 386)

(Canguilhem, 1994, 386). Kant recognizes in the living being the ability to modify itself and its purpose saying:

> The way nature comes, in these forms of life, to her own aid in the case of injury, where the lack of one part necessary for the maintenance of the neighboring parts is made good by the rest; the miscarriages or malformations in growth, where, on account of some chance defect or obstacle, certain parts adopt a completely new formation, so as to preserve the existing growth, and thus produce an anomalous form: these are matters which I only desire to mention here in passing, although they are among the most wonderful properties of the forms of organic life (Kant, CJ §64).

This formulation is similar to what Canguilhem describes as biological normativity (cf. Sect. 3). According to Kant, thanks to the normative power of the living, the latter can resist malformations and disease by way of compensation. The living reacts, as its parts modify their internal purpose, thus creating new ways of life.[23] As Ginsborg (2014) suggests, purpose in this sense can be read as normative. This autopoietic feature of the organism points to a further characterization: it is ruled by an internal purpose. The internal purpose as the organism's self-realization is opposed to an external purpose. For Kant, external purpose is operated by an external agent (such as God or a craftsman) who orientates events. The internal purpose does not presuppose any intelligence, since the origin is the very justification of the purpose. As Canguilhem points out in "Machine and organism," Kant distinguishes the organism from the machine (Canguilhem, 2008, 145–152). His reasoning can be summed up as follows:

1. A machine cannot produce its parts or modify them.
2. A machine cannot produce another machine.
3. A machine cannot be the origin of the purpose it was built for.

A machine always requires a purpose external to it. The difference between organism and machine is not in their *totality*: organism and machine correspond to the definition of totality proposed by Kant in the *CJ* §65, which is often mistakenly confused with the definition of the organism. According to Kant, what distinguishes the organism from the machine is the organism's self-formative capacity, provided by its internal purpose.[24] Similarly, in 1941, Canguilhem argued that "if we conceive purpose as the causality of totality" we are mistaken. The totality modifies what we understand as function or part, through the way each part functions in light of the whole. "The function of a part in fact, does not replace itself as such in the totality," it gives way to a more complex activity that is qualitatively irreducible to a quantitative link (Canguilhem, 1941, 30). This qualitative difference does not allow an analytical-physiological study of the totality, but only its parts and their functions. It is possible to ascertain the organic totality only through an "experimental biological study," says Canguilhem. (Canguilhem, 1941, 32).

---

[23] A new way of life is what Canguilhem names *allure de vie* in NP.

[24] A third way to consider the relationship between mechanism and purpose is that of teleomechanism as proposed by Timothy Lenoir (Lenoir, 1981).

2.3 *The Organism in the* OP

Kant analyzes the relationship between organism and mechanical forces in the *OP*. In advance, it should be noted that Kant does not refer to biology,[25] but to physics.[26] Kant distinguishes between the organic and inorganic in physics, arguing that this distinction is *a priori*.

> One can, in fact, also draw on the concept of organic (as opposed to inorganic) nature, in the consideration of the moving forces of nature, without, [thereby,] transgressing the limits, determined *a priori*, of the transition to physics, or mixing into it what belongs to the material part of physics (thus to the doctrine of experience as a part of it). One can, in fact, define the former as follows: organized beings are those of which, and in which, each part is there for the sake of the other (*propter, non per aliam partem eiusdem systematis*) (Kant OP, 21:185).

According to Kant, only the forces able to internally organize a totality are vital. Indeed, Kant distinguishes between "living force (by impact) (*vis viva*)" from "the vivifying force (*vis vivifica*). The latter, in a separate world-system (and its generation), is perhaps the cause of plants and animals" (Kant, *OP* 22:210). Canguilhem notes that "Kant's theory is not without relations to the conceptions of the vitalist physicians of the 18th century." It does not, however, become a metaphysical vitalism. According to Canguilhem, Kant supports some vitalist theories, particularly that of Blumenbach, and opposes "iatro-mechanistic and desiatro-chemical theories" (Canguilhem, 1941, 1).

This is not to be confused with a metaphysical vitalism. Rather, this idea of a vital force within the organism can be referred to as attitudinal vitalism (Wolfe et al., 2020, 224). Attitudinal vitalism is a type of vitalism focusing on the relationship that the living being has with its environment. The focus is no longer on the nature of life, it is on the organism's relationship with its environment. According to Canguilhem, the living being recognizes that it has a particular "point of view"[27] through which it "reads" the environment. For this reason, Canguilhem defines vitalism as an exigency; the urge of the living to confront the environment, in order to be able to recognize itself as living (Canguilhem, 2008, 62).

---

[25] Under the name "science of nature," *scientia naturalis* is understood as the system of the laws of matter (of the movable in space); which, when it contains only their principles a priori, constitutes its metaphysical foundations. When it contains the empirical as well, however, it is called physics. The latter, as a doctrine of bodies, i.e. of matter in a figure determined according to laws, is divided in turn into general (*physica generalis*) and particular (*specialis*), in which either the formative force acts merely mechanically, or where one body forms another of the same species, in the propagation of its species, i.e. organically. (Kant, *OP* 21:474).

[26] Kant, OP 21: 184. Here Kant conceives of the organism as in *CJ*: as a totality in which every part is an end for the others and reciprocally a means.

[27] Here again Canguilhem's already noted vagueness regarding the living is emphasized. By "point of view" one should not understand an intellectual way of reading the world, but merely the "positioning" of the living according to its preferences in the environment. Canguilhem, echoing the concept of *Umwelt*, considers the "point of view" as what enables the living being to hierarchize, prefer and avoid, its surroundings. Only humans in such hierarchization have intellectual capacity.

Just like Kant, Canguilhem admits that the organism has a capacity – natural purpose – to produce itself by way of its parts. Each part is the cause of the productive capacity of the whole. The principle of purposiveness guarantees a distinction between organic and inorganic, and makes it possible to perceive organisms as natural purposes. Kant argues that the *causa finalis* (final cause) cannot be understood through the laws of matter and mechanics, but only "by assuming an understanding, independent of matter, which is architectonic for these forms, and to represent the moving forces of matter according to the mere analogy with it." In this sense, the purpose precedes "the empirical investigation" (Kant, *OP* 21: 185).

Up until this point, the theorization of the organism follows that proposed by Kant in the *CJ*. In the *OP*, the purpose is no longer only a theoretical *a priori,* but becomes a material *a priori* within the organism. Following the thesis of Ernst Onnasch, we can argue that, according to Kant, an object has to affect us empirically in order for us to have knowledge of it. In addition, the subject must recognize itself as affected by the object. Finally, the object must have already undergone our construction of it (Onnasch, 2014). Since the subject recognizes itself as an object, it must therefore construct itself as a known object. However, this knowledge "by construction" does not apply to the organism. In fact, according to Onnasch, the organism finds its actualization only when it becomes necessary "for the sake of experience" (Onnasch, 2014, 248). In the *OP* Kant demonstrates that experience is organized through the body.

> Empirical representation combined with consciousness is perception. Consciousness of the combination of perceptions into a whole (not as a fragmentary aggregate but as a system) is not, in turn, itself empirical, but a priori knowledge as to its form – that is, experience. This agreement is not derived out of (or from) experience, but is a synthesis of appearances in the subject for experience, and for the sake of its possibility (Kant OP, 22: 322).

He adds that "the amphiboly of concepts" consists in making "a leap from that which comes to us empirically, and is merely appearance, to experience since the latter would be an appearance of an appearance, and experience cannot be received as a representation which comes to us, but must be made" (ibid.). The human organism is understood as a unit capable of adding consciousness to empirical representations that mediate immediate appearance. This mediation creates phenomena for us. The mediation is derived from two characteristics, mentioned in the *CJ*:

1. The organism is part of the unity of nature (Kant, *OP* 22: 301).
2. The organism and its teleological structure constitutes our experience.[28]

The organism is an *a priori,* since experience is only possible thanks to the organism and its structure. This *a priori* is both formal *and* material.[29]

---

[28] Here, the experience is not merely perception: "A fragmentary aggregate of perceptions is not yet experience; rather, the latter takes place only in a system of perceptions which is founded a priori on a certain form [of their connection]" (Kant, OP 22: 457).

[29] Kant, OP 22: 498.

Sensible data are selected based on the intentionality of the organism which is both an organic body, and a part of the whole of nature. The result is that not all the features of an object are part of the sensible data submitted to the intellect in humans. Only the object's features that respond to the organism's purpose are selected to be submitted to the intellect. As a result, the subject acknowledges its teleological form in experience. For this reason, the organism is not given in experience but through experience.[30] According to Kant, the organism becomes the condition of possibility of experience and its own knowledge[31] in human beings, as Canguilhem stated in *Le concept et la vie* (Canguilhem, 1994, 340–343).

3 Biological Individuality and Organism in Georges Canguilhem's Philosophy

Canguilhem makes frequent allusions to Kant's *CJ*.[32] However, it remains unclear how Canguilhem ended up being interested in the *CJ*. When the *CRP* was included in the French curricula of the 1920s and 1930s, especially in the lectures given by Canguilhem's teachers (like Alain and Brunschvicg), the book was mainly studied in relation to its importance in the construction of an aesthetics.[33] Canguilhem started working on the *CJ* during the second half of the 1930s. In 1936–1937, he gave a series of lectures on "Science et technique" in which he analyzed Kant's teleological judgment (Canguilhem, 1937). Through the analysis of Canguilhem's unpublished annotations and thanks to the notes in his personal copy of the *CJ*, we can argue that Canguilhem developed an interest in Kantian purposiveness through René Berthelot's book *Science et philosophie chez Goethe* (1932), in which the author[34] discusses the organism's role in the *CJ* (Berthelot, 1932). Through Berthelot, Canguilhem understood the capital importance of Kant's focus on the organism, starting from the *CPR*. In *EHPS*, he writes that "The Transcendental Analytic had set forth the conditions of possibility of the knowledge of nature in general and found a limit in the fact that life is not only nature in the sense of 'natura naturata' but in the sense of 'natura naturante'" (Canguilhem, 1994, 352).

Canguilhem had clearly appropriated the definition of organism offered by Kant in the *CJ*:

---

[30] See also Van de Vijver, 2006.

[31] Kant, OP 22: 457. In this regard I would like to thank Levi Haeck for our discussion on the *OP*. Here, he stresses the importance for Kant of *the feeling of body:* throughout the feeling of your bodily movement you can conceive you organization (you can form the concept of organization).

[32] Canguilhem, 2008, 121; Canguilhem, 1939, 64–69; Canguilhem, 1994, 343–346.

[33] See also Bianco, 2012.

[34] Berthelot was one of the first to interpret Nietzsche as a pragmatist (See Berthelot, 1909). Berthelot was also a great scholar of Bergson. It was because of Canguilhem's interest in Bergson that he decided to read Berthelot's texts. See also Bianco (2006).

> An organism is a being which is both cause and effect of itself, which organizes itself and reproduces its organization, which forms and gives itself replication, in accordance with a type, and whose teleological structure in which the parts are in relation to each other under the control of the whole, attests the non-mechanical causality of the concept" (Canguilhem, 1994, 345).

The concept of biological normativity resembles that of an autopoietic capacity. According to Canguilhem, the living being produces its own norms, because life is a polarized activity, characterized by the power to choose (to exclude and prefer). In this sense, all organisms turn towards whatever is more desirable; trying to avoid whatever would lead it to perish. Referring to the normativity of health and disease, Canguilhem excludes matter from having normative powers saying: "there is no physical or chemical or mechanical pathology" (Canguilhem, 2011, 127). The idea of normative polarity brings Canguilhem close to the Kantian idea of judgment. "To live," Canguilhem writes, "means to valorize objects and circumstances of one's own experience. It means to prefer and eliminate the means, the situations, and the movements. Life is opposed to the relation of indifference towards the milieu" (Canguilhem, 1989). When Canguilhem mentions the state of wonder that strikes us when we look at an organism, he captures an idea that is already present in *CJ*. Although organisms may seem to "automatically" move and interact with the environment, there is always something that remains irreducible to automatism and mechanism. Canguilhem openly states that as much as it is absurd to assimilate the social field to the biological one, it is equally absurd to reduce an organism to a machine, or to a series of tropisms (Canguilhem, 2008, 119–126). The ability to choose between the polarity of life and normativity, exists even in the simplest living organisms (Canguilhem, 2011, 135–135).

It should be noted that, for Canguilhem, normativity is an internal power proper to the living. "If biological norms exist, it is because life, as not only subject to the environment but also as an institution of its own environment, thereby posits values not only in the environment but also in the organism itself. This is what we call biological normativity" (Canguilhem, 2011, 227). The teleological structure is the basis of the living being's normative power in its relationship with the environment. It guides the interaction between the living being and the environment. Thus, Canguilhem can state the following:

> We have no *a priori* knowledge [of normativity]. These forces that are forms and these forms that are forces are part of nature, they are in nature, but we do not know this through our intellect, we know it through experience. This is why the idea of natural purpose, which is the very idea of the organism that forms itself, is not a category in Kant, but a regulative idea whose application can be made only through maxims (Canguilhem, 1994, 227).

Canguilhem correctly points out that purpose does not have a constitutive role in determinative judgment, but is a maxim for teleological judgment.

In *EHPS,* Canguilhem recalls the importance of purpose – and its regulative form – for Kant's theory of the organism. He argues that the idea of purpose as a formal and structural *a priori* defines a particular relationship between the whole and the parts, namely in terms of a mutual causality (Canguilhem, 1994, 343). However, after the discovery of DNA, a formal *a priori* as in Kant's philosophy is no longer sufficient. From a formal *a priori* we must move to an "objective a priori," or "an a priori that is properly material and not only formal" (Canguilhem, 1994, 362). Canguilhem means that the discovery of DNA has resulted in a new concept of life. Biology is no longer based only on the concepts elaborated by chemistry and physiology but depends on those proper to the theory of information (Canguilhem, 1994, 360). Investigating life means investigating an order of language. Following an Aristotelian tradition, Canguilhem inscribes the *logos* into the living. Though matter returns to the foreground in the new biology, the living does not risk being confused with it. For Canguilhem, only the living is capable of self-reproduction, as was the case for Kant. Reproduction is the transmission and creation of a message. Nonetheless, the living being is not programmed solely by its DNA, since only a part of the information is found in the genetic code. To live, the human being must pass from a material level – genetic – to a cognitive level; producing concepts in order to know(Canguilhem, 2011, 208–2012). According to Canguilhem, the human being provides both of its possible living conditions, material and cognitive,[35] through its *structure* and its way of *functioning*. According to Canguilhem, while investigating genetic information and models, it is necessary to find "the Newton of the living organism"[36] in order to grant biology a legitimate scientific status (Canguilhem, 1994, 360). Although Kant never admitted the possibility of a science of the organisms – a biology – he nonetheless recognized the existence of a material *a priori* in the organism. In the *OP*, Kant emphasizes how the material *a priori* is connected with purposiveness (cf. Sect. 2). As seen in part two, purposiveness is normative. On the one hand, the purpose is not already given. Rather, as Canguilhem argues, it must be sought by looking at the organism and realizing that it is a set of parts held together by a purpose. This is why Canguilhem suggests replacing the term "teleological" with "organismic." On the other hand, purpose is normative as it is not absolute; it allows different interpretations. For example, "To assert that the adrenal capsule is necessary for life is a biological value judgment which does not relieve one from inquiring in detail into the causes through which a useful biological result is obtained" (Canguilhem, 2011, 216–217). Canguilhem stresses that, in biology, the idea of an organism's purposiveness should not exempt research from causal explanations. He suggests that the purposiveness is "the concept of sense, the concept of a possible organization, therefore an organization which is not granted" (Canguilhem, 2011, 179). Because the norm is not already given, it can change. Canguilhem demonstrates this by reflecting on health. Health consists in the ability of the living being to create new norms that replace previous ones.

[35] We can argue that the living can provide for its living conditions materially, while the human being does so both materially and cognitively.

[36] Investigating genetic information allows us to see that "the Newton of the living organisms" is, in Canguilhem's view, Claude Bernard.

4 Conclusion

In this chapter, I first considered how Canguilhem describes the formation of concepts in their relation to life (cf. Sect. 1). I then examined the Kantian theory of the organism, trying to show how the concept of the organism evolved in Kant's thought (cf. Sect. 2). I concluded that, in the *OP*, Kant viewed the teleological structure of the organism as a material *a priori*. Finally, I discussed Canguilhem's reading, and critique, of Kant's concept of purpose (cf. Sect. 3). Here, I tried to demonstrate how the idea of a material *a priori* (introduced by Canguilhem in 1966) correlates with the idea of the organism in the *OP*. In Canguilhem's work, norms have two characteristics that were already present in Kant. First, they are formed by generalization and formalization of data offered by sensibility. Secondly, they are formed by the interpretation and construction that the human beings have of theirs environment.

In Sect. 2, I argued that, according to Kant, knowledge is produced through the filtering action of the organism applied to empirical data and, in a second moment, in humans through generalization/objectification by the intellect. Additionally, Kant's cognitive model has strong affinities and resonance with Canguilhem's cognitive model. In *EHPS*, Canguilhem criticizes Kant for not accounting for his own theory of knowledge in his framework of a theory of life (Canguilhem, 1994, 352). This was presumably because he had not fully studied the *OP*. Here, the possibility of knowledge shifts from the subject to the organism. This shift is the foundation of both vital normativity, and of Canguilhem's theory of knowledge after the discovery of DNA.

Ultimately, I would state that the Kantian idea of the organism is, for Canguilhem, the starting point from which he articulates his historical-epistemological analysis of the living. Historical epistemology involves an investigation of the role that living humans have in different realms, such as history, politics and science. To put it in Canguilhem's words, "the idea of organic totality is a universal norm of judgment or research, of which we must ask ourselves: where is the model?" (Canguilhem, 1941, 88).

References

Beck, L. W. (1981). Kant on the Uniformity of Nature. *Synthese, 47*(3), 449–464.

Bernard, C. (1927). *Introduction to the Study of Experimental Medicine* (L.J. Henderson, Trans.). Macmillan.

Berthelot, R. (1909). Sur le pragmatisme de Nietzsche. *Revue de métaphysique et de morale, 17*(3), 386–412.

Berthelot, R. (1932). *Science et philosophie chez Goethe*. Alcan.

Bianco, G. (2006). La vita nel secolo da Canguilhem a Deleuze passando per Bergson. In G. Bianco (Ed.), *Il significato della vita*. Mimesis.

Bianco, G. (2012). Pacifisme et théorie des passions; Alain et Canguilhem. In Murat, M. et Worms F. (Eds.), *Alain, littérature et philosophie mêlées*. Presses de la rue d'Ulm.

Brilman, M. (2018). Canguilhem's Critique of Kant: Bringing Rationality Back to Life. *Theory, Culture & Society, 35*(2), 25–46.
Canguilhem, G. (1929/2011). À la gloire d'Hippocrate, père du temperament, in Canguilhem, *Œuvres Complètes* vol. I: *Écrits philosophiques et politiques 1926-1939*, ed. J.-F. Braunstein, Y. Schwartz. Vrin.
Canguilhem, G. (1937). Science et technique, lectures held in Toulouse. *Fonds Georges Canguilhem, GC, 11*(1), *1*.
Canguilhem, G. (1939). *Traité de logique et de morale*. Robert et Fils Editeurs.
Canguilhem, G. (1941). Organisme et totalité. *Fonds Georges Canguilhem, GC, 11*(1), 2.
Canguilhem, G. (1945). Nature et valeur du concept, lectures held in Clermont-Ferrand. *Fonds Georges Canguilhem, GC, 12*(1), 1.
Canguilhem, G. (1977a). *Idéologie et rationalité dans l'histoire des sciences de la vie*. Vrin.
Canguilhem, G. (1977b). *La formation du concept de réflexe aux XVIIe et XVIIIe siècles*. Vrin.
Canguilhem, G. (1989). Vie. *Encyclopedia Universalis*.
Canguilhem, G. (1994). *Études d'histoire et de philosophie des sciences*. Vrin.
Canguilhem, G. (2008). *Knowledge of Life* (S. Geroulanos & D. Ginsburg, Trans.). Fordham University Press.
Canguilhem, G. (2011). *The Normal and the Pathological* (C. Fawcett, Trans.). Zone Books.
Daled, P.-F. (2021). *Le concept d'anomalie chez Georges Canguilhem*. Classiques Garnier.
Ginsborg, H. (2014). *The Normativity of Nature: Essays on Kant's Critique of Judgement*. Oxford University Press.
Illetterati, L. (2014). Teleological Judgment: Between Technique and Nature. In I. Goy & E. Watkins (Eds.), *Kant's Theory of Biology* (pp. 81–98). De Gruyter.
Kant, I. (1781). *Critique of Pure Reason*. In I. Kant, P. Guyer, & D. A. Wood (Eds.), *The Cambridge Edition of the Works of Immanuel Kant* (Vol. 1999). Cambridge University Press.
Kant, I. (1790). *Critique of the Power of Judgement*. In I. Kant & P. Guyer (Eds.), *The Cambridge Edition of the Works of Immanuel Kant*. Cambridge University Press, 2000.
Kant, I. (1936–1938). *Opus Postumum*. In I. Kant & E. Föster (Eds.), *The Cambridge Edition of the Works of Immanuel Kant*. Cambridge University Press, 1992.
Kant, I. (1992a). *Inquiry Concerning the Distinctness of the Principles of Natural Theology and Morality*. In K. Immanuel & D. Walford (Eds.), *Theoretical Philosophy, The Cambridge Edition of the Works of Immanuel Kant* (pp. 1755–1770). Cambridge University Press.
Kant, I. (1992b). The Only Possible Argument in Support of a Demonstration of the Existence of God. In I. Kant & D. Walford (Eds.), *Theoretical Philosophy, The Cambridge Edition of the Works of Immanuel Kant 1755–1770*. Cambridge University Press.
Kant, I. (1992c). *Universal Natural History and Theory of Heaven*. In K. Immanuel & D. Walford (Eds.), *Theoretical Philosophy, The Cambridge Edition of the Works of Immanuel Kant* (pp. 1755–1770). Cambridge University Press.
Kant, I. (1992d). In M. J. Gregos (Ed.), *The Conflict of the Faculties*. University of Nebraska Press.
Kant, I. (1992e). In J. M. Young (Ed.), *Lectures on Logic*. Cambridge University Press.
Kant, I. (2007). On the Use of Teleological Principles in Philosophy. In R. Louden & G. Zöller (Eds.), *Anthropology, History, and Education, The Cambridge Edition of the Works of Immanuel Kant* (pp. 192–218). Cambridge University Press.
le Blanc, G. (2008). *Canguilhem et les normes*. PUF.
Lenoir, T. (1981). Teleology Without Regrets. The Transformation of Physiology in Germany, 1790-1847. *Studies in History and Philosophy of Science Part A, 12*(4), 293–354.
Lotfi, S. (2010). The Purposiveness of Life: Kant's Critique of Natural Teleology. *The Monist, 93*(1), 123–134.
Marcucci, S. (1990). Le deduzioni trascendentali della finalità oggettiva. Il trascendentale nella *Critica del Giudizio. Il Cannocchiale. Rivista di studi filosofici, 3*, 105–133.
McLaughlin, P. (1990). *Kant's Critique of Teleology in Biological Explanation*. Edwin Mellen Press.

Montuschi, E. (2018). Constructed Objectivity and Realist Presuppositions: A Kantian Framework in CHESS Working Paper 2018–1, Durham University. https://www.dur.ac.uk/resources/chess/CHESSK4UWP_2018_01_Montuschi.pdf

Onnasch, E.-O. (2014). The Role of The Organism in the Transcendental Philosophy of Kant's *Opus Postumum*. In I. Goy & E. Watkins (Eds.), *Kant's Theory of Biology* (pp. 239–255). De Gruyter.

Vaihinger, H. (1935). *The Philosophy of 'as if': A System of the Theorethical, Practical and Religious Fictions of Mankind*. Routledge & Kegan Paul.

Van de Vijver, G. (2006). Kant and the Intuitions of Self-Organization. In B. Feltz, M. Crommelinck, & P. Goujon (Eds.), *Self-Organization and Emergence in Life Sciences* (pp. 143–161). Springer.

Wolfe, C. T., Penoncelli, M., & Wong, A. (2020). What Kind of Vitalism in 'The Normal and the Pathological'? In P.-O. Méthot (Ed.), *Vital Norms: Canguilhem's the Normal and the Pathological in the Twenty-First Century* (pp. 224–254). Hermann.

Zuckert, R. (2007). *Kant on Beauty and Biology: An Interpretation of the Critique of Judgment*. Cambridge University Press.

Giulia Gandolfi is a PhD student in Philosophy at Cà Foscari University in Venice and at Paris 1 Panthéon Sorbonne University in Paris. During her master she studied at CAPHES (Centre d'Archives en Philosophie, Histoire et Édition des Sciences) where Canguilhem's personal library and manuscripts are stored. She works on Canguilhem's biological normativity concept connected to medical and social fields.

Canguilhem's Divided Subject: A Kantian Perspective on the Intertwinement of Logic and Life

Levi Haeck and Gertrudis Van de Vijver

Abstract By reappraising the biological theory of vitalism, Canguilhem attempted to give pride of place to the idea that acquiring knowledge *about* living beings is an activity *of* living beings. He is indeed credited with the view that knowledge in particular and rationality in general are "tied to a conception of life" whereby "life predominantly manifests itself in organic individuals that act and react within specific environments which, in turn, are defined by the needs and desires of these individuals" (Schmidgen H. Hist Philos Life Sci, 36(2): 4, 2014. doi:10.1007/s40656-014-0030-1). These needs and desires, we are told, constitute "un système de référence irréductible et par là absolu" (Canguilhem G. Le vivant et son milieu. In: La connaissance de la vie (Deuxième édition revue et augmenté). Librairie philosophique J. Vrin, 129–154, 1992c). Canguilhem's legacy is exactly this: rationality is rooted in life, and not the other way around. And yet, in "Le concept et la vie" (from 1966) and "De la science et de la contre-science" (from 1971), Canguilhem seems to tell another story about the complex intertwinement of life and rationality. Not only are we condemned to enter the realm of rationality (i.e., to take part in logical activities such as forming concepts and judgments about the world and about our own condition) because we have needs and desires as living beings, but we also have needs and desires as living beings that depend on the fact we are always already caught-up in the dynamics of rationality, i.e., always already logically active. At this point, Canguilhem's thinking comes closer than ever to a Kantian, transcendental point of view on rationality. Paradoxically, the inscription of human rationality in

L. Haeck (✉)
Department of Philosophy, Ghent University, Ghent, Belgium

The Research Foundation – Flanders (FWO Flanders) – Grant number: 1132621N, Brussels, Belgium
e-mail: levi.haeck@ugent.be

G. Van de Vijver
Department of Philosophy and Moral Sciences, Ghent University, Ghent, Belgium
e-mail: gertrudis.vandevijver@ugent.be

G. Bianco et al. (eds.), *Canguilhem and Continental Philosophy of Biology*, History, Philosophy and Theory of the Life Sciences 31,
https://doi.org/10.1007/978-3-031-20529-3_7

organic life brings with it the idea of a *subject*, divided between two inverse but correlated realms in which it cannot but participate: the singular realm of sensibility and the general realm of logic.

Keywords Canguilhem · Kant · Divided subject · Life · Logic · Need satisfaction · Vitalism

1 Introduction

In "Aspects du vitalisme", Canguilhem introduces a kind of vitalism, which, as Wolfe has it, is neither substantival nor heuristic but *attitudinal* (Wolfe, 2011). Vitalism, it is argued, is an 'orientation of biological thought' rather than one (outdated) biological theory among others (Canguilhem, 2021a, 84). Instead of seeking a justification for the claim that life, as a specific natural object, is in the possession of specific ontological properties, Canguilhem suggests that our (scientific) questions about life result from a *point of view* on experience (Canguilhem, 2021a, 84; Wolfe, 2011, 226). In order to flesh out the import of this idea, we propose to connect it to Canguilhem's many suggestions, scattered throughout his oeuvre, that the very activity of asking questions about life is indicative of the fact that the questioners themselves are always already part of life. There is, in this sense, an intrinsic tie between life and rationality, or, to put it in a philosophically more tractable way: there is an intrinsic tie between life and *logic*. It is however not immediately clear whether one of the two should take precedence over the other, a clear sign of which is the very conjunction in the title of "Le concept et la vie", Canguilhem's 1966 essay written in light of the then recent developments in molecular biology. We will analyze this text as shedding a distinctively Kantian light on what it means and what it has historically meant for logic and life to be rooted in each other, although it appears at times to treat the Kantian point of view as merely one among many options. Then we turn to Canguilhem's 1971 essay, "De la science et de la contre-science," where, once again as befits a good Kantian, the focus is shifted from 'the concept' to 'judgment'.[1] In our view, the originality of Canguilhem's approach in these two essays consists in a depiction of the life of the human subject as to some extent determined by the dynamics of logic: first in relation to 'the concept', then in terms of judgment. Although Canguilhem's oeuvre and legacy arguably tease out the idea that life is to be prior-ranked to rationality, whereby rationality is ingrained

[1] If our analysis is informed to a large extent by Kant's philosophy, this is so because, as argued elsewhere (Van de Vijver & Haeck, 2022), the latter also reveals an intrinsic link between logic and organicity/life. So in Van de Vijver and Haeck (2022) we argue in opposition to Brilman (2018) who maintains that in Kant, life is external to rationality *because* he takes life to be a blind spot to rationality. We think this is a misleading opposition at best.

in life rather than the other way around (cf. Schmidgen, 2014; Brilman, 2018), our close-reading of these two essays adds a significant nuance: if logic is a part of human life, it must have a life of its own.[2]

Throughout our analysis, we are guided by a question raised by Badiou (2012) about his mentor: *Y a-t-il une théorie du sujet chez Canguihem?* Our survey of "Le concept et la vie" and "De la science et de la contre-science" will bring us to respond to this question in the affirmative. Canguilhem's characterization of life as determined by logic, whereby logic is itself to be seen as a living dynamic, puts forward a divided subject in a surprisingly Kantian fashion. We are dealing with a subject who's life consists in thinking about life, and who's logical endeavors—consisting, basically, in forming judgments and concepts—make up its life. At stake is a subject that is, in this process, torn between its sensible impressions and its logical capacities, and therefore structurally unsatisfiable. In this way, our contribution purports to show to what extent Canguilhem himself might have had doubts about his well-known and well-received view that knowledge and rationality are, as Schmidgen has it, "tied to a conception of life" whereby "life predominantly manifests itself in organic individuals that act and react within specific environments which, in turn, are defined by the needs and desires of these individuals" (2014, 4). Perhaps the needs of organic individuals are, in the end, not so easily to be seen as amounting to an irreducible and therefore absolute system of reference, as Canguilhem has it in "Le vivant et son milieu" (2021b, 154). Perhaps the realm of logic—the realm of judgments and concepts—has a certain autonomy in relation to the system of organic needs. There is indeed, as aptly noted by Schmidgen, a more structuralist side to Canguilhem (2014, 250). We do think it is worthwhile to think through this structuralist strand in Canguilhem's oeuvre a bit more radically. We will do this along the lines of Badiou's suggestion that for Canguilhem there is a subject "insofar as there exists in the universe a living being such that, dissatisfied with meaning and capable of displacing the configurations of its objectivity, it always appears, in the order of life [...], as a living being somewhat *displaced* (2012, 79, our translation).

The Canguilhemian subject, then, is "reducible neither to the living [*le vivant*], nor to the knowing [*le savant*]" (Badiou, 2012, 71; our translation).[3] The subject, which occupies a *middle position* between the two poles whose intertwinement we want to investigate, must be taken seriously as the *effect of* their intertwinement. This opens up the space to further substantiate the idea, emerging in the two texts we will analyze here, that we are not only condemned to enter the realm of logic (i.e., to judge, to form concepts about the world and about our own condition)

[2] In this essay, we are solely concerned with *human* life. This has to do with the fact that for Canguilhem vitalism is no longer a classical theory *about* life, but a point of view on experience upheld *by* living, *human* beings.

[3] We translate 'le vivant' as 'the living', in order to remain as close as possible to the original. When we discuss 'the living', then, this always means something like 'the being that lives'.

because we have needs as living beings, but *also* that we have needs as living beings that depend on the fact we are *already* logically active.[4]

2 The Concept *and* Life: From Ontology to Epistemology, via Logic

The complex argument presented in "Le concept et la vie" is launched by a series of interconnected questions that inquire, firstly, whether life is "the *universal* organization of matter" or rather "the experience of a *singular* living being, i.e., man, conscious *of* life" 2015 [1966] (1966, 335; our italics).[5] Although Canguilhem presents it here as if we must choose between one or the other, both options remain at the horizon throughout his text. In line with Wolfe's (2011) characterization of Canguilhem's vitalism as *an attitude*, this opening question appears to suggest that the universal organization of matter, if this is indeed what we mean when we think about life, is nonetheless to be understood in relation to the standpoint of a singular *living* being that tries to come to terms with the fact that it is alive itself. Canguilhem makes quite clear from the beginning that this tricky situation hinges on a distinction between what he calls 'the living' (*le vivant*) and 'the lived' (*le vécu*). The latter, by which is meant the experience of a singular living being, is commissioned by the former, which is more fundamental (1966, 335).[6] So, if one addresses the relations between life and 'the concept' (1966, 335),[7] a qualification must be kept in mind: if *life* is understood to be the universal organization of matter, it must be seen as the form and the power of *the living*, but not of its *lived* experiences. Thus, Canguilhem conveys that the endeavors of the life sciences are to be situated somewhere *in between* (a) an investigation into the universal organization of matter, on the one hand, and (b) a qualification of the experience of a singular living being, i.e., man, as conscious *of* life, on the other. It is clear that Canguilhem leaves 'the lived' out of the equation and identifies '(a)' with 'the living'. But '(b)', which is at first sight left out of the equation as well (because it seems to correspond to 'the lived'), is, however, not so easily detachable from 'the living'. This confounding borderline position, occupied by 'the living', will form the guiding thread of Canguilhem's arguments in "Le concept et la vie", and bring him to develop an account of the subject as *divided*—or so we will argue.

---

[4] Although we cannot develop this idea here, we are inclined to add to this that even what *counts* as a need depends on the fact that we are always already logically active.

[5] All translations of "Le concept et la vie" and "De la science et de la contre-science" are ours.

[6] He explains it as follows: "With life, one can understand either the present participle or the past participle of the verb 'to live', namely the living and the lived. The second meaning is in my view directed by the first, which is more fundamental." (1966, 335).

[7] Canguilhem does not aim to investigate a specific concept, but conceptuality in general, i.e., *the* concept, although he does focus on the concept of life.

What this borderline position designates can be elucidated by considering Canguilhem's second group of questions, which adds an epistemological layer onto these seemingly ontological concerns. He asks: is it possible for the concept to obtain *access to life*, and if so, how? And do we proceed, in the knowledge of life, (a) from intelligence to life or rather (b) from life to intelligence? In regard to the first case (a), Canguilhem wonders how intelligence could *meet* life, while in regard to the second case (b) he asks how intelligence could possibly *lack* life, i.e., how it could possibly not be alive itself.[8] This is indeed the essential predicament: even though the living resists objectification (that is, conceptualization) by an intelligence (that is, by the concept) so that the possibility of the latter *meeting* the former is problematized, we should nonetheless ask ourselves how this very intelligence (that is, the concept) could *lack* life, if it is indeed, itself, alive. This intertwinement of ontology (how can the concept lack life?) and epistemology (how can the concept meet life?) brings Canguilhem to an attention for logic: *what is intelligence?* In considering the idea that the concept might *be* life itself—although this is not to say that the concept *knows* life—Canguilhem acknowledges that we must ask ourselves whether or not we can, in fact, have access *to intelligence*: "[…] if the concept were life itself, we would have to ask ourselves whether or not it [i.e., the concept] is capable of giving us access to intelligence" (1966, 335).

3 Aristotle's Paradigm: The Primacy of the Individual

On the basis of this immensely rich introduction, Canguilhem then proceeds with a historical analysis of the problem. He traces the identification of the living with the concept (or with intelligence more generally) all the way back to Aristotle, who maintained that the nature of the living is the *anima* or soul, which, as the form life, is at once the reality (*ousia*) and the definition (*logos*) of the living. This means that "the concept of the living is therefore the living itself" (1966, 336). Aristotle's ancient epistemology, in which the modern challenge of having to bridge two realms (knowledge on the one hand and the knowable world on the other) is absent, cannot go unmentioned here: "The natural hierarchy of forms in the cosmos commands the hierarchy of definitions in the logical universe" (1966, 336). This dependence of logic (the logical universe) on ontology (the natural hierarchy of forms) is to be taken literally. The fact that a *syllogism* concludes with necessity is due to the

[8] A peer reviewer notes that this amounts to a category mistake. Perhaps the question about 'intelligence *meeting* life' should not be conflated with the question about 'intelligence *lacking* life', because in the first case life takes the position of the (purportedly) known object, while in the second case it seems to be a feature of the knowing subject. This is a category mistake in the sense that two distinct meanings of the concept 'life' are used interchangeably and applied in contexts in which they do not signify the same thing. However, affirming that they have a different meaning seems to beg the question, according to Canguilhem, for his analysis appears to be bracketing the very distinction itself.

natural hierarchy according to which a species dominated by a genus is in turn dominant with regard to a lower species (1966, 336). This means that there is indeed no issue of having to bridge two realms, because logic and ontology cannot be disentangled to begin with: "[...] knowledge is the universe as thought from within the soul [*l'univers pensé dans l'âme*], rather than the soul thinking the universe [*l'âme pensant l'univers*]" (1966, 336). In line with his subtle introductory discussion of Aristotle, Canguilhem lays bare to what extent the logic implicated by Aristotle's ontology dictates the kind of epistemology we should go for:

> If the essence of a being is its natural form, this entails the fact that beings, being what they are, are known as they are and for what they are. The intellect is identified with what is intelligible [*L'intellect s'identifie aux intelligibles*] (1966, 336).

In this sense it is true that the concept of life is life itself (Canguilhem, 1966, 336). Of all beings, moreover, *living* beings seem to Aristotle to be the epitome of intelligibility, because intelligence is not only *in* the world, but also a capacity *of* living beings in the world.[9] By referring to the living, the idea of rationality's proximity to the world (of logic to ontology) is therefore *reinforced* rather than problematized.

However, in anticipation of modern and post-modern ways to situate the place of the living in the ontological, epistemological and logical enterprise we call 'science', Canguilhem immediately takes this Aristotelian approach to trial: "how could knowledge be both mirror and object, the reflection and the reflected?" (1966, 337). In upholding the Aristotelian stance, according to which life and (its) concept stick together, so that both life and rationality are *what knows* and *what is knowable*, one affirms the following: "The definition of man as [...] a reasonable animal [*zōon logon echon*] [...] amounts to making science [...] an activity of life itself" (1966, 337). Canguilhem makes clear that this Aristotelian paradigm revolves around a realist understanding of the notion of the singular *individual* in logic and ontology:

> [A] difficulty of Aristotelianism that has subsisted concerns the ontological and gnoseological *status of individuality in a concept-based knowledge of life*. If the individual is an ontological reality and not merely the imperfection of the realization of the concept, what should be the range attributed to the order of beings represented by the classification by genera and species? (1966, 339; our italics).

In *Categories*, the notion of 'subjecthood' (*hypokeimenon*, 'what underlies') is indeed closely connected to the notion of a primary substance,[10] which is explained

[9] "The world is intelligible, and the living beings in particular are, because the intelligible is in the world." (Canguilhem, 1966, 336)

[10] For instance, when Aristotle says in *Categories* that "it is because the primary substances are subjects for all the other things and all the other things are predicated of them or are in them, that they are called substances most of all" (2002, 7).

as a "this" (*tode ti*), a singular thing, like 'this horse here' or 'this man'.[11] Such 'this-ness' grounds both logic and ontology, because it grounds both syllogistic reasoning and classifications of reality in terms of species and genera.[12] But this raises a question: how could a *concept*, which is by definition a general representation, be *about* a singular thing such as an individual living being?

> If the concept presides ontologically over the conception of the living being, of which mode of knowledge is the *individual* susceptible? A system of living forms, if it is founded in being, has as a correlative the *ineffable* individual. But an ontological plurality of individuals, if it is given, has as a correlative *the concept as fictitious* (Canguilhem, 1966, 340; our italics).

Canguilhem also formulates the problem as follows:

> *Either it is the universal that makes the individual into a living being* [...], whereby singularity is to life what an exception is to the rule: it confirms it [...], since it is by virtue of the rule and against the rule that singularity appears [...]. *Or it is the individual that lends its color, its weight and its flesh to this abstract phantom that we call the universal*—otherwise universality would be to life a way of talking about it, which is, precisely, saying nothing about it (1966, 340; our italics).

If concepts, which are *general* in nature (or universal, as Canguilhem has it), determine the living being as an object of knowledge, it is unclear how this being, conceived precisely as a *singular* individual, could be known through concepts. From the standpoint of the concept, the individual itself has no place; it is *ineffable*. However, if in biology one takes for granted that *individual* living beings are as such given, the concepts we attach to them become secondary, indeed 'fictitious'; they become merely a way in which we can *speak* about something. It is this kind of ontological priority of the individual, capable of constituting a plurality in and of itself—i.e., without the constitutive role of concepts, as Kant would have it (cf. infra)—that seems to be the central tenet of Aristotle's approach. And, as made clear by *Categories*, this approach is fundamentally, for Aristotle, a view on what *logic* is: a tool to reflect on singular living beings used by singular living beings. "Aristotelian logic received, because the forms of reasoning imitated the hierarchy of living

---

[11] Aristotle explains it as follows: "3b10. Every substance seems to signify a certain 'this'. As regards the primary substances, it is indisputably true that each of them signifies a certain 'this'; for the thing revealed is individual and numerically one. But as regards the secondary substances, though it appears from the form of the name—when one speaks of man or animal—that a secondary substance likewise signifies a certain 'this', this is not really true; rather, it signifies a certain qualification, for the subject is not, as the primary substance is, one, but man and animal are said of many things. However, it does not simply imply a certain qualification, as white does. White signifies nothing but a qualification, whereas the species and the genus mark off the qualification of the substance—they signify substance of a certain qualification. (One draws a wider boundary with the genus than with the species, for in speaking of animal one takes in more than in speaking of man.)" (2002, 9).

[12] In chapter three of *Categories*, Aristotle writes: "Whenever one thing is predicated of another as of a subject, all things said of what is predicated will be said of the subject also. For example, man is predicated of the individual man, and animal of man; so animal will be predicated of the individual man also—for the individual man is both a man and an animal" (2002, 4).

forms, a reassurance of the correspondence between logic and life" (Canguilhem, 1966, 344).

4 The Life *of* the Concept: Kant's Logical Horizon and the Regulativity of the Principle of Purposiveness

Kant's transcendental standpoint on logic portrays the exact opposite scheme, where, by virtue of the assumption that logic consist of a hierarchy of concepts in which concepts refer only to other concepts, the singular or individual living being cannot be grasped:

> Transcendental logic fails, in its *a priori* constitution of nature as a system of physical laws, to constitute [...] nature as the theater of living organisms. We understand better the investigations of the naturalist, but we do not manage to understand the procedures [démarches] of nature. We understand better the concept of causality but we do not understand the causality of the concept. [...] Kant does not admit the identification between the logical horizon of the naturalists and what one could call *l'horizon poïétique de la nature naturante* (Canguilhem, 1966, 344; our italics).

Several things happen in this passage. Firstly, Canguilhem correctly asserts that in Kant's transcendental logic, the *a priori* laws of physics drawn from the categories and principles of the understanding do not constitute nature as "the theatre of living organisms". This is Kant's well-known stance on biology, according to which living organisms do not count as objects. The investigation and analysis of the living by way of *concepts*—for instance in terms of *species and genera*—can only count as a 'regulative' and never as a 'constitutive' use of these concepts. Secondly, Canguilhem is right in considering Kant's transcendental logic as an analysis of what it means to investigate nature, rather than of what objectively *happens in* nature. Transcendental logic helps us to "understand the concept of causality" in its constitution of nature as a system of causal laws, but not to understand "the causality of the concept" (Canguilhem, 1966, 344). The 'causality of the concept' refers to Kant's idea in the *Kritik der Urteilskraft* that the concept of a living being depends on a *teleological* judgment. In the "First Introduction" to this work, we read that a teleological judgment "compares the concept of a product of nature as it is with one of what it *ought to be*. Here the judging of its possibility *is grounded in a concept (of the end)* that precedes it *a priori*" (EEKU, AA 20: 240; second italics are ours).[13] It turns out, then, as Ng has it, that the concept *can* contain "the ground for an object's actuality, and that it is only in such cases that we refer to the concept or object as a purpose or end" (2020, 54). Indeed:

---

[13] We always refer to the *Akademieausgabe* (*Gesammelte Schriften*) of Kant's works, either by indicating title abbreviation, volume and page numbers, or, when referring to the first *Critique*, by way of the customary A and B indications. However, we cite from the English translation if available.

> [p]urposiveness of form [required for our conception of the organism] refers to the power or *'causality' of a concept* with regard to its object, and this relationship defined for Kant the unity of diversity and lawfulness of the contingent that provided the horizon of determinability for judgment" (Ng, 2020, 54; our italics).

This indeed seems to be precisely what Canguilhem means, when he says that

> [a]n organized being is a being which is at the same time cause and effect of itself, which organizes itself and which reproduces its organization, which forms itself and which gives itself the replica, in accordance with a type, and whose teleological structure, where the parts stand in a relation to each other under the control of the whole, testifies to the non-mechanical *causality of the concept* (1966, 344–345; our italics).

However, even if Kant gives a place to the causality of the concept (in addition to the concept of causality) in his analysis of living beings, he does *not* give it constitutive power, but only a *regulative* function. Thirdly, then, and seemingly on the basis of this caveat, Canguilhem presents Kant as noticeably resisting Aristotle's paradigm, which identifies the "logical horizon of the naturalists" with "l'horizon poïétique de la nature naturante" (1966, 345).

Canguilhem's reading of Kant is especially subtle and promising when the Appendix to the first *Critique*'s "Transcendental Dialectic" is brought up. In this text, Kant introduces the concept of a 'logical horizon' in order to account for the *regulative* instead of constitutive character of the principles of homogeneity and specification in nature. In its investigation of nature and, more importantly, living beings, the Kantian faculty of reason is guided by two opposing tendencies or 'interests'. Firstly, reason adheres to the principle of homogeneity (i.e., it seeks "sameness of kind in the manifold under higher genera") and secondly, it adheres to the principle of specification (i.e., it seeks "what is same in kind under lower species").[14] Kant explains that the logic behind these two principles is predicated on the notion of a concept and the "logical horizon" that accompanies it (see KrV, A 654/B 683 and further). It is worth quoting Kant in full here:

> One can regard every concept as a point that serves *as the standpoint of an observer and thus has its horizon*, i.e., there is a multitude of things that *can* be represented and—as it were—surveyed from this standpoint. Within this horizon there must be a *multitude of points that can be indicated ad infinitum, each having in turn its own narrower purview*. I.e., every species contains subspecies, according to *the principle of specification*, and the logical horizon consists only of smaller horizons (subspecies), but *not of points having no range* (*individuals*). But for [several] different horizons (i.e., genera) determined by equally many concepts, a common horizon can be thought as drawn, from which those different horizons can one and all be surveyed as from a central point. This common horizon is the higher genus. And so on, until finally we reach the highest genus […]. To this highest standpoint I am led by *the law of homogeneity* […] (KrV, B 686; our italics)

This discussion of the logical horizon and what it implies for our understanding of the notion of the concept is especially interesting for Canguilhem's purposes,

[14] These two laws are moreover guided by the law of affinity, which combines them by offering "a continuous transition from every species to every other through a graduated increase of varieties" (KrV, A 657-8/B 685-6).

because it too is tied to a notion of living beings. According to Canguilhem, it should be stressed that for Kant, a concept is (*i*) a point of view, (*ii*) tied to a logical horizon that consists itself of an indeterminate manifold of viewpoints, (*iii*) from whence is opened up once again a manifold of horizons, be it of lesser purview, whereby it has to be admitted that (*iv*) a horizon can only be comprised of horizons, just as a concept can only be analyzed into concepts, so that (*v*) species can only be divided into subspecies but never into the individuals themselves supposedly contained under them. For to know is to know through concepts, and concepts cannot, unlike sensible intuitions, 'give' the object in its singular individuality.[15] The transcendental-logical analysis expounded here is implicated by and gives direction to the use of concepts in the investigation of nature (1966, 343–345).

This indicates the extent to which logic—the science of the form of thinking in general (KrV, B XXIII)—functions as the transcendental ground for all naturalist endeavors: "Reason itself, according to Kant, prescribes this procedure; and to prescribe it is to *proscribe the idea of a nature in which no resemblances would appear*, since in this case the logical law of species as well as the understanding itself would be simultaneously annihilated" (Canguilhem, 1966, 344; our italics). Reason *must* seek homogeneity among the diverse and diversity among the homogenous, by virtue of its logical nature. In reading Kant, Canguilhem insists on the idea that this 'interest of reason' manifests itself above all in relation to the phenomenon of life: "[...] where the knowledge of life pursues its heuristic task of determination and classification of species, reason becomes the interpreter of the exigencies [*exigences*] of the understanding. These needs define a transcendental structure of knowledge" (1966, 344; our italics). This is to say: the scientific investigation of life, pursuing a heuristic project of classification, invites reason to fold back onto itself (see Van de Vijver & Haeck, 2022). In folding back onto ourselves as thinking and knowing instances, we ask: what is this 'I', this 'thing' or this 'instance' that thinks, knows, and thinks it can know? While for Kant reason is unable to objectively investigate organized beings, reason's conceptualization of them in terms of a reciprocal causality displays an *internal* purposiveness that is akin to the purposiveness of the faculty of reason *itself*. "One of Kant's greatest services to philosophy was" indeed, as Hegel has it, "in drawing the distinction between relative or *external* purposiveness and *internal* purposiveness; in the latter he opened up the concept of *life*, the *idea* [...]" (WL, 654/12.157).[16] Both living organisms and the faculty of reason are internally purposive, self-organized systems. Or, as Mensch has it: "When reason saw organic activity in nature, according to Kant, what it was really looking at was itself" (2013, 144).

[15] What is given in intuition, is the object's individuality and not the individual object. One can only account for an individual object if the connection of a concept (which thinks the object) with an intuition (which gives the object's individuality under the name of the manifold of intuition) is presupposed (KrV, A 320/B 376-7; KrV, B 137).

[16] See Ng (2020) for interesting analysis of Hegel's reading of Kant in this respect.

5 From Kant's Critique to Hegel's Aristotelianism

Canguilhem reveals quite well that the Kantian subject (throughout this text we have referred to the subject under the name of 'rationality', 'logic' 'intelligence', etc.), if it is organized and self-organizing, is also structurally unsatisfiable: *it will never cease* to scientifically investigate life, *nor will it ever seize* life's true essence objectively. This situation is problematic, according to Canguilhem, and it occasions him to shift his attention to Hegel, who supposedly did not refuse what Kant had forbidden himself. That is, who has not, on the grounds that life cannot be an object graspable through concepts, refused to connect in a fundamental way life to the concept (1966, 345). Hegel, as presented here by Canguilhem (who in turn draws on Hyppolite's reading of Hegel), emerges as offering a promising middle ground between Kant and Aristotle. Hegel's point of view seems to combine *the gap* between life and the concept with *the tie* between them. The crucial Hegelian move appears to be the following:

> [T]he movement of life betrays—betrays because it tries to translate—the infinity of life, which, rising in man to self-consciousness, [thus] inaugurates spiritual life. However, since the multitude of the species forms an obstacle to the universality of life, one could not, at risk of error, conclude in turn that spiritual life is biological life (Canguilhem, 1966, 346).

This means that if, for Hegel, life and the concept are tied together (in line with Aristotle), there is still a fundamental gap between both (in line with Kant). The attempt to remedy for the infinite manifoldness of life's forms inaugurates what we can call 'spiritual life', which is from that moment on opposed to 'biological life' (i.e., the infinite manifold of life's forms). The reasoning here is subtle: even if intelligence *cannot lack* life (as there is no spiritual life without presupposing biological life as originary in the first place), this is not to say that intelligence *could meet* life (as spiritual life is always distinct from biological life).

This complex exposition of the Hegelian point of view is, to us, much more Kantian than Aristotelian. It could however become stronger if its Kantian background were to be recognized and made explicit. Kant's question is what it means, *logically*, for the intellect to generalize what is singular (by bringing the general *to* the singular), and in line herewith what it means to seek diversity among the homogeneous while affirming the homogeneity of the diverse. The answer Kant gives to this question is a highly sophisticated attempt to conceive of rationality (intelligence, logic) as what Canguilhem ascribes to Hegel under the name of 'spiritual life'. Canguilhem's Hegel is less sharp than his Kant, however. In spite of its more appealing exposition of *the life of the concept*, the Hegelian point of view seems less systematically critical of the possibility of *a concept of life* (as in: knowledge about life) than the Kantian one. So indeed, the fact that "Hegel did not refuse what Kant had forbidden himself" (1966, 345) has another meaning as well. Although Hegel's analysis of *the life of the concept* is deeply inspired by Kant's philosophy, he supposedly also proclaimed in a rather Aristotelian vein that life and the concept *do* stick together ontologically: "'Life, Hegel says, is the immediate unity of the concept and its reality, without this concept being distinguished from it"" (1966, 345), which inevitably has epistemological consequences:

> In any case, we must ask Hegel the question of *how, if it is true that the concept and reality immediately coincide in life, a knowledge of life through concepts at the level of science is possible*. The answer is, of course, that knowledge can only organize itself through the life of the concept itself. 'I place that by which science exists in the self-movement of the concept, Hegel says' (1966, 346; our italics).

Hegel is presented here as a less rigorous *transcendental* thinker than Kant. Or, put more neutrally, he is presented as returning to Aristotle's ancient worldview, be it from within a modern framework. Rationality appears not only as a capacity *of* living beings, but ipso facto as something that enables us to *know* them. But this confuses the distinction between (*i*) *the life of the concept* on the one hand and (*ii*) knowledge about life ('*the concept of life*', one could say) on the other. This would not be so problematic if it did not amount to a confusion of (*i*) a reflexive presupposition with (*ii*) a predicative knowledge claim. Kant, on the other hand, by holding on tightly to the distinction between regulative and constitutive principles, did not fall prey to such confusion. If we reconstruct the citation above from a Kantian point of view, we would be forced to say that taking into account, rightfully so, "the life of the concept itself" or "the self-movement of the concept" still does *not* lead to the assertion that "the concept and reality immediately coincide in life" such that we would have something like "a knowledge of life through concepts".

The passage quoted above conveys a crucial dual sense: on the one hand, it represents Hegel's insightful take on the Kantian idea that reason will attempt to objectify life (even if it will never achieve this goal) in terms of the logical self-movement of the concept—that is, in terms of the fact that reason is itself fundamentally alive. On the other hand, however, the assumption that conceptuality *is* life (i.e., the self-movement of the concept) is presented as securing the possibility of knowledge *about* life. Was this not Aristotle's paradigm? Canguilhem's Hegel fails indeed to resist the temptation of the Aristotelian ideal. Presumably, as it is presented here, Canguilhem seems to recognize in Hegel a Kantian and therefore more modern and *critical* reappraisal of the otherwise non-Kantian, but Aristotelian promise (which consists in making the concept and life stick together ontologically and epistemologically). To the extent that Canguilhem is developing a critical viewpoint on the concept and life, his main target remains the *gap* between life and the concept presupposed by Kant's transcendentalism: there is "[...] resistance of the thing, not to knowledge, but to a theory of knowledge that proceeds from knowledge to the thing" (1966, 351). If he grants that there is no scientific meta-perspective on life, he still concedes, in line with Hegel's revised Aristotelianism, that there is nonetheless *a* perspective, which is a perspective *of* life *on* life. The core issue is framed as follows:

> We can [...] ask ourselves how life is inclined to [disposée à] outline in its products precisely what one of its products, man, will perceive, erroneously and correctly at the same time, as an invitation of life to be conceptualized by man (Canguilhem, 1966, 352).

If there is a meta-perspective on life, it must have emerged from life itself. Yet in criticizing, and rightfully so, the idea of a meta-perspective, Canguilhem mistakenly associates it with a Kantian transcendental subject: "[i]t is not because I am thinking, it is not because I am a subject, in the transcendental sense of the term, but *it is because I am living that I must seek in life the reference of life*" (1966, 252; our italics). This joins Canguilhem's attitudinally vitalist and famous proclamation, in the "Introduction" to *La connaissance de la vie*, that "[t]he thought of the living must take from the living the idea of the living" (2021c, 13).

6 Is Logic Intrinsic to Life, or Does Life Disconcert Logic? Towards Life as Dissatisfaction

Canguilhem's thinking is most pressing when it conceives of the living as standing *between* the universal organization of matter, on the one hand, and the experience of a singular, living being that is conscious of life, on the other. Indeed, the very opposition between these two poles is, as will be shown below, the consequence of the fact that there is a fundamental involvement of logic in human life.[17] What, then, should we make of Canguilhem's earlier, rather Kantian idea, expressed in *La formation du concept de réflexe aux XVIIe et XVIIIe siècles* (1955), that life *disconcerts* logic? (2021d, 1). Or, put in a more epistemological vein, that life cannot by grasped by logic? This principle is clearly pushed to the background in "Le concept et la vie". In the latter, Canguilhem dwells on the new promises of molecular biology as prepared by figures as diverse as Aristotle, Hegel, and Bergson. The new paradigm of molecular biology and its scientific promise, exemplified by the Nobel Prize-winning research of Jacob, Monod and Lwoff just a year before the publication of "Le concept et la vie", must have excited Canguilhem—and understandably so.

> So when we say that biological heredity is a communication of information, we reencounter in a way the Aristotelianism from which we started. In presenting the Hegelian theory of the relationship between the concept and life, I wondered whether, in a theory that was so strongly connected to Aristotelianism, we might not find a more faithful way of interpreting the phenomena discovered by contemporary biologists and the explanatory theories they propose than in an intuitivist theory like Bergson's. Saying that biological heredity is a communication of information is, in a certain sense, to return to Aristotelianism, if it is to admit that there is in the living a *logos*, inscribed, preserved and transmitted (1966, 362).

---

[17] This contribution focuses on *human* life and does not purport to analyze non-human life. And yet we would be inclined to say that *all* life forms (including plants, mushrooms, sea urchins, amoeba, etc.) are unavoidably approached by us humans by means of a conception of life which ultimately originates in what we take to be human life. So, although we do not analyze non-human life, our contribution does concern the *knowledge* of non-human life (as part of our analysis of the relation between life and rationality). Although we must reserve the interesting issue of non-human life 'as such' for another occasion, we would like to thank the peer reviewer that pressed us on this point.

But these options—exemplified by Aristotle, Hegel, and Bergson—all testify to the potential annulment of the Kantian credo, echoed in Canguilhem's own œuvre, that life disconcerts logic, because indeed: "[t]o define life as a meaning inscribed in matter is to admit the existence of an objective *a priori*, of a properly material and not only formal *a priori* [...]", whereby "[d]efining life as meaning is to oblige oneself to a work of discovery" (1966, 362). If life is to be found 'in matter' while simultaneously being a 'logical issue', then our *concepts* can grasp life as an object; then biology can be a proper science after all. So even if "Le concept et la vie" begins as a historical and critical analysis of the relation between life and the concept, it culminates in a cautious yet programmatic defense of scientific optimism vis-à-vis the project of (molecular) biology.

Canguilhem's writing is nonetheless quite suspicious (hence his caution) of such a renewed scientific optimism, for he ends his text by addressing once more the epistemological issue of what can legitimately count as knowledge, given the fact that we are alive, or better, given 'the fact of life'. His final remark on knowledge hinges on the notion of a divided subject: "To be the subject of knowledge [être sujet de la connaissance], if the a priori is *in* the things, if the concept is *in* life, *is only to be dissatisfied with the meaning found. Subjectivity, therefore, is only dissatisfaction.* But perhaps *this* is life itself" (1966, 364; our italics). This suggestive conclusion is crucial. Firstly, the conception of life as the universal organization of matter is once again juxtaposed to the conception of life as something about which the subject must become conscious. However, the *juxtaposition itself* is now presented as much more fundamental than the juxtaposed elements, thus marking a division that is *intrinsic* to rationality. Secondly, the point of reference in relation to which the concept and life can be seen as disconnected, is the notion of the *subject*. Thirdly, and in the opposite direction, an element common to both life and the concept is introduced: the subject's *(dis)satisfaction*. If we succeed in scientifically presenting and investigating life in terms of a universal organization of matter (namely by forming concepts *about* life, guided by the assumption that these concepts *belong* to life as predicates to a subject),[18] we will still be unsatisfied with what we discover. Seemingly, we will be left with a sense of inadequacy, of lack, on account of which the formation and usage of concepts (in view of the investigation of life) is *self-moving*: it reinstalls itself over and over again. On this ground, Canguilhem appears indeed to conclude quite fundamentally that the concept *is* life. Put differently, we could say that the concept is not so much a component *of* (human) life but rather that it *is* alive. We find concepts in life, but we also find life in concepts. This means that if we see living activities in nature, whether or not this activity is rational, we will see something that has a striking resemblance to rationality, to the concept. Life and the concept, then, are not intertwined with each other as, respectively, a known object and a knowing subject. They are intertwined by virtue of the fact that they simultaneously take the position of both the knowing subject and the known object.

[18] We are of course invoking Aristotle's logical lingo here, according to which a predicate *belongs* (*huparchei*) to a subject if the predicate is said to assert something *about* or affirm something *of* the subject.

But this ultimately means that the clear divide between the knowing subject and the known object does not apply to life. *Saying* that the concept *is* life entails, contrary to the Aristotelean paradigm, that life *disconcerts* logic (i.e., the concept). If this is Canguilhem's true, yet hidden conclusion, than "Le concept et la vie" is an unknowingly Kantian treatise.

However, Canguilhem is still adamant that he has no use for Kant's transcendental subject: "[i]t is not because I am thinking, it is not because I am a subject, in the transcendental sense of the term, but it is because I am living that I must seek in life the reference of life" (1966, 252). There is truth to this, but it seems to play on a false dichotomy if we take Canguilhem's own conclusion more seriously than he did. This conclusion leads the way to connect more fundamentally subjectivity to conceptual dissatisfaction and the living. For being alive as a human being *is* to be a subject that is structurally condemned to think; *is* to account for the manifold of sensibility in relation to the demands of reason and the understanding; *is* to be proceed conceptually.[19] We read an echo of this in Kant's analysis of reason's conceptual investigation of life, and of nature more generally:

> When merely regulative principles are regarded as constitutive, then they can, as objective principles, be in conflict with each other. But if they are regarded merely as *maxims*, then there is no true conflict but merely a diverse interest of reason that causes *the splitting of the way of thinking* [die Trennung der Denkungsart] (KrV, A 666/B 694).

Why there appears within reason such a 'splitting in the way of thinking' can be explained by reference to the subject's conceptual condition: reason is never satisfied with the understanding's conceptual progressions because concepts never really grasp the thing in itself. It is never grasped, because the human being has a *discursive* understanding, rather than an *intuitive* one, according to Kant, i.e., it will forever be faced with the task of bridging the realms of intuitions and concepts (KrV, A 256/B 311-2; KU, AA 05: 406). In the *Kritik der reinen Vernunft* (but also in the third *Critique*), Kant makes clear that the thinking subject is indeed *condemned* to

[19] Again, one might object that this is a very anthropomorphic conception of life, but some form of anthropomorphism is perhaps unavoidable if we take seriously the idea that vitalism and its conception of life amount to an orientation of biological *thought*. Canguilhem himself, however, would most probably disagree with us here. In most cases (with the exception, at least according to our reading, of "Le concept et la vie" and "De la science et de la contre-science"), he claimed that human life is grafted on life as such. In "La pensée et le vivant" (i.e., the introduction to *La connaissance de la vie*), for instance, knowledge is connected to life by being defined as "a general method for the direct or indirect solution for the tensions between man and milieu" (2021c, 10). This conception of knowledge as in a sense merely a *method of* life, could, according to Schmidgen (2014, 247), shed a light on why Canguilhem and Planet, in their *Traité de Logique et de Morale*, specified what a concept is in terms of "[an] enunciation of the problem to be solved" (1939, 724/96). Still, what follows from this is an open question. Does it entail that human life is at bottom reducible to life as such? Or does it entail that human life, afflicted with its conceptuality, serves as the model for all other life forms in the sense that plants, amoeba, etc., in resolving the tension between themselves and their milieu, could very well be utilizing concepts too? Or does it entail the view, as endorsed by us in this contribution, that life, insofar as it is conceived to be a known object/concept, always presupposes a knower, such that the knower is unavoidably implicated even in what is rightfully called non-human or non-rational life?

judge, to conceptualize, and to form ideas, as it never attains the object as a thing in itself, although this is precisely what it *structurally* aims at. All thinking is representational, even if it fails precisely at this point. On that account, the subject is structurally caught in the activity of judgment. Thus, what Canguilhem gathers under the heading of 'the concept'—i.e., what he calls 'intelligence' and 'rationality'—refers to a self-organized, internally self-moving and purposive system, that is, a *living* system. Life disconcerts logic as an object, yes, but in a way, logic *is* life. In "Le concept et la vie", however, the consequences of this crucial point are not yet followed through. In view of this we must now turn to "De la science et de la contre-science".

7 The Divided Subject's Broken Judgment

Canguilhem's meditation on the living vis-à-vis the concept brings to the fore a divided subject. Although Canguilhem does not explicitly come to this conclusion himself, the subjective division emerging from his thinking has a Kantian flavor. This brings us to the second element of logic that deserves a juxtaposition with life: judgment. In Kant's philosophy, this central power of the intellect occupies a place *between* life and the concept, although in its operations it is 'alive' and 'organized' itself. In this capacity, the figure of judgment reasserts the subjective division alluded to in "Le concept et la vie". In Kant's philosophy, moreover, judgment is subjective and objective at the same time. In the first *Critique*, it is a subjective condition of objectivity, while in the third *Critique*, where, as is well-known, objectification fails, judgment is presented as a condition for subjectivity itself. Across Kant's oeuvre, both the objective and the subjective, the necessary and the contingent, the general and the singular, are opposed to one another by reference to the operations of this fundamental and basic function of our intellect. In Kant, the importance of judgment becomes especially clear in relation to the issue of the living. It is therefore quite striking that in "De la science et de la contre-science", a dense philosophical text about the paradoxical dialogue between knowledge and life, as Machery (2008) has it, Canguilhem too turns his attention to judgment. Certainly for Kant, and perhaps also for Canguilhem, there is no concept without judgment. A concept, for Kant, is a predicate of a possible judgment (KrV, A 69/B 94), so that in every concept a judgment is silently at work (cf. Steckelmacher, 1879, 22). Something very similar speaks from Canguilhem's transition from "Le concept et la vie" to "De la science et de la contre-science".

In the latter, Canguilhem sets the stage with the following opening question: "[s]hould we say, like La Fontaine, 'when water bends a stick, my reason straightens it', or should we say that my reason confirms [...] the broken stick in its brokenness [*brisure*]?" (1971, 173). In his concise answer, the core of his theory of judgment (and, by extension, of his theory of subjectivity) is proclaimed:

> [...] on the one hand, one continues to perceive a stick as broken of which one knows that it is straight, without succeeding in substituting the knowing for the perceiving, like one observation for another; on the other hand, the admission of the necessity of the illusion forces to stop identifying being with appearance. An opposition breaks the unity of the affirmation. Although appearing as broken, the stick is no longer broken, but it is *judgment* that is broken (1971, 173; our italics).

One might object to the weight we attribute to Canguilhem's use of this classical example of ancient skeptical epistemology.[20] And yet we cannot ignore that it immediately summons a comparison with the opening passage of "Le concept et la vie", where a careful distinction was made between the experience of a *singular* living being and the *universal* organization of matter. Both essays appear to be set off by an opposition between a '*me*-perspective' and a '*reality*-perspective'. In "De la science", however, the focus shifts from the concept and life to judgment and life. The problem in relation to the objectivity of the stick's brokenness cannot be explained by reference to the empirical *concept* of a stick only. The problem rests on our *seeing* of the stick as broken, while we know, conceptually, that it is not. It is here that, in a rather Kantian way, judgment is invoked. Judging is primarily the activity of subsuming particulars under general representations (and of *seeking* the general for the particular). As such, judging is what enables objective knowledge, according to Kant, since it enables the co-operation of our two distinct sources of knowledge: *sensibility*, which represents the manifold of sense-impressions through singular intuitions, and *discursivity* (the understanding), which represents the unity of a manifold through general concepts. Both are necessary but insufficient. In an act of judgment, a manifold is *determined* by means of concept, on account of which an object is constituted and objective knowledge becomes possible (see, most famously, KrV, B 141).

However, not all particulars presented to us in intuition can be determined by a concept. There are cases in which a concept is simply *unavailable*. In these cases, as Kant has it, judgment is merely reflective, instead of determinative: it must *find* the general for the particular (EEKU, 20: 209–210; our italics). Quite often, for instance in the aesthetics of beauty and the sublime, this situation results in a contemplation on the particular *as a mere particular*, and not as something that can be grasped objectively. This goes not only for aesthetics, but also for nature in its vast manifoldness. The diversity of natural phenomena will indeed elude to a large extent our conceptual procedures. Living organisms, moreover, are the objectification-resisting natural phenomena *par excellence*. They are, in a sense, formally unexpected, because they exceed the mechanical causal order. Indeed, we know already from "Le concept et la vie" that the categories and principles of the understanding do not constitute nature as "the theatre of living organisms" (1966, 344). However, while these principles and their corresponding concepts of mechanical science are blind for living organisms, the power of judgment is *not*. When the conceptual determination of sensible intuitions fails or becomes contentious, there is a sense in which judgment keeps track of the elements involved in *any* conceptual

20 We would like to thank Massimiliano Simons for voicing this objection.

determination of sensible intuitions, failed or successful. The activity of judgment is indeed the only activity from within which a distinction can be made between sensible representations, which hold only 'for me' because they relate to the receptivity of my subject in a contingent time and place, and determining conceptual elements, which must, as necessary rules of the understanding, always hold 'for us'.

In this sense, we *must* hold on to the stick's brokenness, because this is how the stick is presented to us in sensibility, and because judgment keeps track of every element involved in its activities. The reason why we must hold on to the stick's brokenness, then, is twofold: it is both because we receive sense impressions *and* because we judge (by means of concepts). On the other hand, the brokenness of the stick *must* be foreclosed, because it adheres to a me-perspective incapable of possessing the universal (or general) normativity of concepts. But these *universal* concepts depend on the same capacity to judge that makes us hold on to the *singular* me-perspective. Two imperatives, then, are to be accounted for from within the activity of judgment: "one continues to perceive a stick as broken of which one knows that it is straight, without succeeding in substituting the knowing for the perceiving" (i.e., the first imperative), and "the admission of the necessity of the illusion forces to stop identifying being with appearance" (i.e., the second imperative) (Canguilhem, 1971, 173). But these imperatives, although they belong to the same activity of judgment, are conflictuous. Our concern should therefore be the question why *judgment* is broken, rather than whether or not, and if so, why, the *stick* is to be seen as broken. Canguilhem specifies:

> At first, judgment retained two affirmations in one: there is a reality given to me from where I am and from where I perceive it. Now two affirmations diverge: there is a reality that I do not grasp as such from where I am and from where I perceive it. The real is affirmed as a vection of *my* judgment. And this vection is the opposite of the vection according to which a judgment can be *said* to be my judgment (1971, 173; our italics).

In these passages, Canguilhem is at work in a Kantian universe: there is a sense in which reality is, at first, presented to me 'here and now', i.e., with regard to the necessary contribution of my sensible intuitions. And yet reality has nothing to do with 'me, here and now'—reality is a normative issue, governed by categories and principles of the understanding that serve as universal rules. This universal order runs counter what can be called the singular order of *my* judgment. The real is, moreover, a 'vection of my judgment', according to Canguilhem. The term 'vection' comes from medicine and signifies the transmission of a pathogen from one organism to another. Just as if someone would infect someone else with a virus, the relation between the reality-perspective and the me-perspective is characterized by a certain *transmission* from the latter to the former. But then Canguilhem adds that the vection which appears to install the real is the inverse of the vection that installs the me-perspective. As a consequence, the transmission has two directions. That something can be *called* my judgment, depends on the affirmation of the real, although the affirmation of the real results from a transmission from the me-judgment: "Appearance is born as appearance, at the same time as reality, at the moment when it is placed next to reality and against it, that is to say outside of it" (Canguilhem, 1971, 174).

In Kant's epistemology, a similar dynamic is upheld. On the one hand, sensible impressions 'awaken' my discursive faculties "into exercise" (KrV, B 1), whereby the power of judgment makes possible the transmission from singular representations presented in intuition to general representations in concepts, serving as 'marks' of the former (KrV, A 320/B 376-7; Log, AA 09: 91; Longuenesse, 1998, 107–127). On the other hand, however, it is only because we are always already at work within the realm of generality (through pure categories of the understanding), that we can identify something *as* being non-general and non-conceptual (or non-categorial), i.e., *as* purely singular or sensible. In judging, we are torn between two poles that proceed in different directions and are equally 'prior' to one another, be it in a different sense: the general pole is *transcendentally* prior because it allows, in one fell swoop, for the identification of the singular pole as *genetically* prior. This is why Kant notes that "all our cognition commences with experience,yet it does not on that account all arise from experience" (KrV, B 1). Much more poignantly than Kant, however, Canguilhem shows to what extent these two poles structurally run *counter* each other:

> From the moment that the initial affirmation has split into two, simple, *inverse and correlative*, one of the two is not the other. And they can only be sustained together and in the same way by forgetting *their reciprocal exclusion*. One cannot be simultaneously naive and warned, credulous and critical, presumptuous and lucid, ignorant and learned. *Science is the explicit negation, by positive exclusion, of that negation of science*, implicit by confusion, which was ignorance. [...] Ignorance, here, is the innocence, the illusion, the initial presumption that made one take for resolved a question not yet asked, [that made one take] for an affirmation of reality *a judgment not yet divided* (Canguilhem 1971,174; our italics).

The two poles are indeed inverse only insofar as they are correlated, but, since the exclusion is reciprocal, they correlate because they are inverse. As soon as the genetically prior moment of appearance is opposed to the one of reality, we reason as if the moment of reality is prior: "'The stick is not broken' coming after 'the stick is broken' means: 'the stick was never really broken'" (1971, 176). He adds a bit further that "[t]he false is therefore never a moment of the true" (1971, 174) because "the true is immediately retroactive, timeless" (1971, 174). It is clear that the epistemological relation between the true and the false, as well as between the real and the appearance, is governed by the logical principle of non-contradiction. Canguilhem asserts "that the same judgment cannot be sometimes true, sometimes false, but that it is one or the other, once and for all" (1971, 175). Moreover, "[i]n the order of the true it is impossible to qualify that which is different from the true other than as vain, illusory, *nul*" (1971, 176). The true itself is indeed not "a pro-position but a pre-supposition" (1971, 176). This is why science, according to Canguilhem, is the explicit negation of a negation of science. It operates from within the reality-perspective and thus reverses the order of the progression of knowledge altogether, so that our initial moment of *affirmation* is considered as already a *negation* of something else, within which, however, this 'something else' remains unqualified. The initial moment of affirmation is indeed an 'ignorance': it 'considers resolved a question not yet asked', and it affirms a reality through a still *undivided* judgment. But this initial moment can only be identified as such by leaving it behind and attributing to it, from then on, a certain nature. In this way, the *division* proper to judgment, to science, between appearance and reality, between the me- and the reality-perspective, i.e., between the "two references of judgment" (1971, 174), is in a most fundamental sense prior to the non-division. This is why "the search for truth is the effect of a choice that does not exclude what is inverse to it" (1971, 177).

8 The Desire to Find the True Is What Invents It. On Logic's Needs

What can this tell us about the intertwinement of the living with logic? Canguilhem follows Nietzsche here in claiming, on the basis of this analysis of judgment, that logic must be taken as "a will to find the truth, and therefore, in the final analysis, as a pretext [expédient] to invent it" (1971, 175). The desire to find the true, then, is what invents it. It is at this crucial point that life and logic enter the scheme *together*: "[l]ogical thinking can only function under the presupposition of the myths of being, substance, thing, identity", whereby these myths are institutions that have a "sécurité vitale" (1971, 178). True, we might also say that it is "for the peacefulness of the living, as a preventive defense reaction against the unexpected" (1971, 178), that the realm of the reality-perspective is opened and entered. But this is to miss the subtlety of the intertwinement of logic and life. In the Nietzschean universe, it is clear that logic is a way *of* life to annul itself:

> It is logic itself that is error. [...] This error that is logic is not an error in the logical sense. Otherwise, Nietzsche would have oriented his philosophy according to the magnetic needle of logic. On the contrary, he situated the logic in his philosophy according to the wind rose of life. The truth is a kind of error, in the sense of a vital illusion, without which a certain kind of living being, man, would not be able to live (1971, 178).

Nietzsche, of course, is correct to highlight that it is not because logic is an error *of* life, that there should be solidarity between life and logic: "[k]nowledge, a negation of life, itself a condition of knowledge, would thus be a perversion of life, or perhaps only an expression of its *fatigue*" (1971, 179). And indeed, if logic is a way of life to get control over itself,[21] then logic, when it becomes knowledge, must *fear*

[21] Instead of speaking in terms of control, Jacques Lacan will say that logic is an attempt at meta-language, by which he means that logic attempts, but always structurally fails, to capture through formal means what it is that cannot be grasped through language (2006, 30). No reason for him, though, not to continue to stress the importance of *saying*, the importance of enunciation, however much it is inscribed in a logic of structural failure. In line with this idea, Marjorie Grene, in discussing the issue of reductionism in the context of biology and philosophy of biology, quite similarly stresses the importance of discourse: how could a reductivist viewpoint, stating that, for instance, biological systems could and should be reduced to constituent particles and their workings, account for the fact it is a theory, a text, a discourse in which it is expressed and has to be expressed? "In short, were reductivism true, knowledge would be impossible, including the knowledge that reductivism is true. And were reductivism true, language would be impossible, including the formulation of the reductivist's thesis" (Grene, 1974, 56).

life and seek death: "[k]nowledge, anxious of a stable object that is identical to itself, would be afraid, not of death, but of life, which is power, struggle, invention, risk" (1971, 179). But while he is clearly inspired by it, Canguilhem appears to be discontented with the way in which Nietzsche allegedly treats human life as a unified force that comes to the point of the search for truth, science, and knowledge while its essence is, at the same time, being obstructed by it. Canguilhem wonders how, from within human life, something like logic could come to the fore and be developed into science, if the latter is ultimately a perversion of life, its destruction? Indeed, his question is not so much philosophical as it is biological: "[i]f life were nothing but life, force, and will to power, its drop of tension [*chute de tension*] would be unintelligible" (1971, 180). That is,

> If life contains its own limitation, why should science, which theorizes about it [i.e., about life] in taking it as its object, be only an 'error' of life? Why could science, the daughter of the fear of life, as a determination of the limits of life, not be accepted by life and be courageously used by life? (Canguilhem, 1971, 180).

But both Canguilhem's Nietzsche and Canguilhem himself seem confused. In view of remedying this confusion and offering some clarity to the argument, we need to connect the analysis of logic in terms of a broken judgment to the analysis of logic as a desire to find the true that thereby *invents* the true. In *Träume eines Geistersehers*, an early text of Kant's, reason is described as "halb dichtende und halb schließende" (*Träume*, AA 02: 348). Lachterman takes this to mean that reason is "half inventive, poetizing, half inferential, syllogistic" (1990, 197). This duality is the clue to the riddle. Let us take a look at the central description of the operation of the faculty of reason, found at the beginning of the first *Critique*'s "Transcendental Dialectic". On the one hand, there is the so-called 'logical principle of reason' according to which "the proper principle of reason in general (in its logical use) is to find the unconditioned for conditioned cognitions of the understanding, with which its unity will be completed" (KrV, B 364/A307). On the other hand, closely connected to the former, there is what we could call the 'metaphysical principle of reason': "when the conditioned is given, then so is the whole series of conditions subordinated one to the other, which is itself unconditioned, also given (i.e., contained in the object and its connection)" (KrV, B 364/A307-8). In reason, then, the unavoidable *tendency towards x* amounts, necessarily, to an unjust but perhaps equally unavoidable *affirmation of x*. Kant describes this overflow as a "need of reason" (B 365). Indeed, "if a cognition is regarded as conditioned, reason is *necessitated* to regard the series of conditions in an ascending line as completed and given in their totality" (KrV, B 388). Kant similarly suggests in *Was heißt: Sich im Denken orientiren?* that through its drive for knowledge (*Erkenntistrieb*), which is connected to the acknowledgement of a certain lack (*Mangel*), reason installs a feeling of need (*Gefühl des Bedürfnisses*) for the subject (WDO, AA 08: 139n). This is why a 'critique of pure reason' is required: reason's formal procedures lead, enthusiastically, to the ideas of the soul, the world-whole or -sequence, and God as the central 'concepts' of dogmatic metaphysics. More crucially, this means that reason is indeed half inventive, half inferential. Or, that logic is a desire to find the true that invents it, even if it is unjustified in doing so. The motor behind these dynamics of reason, says Kant, are the activity and the form of judgment (KrV, A 405-406/B 432-433, but also A 340/B 398).

Now, if the *division* proper to judgment mentioned earlier, i.e., the division between appearance and reality, between the me- and the reality-perspective, as the "two references of judgment" (1971, 174), is prior to the non-division, and if this division installs the search for truth, then it is indeed a matter of the dynamics of *judgment* when we assume, with Canguilhem and Nietzsche, that logic must be taken as "a will to find the truth, and therefore, in the final analysis, as a pretext [expédient] to invent it" (1971, 175). If we read that "[l]ogical thinking can only function under the presupposition of the myths of being, substance, thing, identity", such that these myths have a "sécurité vitale" (1971, 178), then this means, when we approach things with a Kantian state of mind, that we are dealing here with a "sécurité vitale" *of logic*. If the ideas of the soul, God, and the world-whole result from needs of reason that derive from its logical nature and impose themselves on the human subject, then logic is not only a need *of* the human, living subject, enabling it to lead its life by installing safety and controllability (which, paradoxically, is death).[22] If this is the case, then logic is *itself already alive*. This is what ultimately constitutes subjective division: as a living organism that has needs that must be relieved, the human being can, to that end, do nothing but take recourse to a realm in which it already takes part and that already has certain needs of its own. In this way, the living organism we call the human being *becomes* a subject—subject to the division it cannot but reestablish time and again for itself.

9 Conclusion

We cannot be accused or praised for simply having placed some kind of 'grid' on Canguilhem's oeuvre in order to find what is there and what is not. Instead, we reflected on what was being said in view of taking it beyond itself. And indeed: although we have wrung out his thinking like a sponge, we cannot ignore the water that came out. Three interrelated ideas were intimated and expounded in carrying out this analysis:

1. There is a hidden theory of the subject ingrained in "Le concept et la vie" and "De la science et de la contre-sience", because in these writings the intertwinement of logic and life is implicitly thematized from a Kantian point of view.
2. According to this theory, the heterogeneity between sensibility and logic constitutes a divided subject that is structurally unsatisfiable.

[22] It would take us too far to expand upon the Freudian death drive and the way Lacan translates it: not as a drive for destruction, but as enjoyment (*jouissance*). Although this would be an interesting parallel, we reserve this for another moment.

3. This revealed that logic is not only a part of life, but also has a life of its own. More fundamentally, this realization was taken to entail (rather than to thwart) the view that life also *disconcerts* logic.

There are basically two options when it comes to assigning a place to the subject within the intertwinement of logic and life. If logic is only a part of life, whereby the subject merely makes use of its rational capacities in order to relieve its organic needs, then we must assume the subject to be the *ground* of the intertwinement. Canguilhem's legacy seems to boil down to exactly this option. If, on the other hand, logic has a life of its own, then we must assume the thinking subject to be an *effect* of it. Now this option, although it is at odds with his legacy, seems to be upheld by Canguilhem too, be it in his late and much-ignored, yet surprisingly rich reflections on logic.

References

Aristotle. (2002). *Categories and De Interpretatione*, trans. J. L. Ackrill. Oxford University Press.

Badiou, A. (2012). Y a-t-il une théorie du sujet chez Canguilhem ?. In *L'aventure de la philosophie française* (pp. 65–79). La Fabrique.

Brilman, M. (2018). Canguilhem's Critique of Kant: Bringing Rationality back to Life. *Theory, Culture & Society, 35*(2), 25–46. https://doi.org/10.1177/0263276417741674

Canguilhem, G. (2015 [1966]). Le concept et la vie. In *Etudes d'histoire et de philosophie des sciences*. Vrin.

Canguilhem, G. (2018 [1971]). De la science et de la contre-science. In *Hommage à Jean Hyppolite* (pp. 173–180). PUF. Reprinted in Canguilhem, *Œuvres complètes* vol. 5 : *Histoire des sciences, épistémologie, commémorations: 1966-1995*, ed. C. Limoges (pp. 403-410). Vrin

Canguilhem, G. (2021a [1947]). Aspects du vitalisme. In C. Limoges (Ed.), *Œuvres complètes* (Vol II). Vrin.

Canguilhem, G. (2021b [1947]). Le vivant et son milieu. In C. Limoges (Ed.), *Œuvres complètes* (Vol II). Vrin.

Canguilhem, G. (2021c [1952]. La connaissance de la vie. In C. Limoges (Ed.), *Œuvres complètes* (Vol II). Vrin.

Canguilhem, G. (2021d [1955]). *La formation du concept de réflexe aux XVIIe et XVIIIe siècles*. In C. Limoges (Ed.), *OEuvres complètes* (Vol II). Vrin.

Canguilhem, G., & Planet, C. (2011 [1939]). *Traité de logique et de morale.* In C. Limoges (Ed.), *Œuvres complètes* (Vol. I). Vrin.

Grene, M. (1974). Reducibility: Another Side Issue? In *The Understanding of Nature. Essays in The Philosophy of Biology* (53–73). D. Reidel.

Hegel, G. W. F. (2010). *The Science of Logic* (G. Di Giovanni, Ed. & Trans.). Cambridge University Press. https://doi.org/10.1017/9780511780240

Kant, I. (1900). *Gesammelte Schriften* (Vol. 1–23). Deutsche Akademie der Wissenschaften.

Kant, I. (1998 [1781/1778]). *Critique of Pure Reason* (P. Guyer & A. W. Wood, Eds. & Trans). Cambridge University Press. https://doi.org/10.1017/CBO9780511804649

Kant, I. (2000[1790]). *Critique of the Power of Judgment* (P. Guyer & E. Matthews, Trans.). Cambridge University Press. https://doi.org/10.1017/CBO9780511804656

Lacan, J. (2006 [1965-66]). *Le séminaire XIII: L'objet de la psychanalyse: 1965–1966*. M. Roussan.

Lachterman, D. (1990). Kant: The Faculty of Desire. *Graduate Faculty Philosophy Journal, 13*(2), 181–211. https://doi.org/10.5840/gfpj19901329

Longuenesse, B. (1998). *Kant and the Capacity to Judge: Sensibility and Discursivity in the Transcendental Analytic of the Critique of Pure Reason* (C.T. Wolfe, Trans.). Princeton University Press.

Machery, P. (2008). *Canguilhem et le concept d'idéologie scientifique*. Groupe d'études « La philosophie au sens large ». https://philolarge.hypotheses.org/files/2017/09/14-05-2008.pdf

Mensch, J. (2013). *Kant's Organicism*. The University of Chicago Press.

Ng, K. (2020). *Hegel's Concept of Life: Self-Consciousness, Freedom, Logic*. Oxford University Press.

Schmidgen, H. (2014). The Life of Concepts. *History and Philosophy of the Life Sciences, 36*(2), 232–253. https://doi.org/10.1007/s40656-014-0030-1

Steckelmacher, M. (1879). *Die formale Logik Kant's in ihren Beziehungen zur Transcendentalen*. Verlag von Wilhelm Koebner.

Van de Vijver, G., & Haeck, L. (n.d.). Judging Organization: A Plea for Transcendental Logic in Philosophy of Biology. In M. Mossio (Ed.), *Organization in Biology. Foundational Enquiries into a Scientific Blindspot* (forthcoming). Springer.

Wolfe, C. T. (2011). From Substantival to Functional Vitalism and Beyond: Animas, Organisms and Attitudes. *EIDOS, 14*, 212–235.

Levi Haeck is an FWO doctoral candidate at Ghent University (Belgium), specializing in Kant's transcendental logic. He is especially interested in the relationship between logic and metaphysics, as well as between logic and life in Kant's philosophy and in the Kantian tradition.

Gertrudis Van de Vijver is full professor in philosophy at Ghent University. Her research deals with the issue of complexity, self-organization and teleology in the life sciences and with issues around the system of the psyche from within the psychoanalytical theories of Freud and Lacan. For both, she contributed to the articulation of a transcendental epistemology, centring on the idea of co-constitution. From 2016 till 2018 she was the director of the Centre for History of Philosophy and Continental Philosophy (HICO) at Ghent University. Between 2005 and 2010, she led an interdisciplinary research project on Complexity and Systems Biology that involved philosophers, biologists, bio-engineers and communication scientists.

Knowledge, Life, and Error. Nietzschean Themes in the Work of Georges Canguilhem

Henning Schmidgen

Abstract It is well known that authors such as Emile Chartier, i.e. Alain, Henri Bergson and Kurt Goldstein impacted importantly on the work of Georges Canguilhem. This paper argues, however, that it is the philosophy of Friedrich Nietzsche which gave Canguilhem's work its most distinctive traits. Themes such as the relationship between health and disease, the influence of language on perception and knowledge, or the conception of philosophy as a philosophy of values, are obviously Nietzschean. The paper shows that in Nietzsche as well as in Canguilhem these themes rely on and refer to the research direction of "General Physiology," which since the 1860s investigated the relation of the organism to its environment with regard to phenomena such as assimilation, nutrition, and orientation. At the same time, the paper shows that Canguilhem's reliance on Nietzsche remains intact even as General Physiology is being eclipsed by the emergence of Genetics and Molecular Biology, and as the life sciences as a whole reconfigure themselves with respect to the concept of information.

Keywords Philosophy of Values · General Physiology · Molecular Biology · Assimilation · Information · Error

Canguilhem is said to have once remarked to a student, "*Je suis un nietzschéen sans carte*" (Stiegler, 2000, 99),[1] which might be translated as, "I am a Nietzschean without any official membership card." Michel Foucault seems to cautiously confirm the content of this statement when he speaks in the 1980s of Canguilhem being "very interested in Nietzsche" (Foucault, 1983, 199) and of his works displaying a "certain affinity with Nietzsche" (Foucault, 1991, 69).

[1] Unless otherwise stated, all translations are my own (H. Sch.).

H. Schmidgen (✉)
Department of Media Studies, University of Weimar, Weimar, Germany
e-mail: henning.schmidgen@uni-weimar.de

G. Bianco et al. (eds.), *Canguilhem and Continental Philosophy of Biology*, History, Philosophy and Theory of the Life Sciences 31,
https://doi.org/10.1007/978-3-031-20529-3_8

However, one looks in vain for explicit committments to Nietzsche's philosophy in Canguilhem's work. At the same time, references to Nietzsche's works, especially to the edited volume *La volonté de puissance* (Nietzsche, 1935/37), are an almost constant feature of Canguilhem's writings, from the early *Traité de Logique et de Morale* (Canguilhem & Planet, 1939), the 1952 essay collection *Knowledge of Life* (Canguilhem, 2008) and his 1958 critique of psychology to the remarkable essay "De la science et de la contre-science" (Canguilhem, 1971), published side by side with Foucault's (1977) famous piece "Nietzsche, Genealogy, History," and his late writings on the philosophy of medicine and health (Canguilhem, 2012).

As is well-known, authors such as Emile Chartier, i.e. Alain, Célestin Bouglé, Henri Bergson, Kurt Goldstein, and Gaston Bachelard crucially impacted on Canguilhem's thought. But it is the ongoing dialogue with Nietzsche that gives his philosophy its most characteristic grounding. Indeed, the main themes of his philosophy can be readily identified as Nietzschean: philosophy as a philosophy of values and meaning, as an activity of critique and interpretation; the understanding of history as a "useful" element of life, the interpretation of the past from the "supreme power of the present" (see also Bachelard, 1951, 24); the question of the relationship between sickness and health, the fight against the "utilitarian" equation of the normal and the average, the consideration of health from the perspective of illness and vice versa; the assumption of the language-bound nature of all knowledge of nature and the associated critique of language with its particular focus on the lineages and genealogies of concepts; finally, the preference for polemical, limited interventions and quasi-experimental forms of writing, the resistance to any system of philosophy.

Above all, however, Nietzsche appears as a decisive point of reference for Canguilhem, insofar as the author of *Ecce homo* got involved with the sciences and especially the life sciences of his time – not only through reading, but equally through visits to some laboratories, e.g. the marine station in Villefranche-sur-Mer, not far from Nice. Against this background, Nietzsche contributed to the founding of a new kind of philosophy after the end of metaphysics (see Krummel, 1988; Müller-Lauter, 1974, 1978; and Stiegler, 2001). It is therefore no surprise that even where Canguilhem does not refer directly to Nietzsche or, as in *Knowledge of Life,* lets a passage from Thomas Mann's novel *Doctor Faustus* speak for his themes (Canguilhem, 2008, 133), the trace of precisely this philosopher is often clearly recognizable.[2]

Remarkably, this is also true for the 1940s, i.e. the famous *Essay on Some Problems Concerning the Normal and the Pathological*, Canguilhem's medical doctoral thesis. To conceive of life as essentially "normative," i.e. value-setting, to see the living being as endowed with a "spontaneous effort" to "dominate the environment and organize it according to his values" (Canguilhem, 1989a, 228–229), to describe health as a state in which "one feels that one is not only possessor or bearer

[2] See also Brusotti (2021), who presents and discusses an unpublished essay, written by Canguilhem in 1947/48, that deals with Nietzsche's maxim „Become who you are!"

but also, if necessary, creator of value, establisher of vital norms" (ibid., 201) and to translate "to be cured" without hesitation as "to be given new norms of life, sometimes superior to the old ones" (ibid., 228), to speak at all of men "for whom it is normal to break norms and establish new ones" (ibid., 165) and finally to tie the knowledge of norms to the experience of "breaking," i.e. transgressing them (ibid., 209) – all these are not only topics of resistance per se, but also connections to Nietzsche's physiologically grounded theory of the will to power, connections that at this time, i.e. in 1943, when he defended his thesis, are obviously directed against a reception of this theory by an exactly opposite politics. One might even say that, in the *Essay*, Canguilhem turns Nietzsche against the appropriation of precisely this philosopher by the National Socialists.

Comparably implicit but nevertheless clear is the presence of Nietzsche in Canguilhem's philosophical *thèse* on the formation of the concept of reflex in the Seventeenth and Eighteenth centuries, which appeared in 1955, twelve years after the defense of the *Essay* (on this point, see already Schmidgen, 2008). On the one hand, it is the relation between language and assimilation that makes one think of Nietzsche. The formation of the reflex concept by Thomas Willis may be a result of "speculative changes" of older theories, as Canguilhem states; at the same time, it is based on the more or less successful "assimilation" of biological phenomena to technological ones. Thus, the formation of concepts is on the one hand a question of the "imaginative and analogical talent" (Canguilhem, 1977, 81), of the "analogical imagination" (ibid., 170) or, to speak with Paul Valéry (whom Canguilhem also frequently quotes): "our faculty of changing images, of combining them, of making part of one coexist with part of another, and of perceiving, voluntarily or involuntarily, the connections in their structure" (Valéry, 1972, 11).[3]

On the other hand, the analogizing access to preliminary forms of knowledge is bound to specific material cultures, to certain forms of technology. In his book on the formation of the reflex concept, Canguilhem identifies a striking difference between Willis and Descartes. In the latter, the assimilation of physiological functions to mechanical and hydrodynamic models dominates: levers, winches, pulleys, clocks, organs, fountains; in Willis, in contrast, it is light machines and fire devices that serve as explanatory comparisons: glowing mirrors, Greek fire, cannons, gunpowder. It is these comparisons that lead to the reflex concept becoming an intellectual constant in Willis, rather than being reduced, as in Descartes, to two times using a word (Canguilhem, 1977, 46; 66).

To speak of "assimilation" in this context is not an arbitrary choice. Like his sociological teacher Célestin Bouglé, Canguilhem considers the recourse to life necessary in order to adequately grasp the emergence and evolution of facts that ultimately manifest themselves as social (technology, science, language, etc.). In this sense, he already poses the rhetorical question in the 1943 *Essay*: How should

[3] Canguilhem cites another passage from Valéry's "Introduction" in the reflex book and other Valéry texts elsewhere in his work (e.g., Canguilhem, 2008, 140). In contrast, he hardly refers to considerations in the philosophy of science concerning the relationship between concept and analogy as, for example, spelled out in Metzger (1926).

the normativity essential to human consciousness be explained "if it did not in some way exist in embryo in life" (Canguilhem, 1989a, 127)? Similarly in his article on the history of cell theory: How should one be able to understand that "stupid humanity" has become intelligent one fine day, if one radically devalues "old intuitions" (Canguilhem, 2008, 56)? At least implicitly, in the reflex book, the formation of concepts is also traced back to biological phenomena. The process of assimilation, which is decisive for the formation of concepts, is not only to be understood as linguistic assimilation or comparison, but also as living appropriation, as incorporation and transformation.

"Even for an amoeba, living means preference and exclusion," writes Canguilhem (1989a, 136) in his medical *thèse*. The human judgment function thus sees itself referred back to a polarity anchored in the elementary expressions of organic individuality. The trace of such a polarity can also be recognized in the reflex book: here hydrodynamics as central analogy, there pyrotechnics; here the mechanism of Descartes, there the vitalism of Willis. These theoretical choices also imply 'preferences and exclusions.'

In addition to this, in 1966, in an essay on the reflex concept in the Nineteenth century, Canguilhem states: "In the *species* of movements, the reflex concept delimits a certain *type*" (Canguilhem, 2002, 295; italics added). Here it seems as if it was life itself that had expressed itself conceptually before it enters the conceptual expressions of humans.

Something similar, perhaps one should say *very* similar, can be found in Nietzsche. This is especially true with regard to the relationship between assimilation and language. In "On Truth and Lie in the Extra-Moral Sense" Nietzsche observes: "What the researcher [...] is seeking is only the metamorphosis of the world into man. He strives to understand the world as a human-like thing, and at best he achieves by his struggles the feeling of an assimilation" (Nietzsche, 1968, 105; on this point see also Emden, 2004). In contrast to the neurophysiological theory of metaphor that Nietzsche also develops in this context ("A nerve stimulus, first transposed into an image! First metaphor," etc.; Nietzsche, 1968, 101), his comparable considerations from the 1880s are derived from general biology.

More precisely, it is the research tradition of General or Comparative Physiology, which in turn prepares Theoretical Biology, that forms the framework for Nietzsche's considerations – similar as in Canguilhem at a later point in time. Since the 1860s, unicellular organisms, especially amoebae, became important model organisms for physiological research. Situated halfway between cell and organism and oscillating in their status between plant and animal being, Ernst Brücke recognized in them "elementary organisms" and thus opened the perspectives of their broad reception as theoretical models: from Freud to Bergson and Goldstein (see Johns Schloegel & Schmidgen, 2002).

Informed by the same research tradition, Nietzsche stated in the spring of 1884: "[I]n the highest functions of the spirit I recognize only a sublime kind of organic function (assimilation selection secretion, etc.)" (Nietzsche, 1980a, 106). Shortly thereafter, in the summer of 1886/fall of 1887, he adds: "All thinking, judging, perceiving as *comparing has* as a prerequisite a 'equating,' even earlier a 'making equal.' Equating is the same as the incorporation of appropriated matter into the amoeba." (Nietzsche, 1980b, 209).

Canguilhem will have been aware of the fact that such reflections concerning the biology of knowledge transition smoothly into the theory of the will to power. The assimilation, the primitive nourishment, the, as Nietzsche writes, "original tendency of the protoplasm, when it sends out pseudopodia and gropes around itself" (ibid., 424), is, according to this view, not to be understood simply as a response to a need such as "hunger," but as an attempt of overcoming and incorporation, or with Canguilhem as "an activity of information *and* assimilation" (Canguilhem, 1989a, 130; italics added). At this point it seems no longer important to distinguish between knowledge and power, reason and control (in a similar vein, see Latour, 1988, 153).

Even conceptual knowledge, i.e. the identification, naming, and definition of circumscribed entities or phenomena, appears in this comparative-physiological perspective as the consequence of a will to become stronger. Unlike Nietzsche, however, Canguilhem assumes that the anthropomorphism of judgment and concept formation can be overcome. This is another reason why he emphasizes the difference between life, technology, and science. As constructive as the technical analogies are (or at least can be) in the description and development of biological phenomena, only when they have been worked through, i.e. modified or even abandoned, only when the difference between technical and biological notions has been brought out, one arrives at an athentic knowledge of life.

In the reflex book, Canguilhem's affinity with Nietzsche is underscored in yet another way. Just like the author of *The Gay Science,* Canguilhem distinguishes with regard to the history of physiology "two kinds of causes" (Nietzsche, 2001, 225),[4] more precisely: two configurations of the interval between cause and effect. According to him, Nineteenth century organic physics aimed at a mechanics, but its antecedent counterpart, vitalistic physiology, aimed at an energetics of life. While the mechanistic reflex theorists from Descartes to du Bois-Reymond and Pavlov therefore assumed a linear, as it were continuous relationship between stimuli and reactions, causes and effects, the reflex energeticists from Thomas Willis to Keith Lucas and William Bayliss focused on the discontinuous and sudden and above all on the disproportional relation between *causa* and *effectus*. It was Willis who introduced ballistic analogies for the explanatory description of the reflex: fuses, gunpowder, cannons, explosions. But still in the physiology of the early Twentieth century the effectiveness and fruitfulness of these analogies is obvious, for instance when talking about "firing neurons" or about the nervous function as a combustion along a "train of gunpowder" (Bayliss, 1915, 397).

In order to characterize the peculiarity of the conception of the living which is connected with it, Canguilhem refers to Bergson's essay "Life and Consciousness," in which it is said, among other things: "To execute a movement, the imprisoned

---

[4] "*Two kinds of causes that are often confused.* [...] The first kind of cause is a quantum of dammed-up energy waiting to be used somehow, for something; the second kind, by contrast, is something quite insignificant, mostly a small accident in accordance with which this quantum 'discharges' itself in one particular way: the match versus the powder keg."

energy is liberated [by the animal organism]. All that is required is, as it were, to press a button, touch a hair-trigger, apply a spark: the explosion occurs, and the movement in the chosen direction is accomplished" (Bergson, 1920, 19; see Canguilhem, 1977, 74–75).

Canguilhem also points out that in *Creative Evolution,* in a similar context, Bergson talks about causality in terms of "releasing" (*déclenchement*) (Bergson, 1911, 73; see Canguilhem, 1977, 192). Neither Bergson nor Canguilhem explicitly mention Nietzsche. Nevertheless, "releasing," or *Auslösung*, seems to mark with sufficient clarity a nexus in Nietzsche's thinking, where again the physiological and the philological, the biological and the historical-conceptual are connected with each other.

It is well known what importance Robert Mayer's study *Ueber Auslösung* (1876) had for the thinking of the late Nietzsche (Mittasch, 1952; Abel, 1998, 43–99; 93–95). Nietzsche had read Mayer's short text in the spring of 1881 and considered it "the most essential and useful" of Mayer's writings.[5] Indeed, his unpublished notes show that the motif of trigger, discharge, and explosion becomes for Nietzsche a kind of passe-partout with which biological functions, social phenomena, and historical processes can be compared and related to each other: from the explosive reactions of the simplest living beings to the sudden effects of music on a hysterical audience to the solitary founder of religion who can abruptly fascinate and mobilize entire crowds.

One might even argue that it was the reception of Mayer's writings that led Nietzsche to combine a physiological and psychological economy of forces with a *historical energetics*, a historical thinking in terms of wound-up potentials and their discharges. Historiography would thus be transferred into a form of genealogy, which in turn might be seen as an experimental handling of triggers, pretty much in line with Foucault's later understanding of genealogy: "We have to crack open words and things, experiences and concepts, in order to free the knowledge encapsulated in them, the embedded affects and values, the active and reactive forces" (Schäfer & Vogl, 2004, 199).

Canguilhem, however, does not describe the emergence of the reflex concept as a sheer discontinuity, a simple explosion, and to that extent he is not interested in specifying the conditions for such triggering processes and uncovering a corresponding genealogy. Instead, he even emphasizes that the "revolution" associated with the emergence of the reflex concept did not occur "all at once, not without timidity, and not without concessions to the traditional conception of animal movement" (Canguilhem, 1977, 127). In other words, as suggestive as the formula seems,

---

[5] For Canguilhem, the Mayer reception of Nietzsche has not been a secret, as he included a fragment from Nietzsche's manuscripts, in which Mayer is quoted, in an edited volume anthology in the early 1950s. See Canguilhem, 1952, 58–59. In the precise form used by Canguilhem, this fragment can be found in Würzbach, 1940, 179–182. The part concerning Mayer, however, is omitted in Canguilhem's selection. For this particular fragment, see Nietzsche, 1980c, 451, for the other parts, see Nietzsche, 1980b, 89; 300.

Canguilhem does not consider that the concept is a reflex before "reflex" becomes a concept.

Nevertheless, he is concerned with the affects and values that the reflex concept includes and implies, precisely those active and reactive forces that have seized and still seize certain things with the help of this word. And seen in this way, the reflex book certainly offers a genealogy in Nietzsche's sense, a genealogy of scientific knowledge that is indissolubly linked to the problem of the will and thus to that of normativity.

Canguilhem only alludes to this, but the allusion is clear enough: "[I]f the distinction between voluntary and involuntary movement has become a problem of physiology, it is only because of the meaning that this distinction first derives from its religious, moral, and juridical aspects. Before it becomes a scientific problem, it is a question concerning the practical handling of guilt and responsibility" (ibid., 148–149).

The connection between will and knowledge, between power and reason, is therefore by no means a merely natural-historical one, but also an eminently cultural-historical one. The debates that even in recent times flare up around the question of free will between brain scientists, philosophers, politicians, lawyers, etc., seem to confirm this rather impressively (as one example, see Massumi, 2002, 23–45).

In 1971, in his essay "De la science et de la contre-science," Canguilhem underscores his proximity to Nietzsche. It appears all the greater there because in this text a kind of balance sheet is drawn concerning the changes that took place in his work under the impression of the developments in biology during the 1960s. As can be seen from "New Reflections on the Normal and the Pathological" (Canguilhem, 1989b) and the section on "La nouvelle connaissance de la vie" in *Etudes d'histoire et de philosophie des sciences* (which contains the remarkable essay "Le concept et la vie"), at the center of these changes is the concept of *error*.

Certainly, this concept was already present in Canguilhem's earlier writings, in the early essay "Descartes et la technique," for example, as the failure of technical action, as the "resistances" encountered by human art; then, in the 1943 *Essay*, as the "failures of life," the revealing obstacle of illness; finally, in the reflex book, as the contents of scientific consciousness, which can be grasped in their creative potential only by regressing to a scientific unconscious. In the mid-1960s, however, under the impression of the progress of molecular biology, Canguilhem begins to define error anew and differently (as comprehensive studies on this topic, see Talcott, 2019 and Cammelli, 2022)

He inserts it into the basic structure of life, thus taking the very step that Giorgio Agamben ascribes to the late Foucault, namely to transplant "the subject taken from the field of *cogito* into the field of *life* […] – a life, however, that is understood as the essential region of error" (Agamben, 2004a, 7–8). In the 1970s, Canguilhem quotes Nietzsche to clarify and locate the underlying thought: "[O]ur *organs* (for *life*) are devised for erring. […] Life is the condition of knowledge. To err is the condition of life, and indeed to err at the deepest level. […] We must love and cultivate error; it is the mother's womb of knowledge" (Canguilhem, 1971, 178; see Nietzsche, 1980c, 504).

Canguilhem's error statements from the 1960s and 1970s can be understood as the result of a rigorously "untimely" reading of this Nietzsche fragment. Eventually, these statements were inspired by the contemporary success of molecular biology, epitomized in 1965 when the Nobel prize for physiology and medicine was awarded to François Jacob, Andre Lwoff and Jacques Monod. Against this background, it is contemporary information theory which has grasped the 'deepest level' of life and has thus also conquered the 'womb of knowledge.'

This is obviously not to be understood in the sense of the simple empiricism that Canguilhem has repeatedly criticized, but as a redefinition of the starting point for a critique and interpretation of the life sciences. As in his study of the concept of reflex, the point here is to trace and challenge those forces that have taken hold of a particular thing by means of a particular word – in this case, *information*.

In this connection, Canguilhem counters the cybernetic discourse on information and communication technology with the thesis that the theory of information is a general one: it applies not only to the objects of biological knowledge (organisms, cells, genes, etc.), but also to knowledge itself. And what is valid for information is also valid for error, noise and chance. For Canguilhem, there is no difference between the evolutionary error of life and the error of knowledge: "The first furnishes the key to the second" (Canguilhem, 1989b, 277).

The result of this updated bending back of the epistemological to the biological is a new theory of the knowledge of life, a "Nouvelle connaissance de la vie." The problem of knowledge is no longer related only to an individuality of behavior and experience, as still in the example of the amoeba, but also to a collectivity of heredity, of biological transmission, transcription and expression of genetic information. Knowledge now no longer means analyzing in the sense of measuring and calculating (as in the 1950s, in *Knowledge of Life*). In 1966, Canguilhem states: "to know is to be informed, to learn to decipher or decode" (Canguilhem, 1989b, 277). Conversely, error no longer resides just in the wrong *form*, an inappropriate, rudimentary analogy for life, but also in faulty *information*, in human and non-human misunderstandings and misreadings.

Even to this new epistemology Canguilhem gives a Nietzschean twist. Recalling Nietzsche's discourse about man as the animal that is not yet fixed or defined, he writes: "In fact, human error is probably one with human errancy. Man makes mistakes because he does not know where to settle. He makes mistakes when he choses the wrong spot for receiving the kind of information he is after" (Canguilhem, 2000, 319). At the same time, it is precisely this ability to err, to move and to relocate – of oneself, but equally of things – that allows human beings to know, that is, to decipher and decode.

Canguilhem has repeatedly returned to the fact that the preforms of knowledge are to be found in life itself. In contemporary thought, he thus marks a position that situates itself not beyond, but beneath the 'machinic' divisions that, according to Agamben (1998), have led to the emergence of *bare life* in the political, religious, juridical, and scientific realm. For Canguilhem, there is no bareness, nakedness or purity of life, however. For him, life is always already the *living* (*le vivant*), that is, always already organized in forms, and it is these forms – from genetic information to the morphology of bodies (and technology) to the images of thought and speech – in which a knowledge of life is formed long before it is put into scientific terms.

The theory and history of the life sciences are in this respect connected with life as such. Pierre Macherey pointed this out early on: "Along the path of a history of biology, what develops [in Canguilhem] is not a biology of knowledge in the traditional sense of the word, that is, a mechanistic explanation of the production process of knowledge, but a reflection on the knowledge of biology, which in turn is illuminated by biology" (Macherey, 1964, 69). And Foucault says something similar when he writes: "Georges Canguilhem, through the elucidation of knowledge concerning life and the concepts which articulate this knowledge, wants to rediscover in how far the *concept belongs to life*" (Foucault, 1989, 20–21; amended).

What Canguilhem is concerned with in his discussion of the (life) sciences is thus the uncovering of a knowledge that is immanent to life – which includes, of course, understanding science as an integral part of life. When Agamben says: "[E]verything happens as if, in our culture, life were *what cannot be defined, yet, precisely for this reason, must be ceaselessly articulated and divided*" (Agamben, 2004b, 13; italics in the original), we can read this in the sense of a continuation of the philosophical projects of Foucault and Deleuze, which were ultimately oriented towards the question of life and immanence.[6] But, as we have seen, it is also an echo of Canguilhem's view of life as a peculiar object of knowledge.

Vitalism, which takes this peculiarity into account, is consequently distinguished from mechanism by being a "cautious positivism" that does not rush to define life. Vitalism in this sense is, as Canguilhem specifies, "perhaps only the sense of an ontological, that is, chronologically irreducible anticipation of life vis-à-vis mechanical theory and technique, vis-à-vis intelligence and the simulation of life" (Canguilhem, 1977, 123).

Accordingly, authentic biological research is driven in its articulations and analyses by a vital foresight of an almost inevitable but always unexpected missing of its object, the constructive-anticipatory acceptance of a necessary failure of its concepts. In other words, the preforms and the forms of knowledge are not simply informations. They are always also misforms (*Fehlformen*) of knowledge, even if this can only become apparent *après coup*. Thus, even in the age of information, life in our culture "is" not truth, but without life there would not be that distinction which is actualized with every scientific judgment: that between truth and error. Highlighting this point, Canguilhem's philosophy continues to develop Nietzschean themes – *sans carte*, as it were.

[6] See the parallel reading of the two late texts "Life: Experience and Science" (Foucault) and "Immanence: a life..." (Deleuze) in Agamben, 2003. On Deleuze and Canguilhem, see Bianco, 2006.

References

Abel, G. (1998). *Nietzsche: Die Dynamik der Willen zur Macht und die ewige Wiederkehr* (2nd ed.). de Gruyter.

Agamben, G. (1998). *Homo Sacer: Sovereign Power and Bare Life* (D. Heller-Roazen, Trans.). Stanford University Press.

Agamben, G. (2003). Absolute Immanence. In J. Khalfa (Ed.), *Introduction to the Philosophy of Gilles Deleuze* (pp. 151–169). Continuum.

Agamben, G. (2004a). Wahrheit als Irrsal (Andreas Hiepko, Trans.). In R. M. Kiesow & H. Schmidgen (Eds.), *Das Irrsal hilft* (pp. 7–14). Merve.

Agamben, G. (2004b). *The Open: Man and Animal* (K. Attell, Trans.). Stanford University Press.

Bachelard, G. (1951). *L'activité rationaliste de la physique contemporaine*. PUF.

Bayliss, W. M. (1915). *Principles of General Physiology*. Longmans & Green.

Bergson, H. (1911). *Creative Evolution* (Arthur Mitchell, Trans.). Holt.

Bergson, H. (1920). Life and Consciousness. In *Mind-Energy. Lectures and Essays* (F. Wildon Carr, Trans.), (pp. 3–26). Holt.

Bianco, G. (2006). Introduzione. In G. Bianco (Ed.), Georges Canguilhem and Gilles Deleuze, *Il significato della vita: Letture del III capitolo dell'*Evoluzione creatrice *di Bergson* (pp. 7–51). Mimesis.

Brusotti, M. (2021). 'Werde, der du bist!' Selbsterkenntnis, Handeln und Selbstgestaltung bei Nietzsche in einem Ineditum von Georges Canguilhem. *Nietzsche Studien, 21*, 181–216.

Cammelli, M. (2022). *Canguilhem philosophe. Le sujet et l'erreur*. PUF.

Canguilhem, G. (1971). De la science et de la contre-science. In S. Bachelard et al. (Eds.), *Hommage à Jean Hyppolite* (pp. 173–180). P.U.F.

Canguilhem, G. (1977). *La formation du concept de réflexe aux XVIIe et XVIIIe siècles* (2nd ed.). Vrin.

Canguilhem, G. (1980). What is Psychology? (H. Davies, Trans.). *Ideology and Consciousness, 7*, 37–50.

Canguilhem, G. (1989a). Essay on Some Problems Concerning The Normal and The Pathological. In *The Normal and the Pathological* (C. R. Fawcett, Trans. in collaboration with R. S. Cohen) (pp. 25–229). Zone Books.

Canguilhem, G. (1989b). New Reflections on The Normal and The Pathological (1963–1966). In *The Normal and the Pathological* (C. R. Fawcett, Trans. in collaboration with R. S. Cohen) (pp. 231–289). Zone Books.

Canguilhem, G. (2000). The Concept of Life (A. Goldhammer, Trans.). In F. Delaporte (Ed.), *A Vital Rationalist. Selected Writings from Georges Canguilhem* (pp. 303–319). Zone Books.

Canguilhem, G. (2002). Le concept de réflexe au XIXe siècle. In *Etudes d'histoire et de philosophie des sciences,* 7th, enlarged ed., 2nd printing, (pp. 295–394). Vrin.

Canguilhem, G. (2008). *Knowledge of Life* (S. Geroulanos & D. Ginsburg, Trans.), eds. P. Marrati & T. Meyers. Fordham.

Canguilhem, G. (2012). *Writings on Medicine* (S. Geroulanos & T. Meyers, Trans.). Fordham.

Canguilhem, G. (Ed.). (1952). *Besoins et tendances*. Hachette.

Canguilhem, G., & Planet, C. (1939). *Traité de Logique et de Morale*. Robert.

Emden, C. J. (2004). Metaphor, Perception and Consciousness: Nietzsche on Rhetoric and Neurophysiology. In G. Moore & T. H. Brobjer (Eds.), *Nietzsche and Science* (pp. 91–110). Ashgate.

Foucault, M. (1977). Nietzsche, Genealogy, History (D. F. Bouchard & S. Simon, Trans.). In D. F. Bouchard (Ed.), *Language, Counter-Memory, Practice: Selected Essays and Interviews* (pp. 139–164). Cornell University Press.

Foucault, M. (1983). Structuralism and poststructuralism [Interview by Gérard Raulet, 1983] (J. Harding, Trans.). *Telos* 55: 195–211.

Foucault, M. (1989). Introduction. In *The Normal and the Pathological* (C. R. Fawcett, Trans. in collaboration with R. S. Cohen), (pp. 7–24). Zone Books.

Foucault, M. (1991). The Subject, Knowledge, and the 'History of Truth.' [Conversation with Ducio Trombadori, 1978]. (R. James Goldstein & J. Cascaito, Trans.). In *Remarks on Marx* (pp. 43–82). Autonomedia.

Johns Schloegel, J., & Schmidgen, H. (2002). General Physiology, Experimental Psychology, and Evolutionism: Unicellular Organisms as Objects of Psychophysiological Research, 1877–1918. *Isis, 93*, 614–645.

Krummel, R. F. (1988). Joseph Paneth über seine Begegnung mit Nietzsche in der Zarathustra-Zeit. *Nietzsche-Studien, 17*, 478–495.

Latour, B. (1988). Irreductions. In *The Pasteurization of France* (pp. 152–236). Harvard University Press.

Macherey, P. (1964). La philosophie de la science de Georges Canguilhem: Épistémologie et histoire des sciences. *La pensée, 113*, 50–74.

Massumi, B. (2002). *Parables for the Virtual. Movement, Affect, Sensation*. Duke University Press.

Metzger, H. (1926). *Les concepts scientifiques*. Alcan.

Mittasch, A. (1952). *Friedrich Nietzsche als Naturphilosoph*. Kröner.

Müller-Lauter, W. (1974). Nietzsches Lehre vom Willen zur Macht. *Nietzsche-Studien, 3*, 1–60.

Müller-Lauter, W. (1978). Der Organismus als innerer Kampf: Der Einfluß von Wilhelm Roux auf Friedrich Nietzsche. *Nietzsche-Studien, 7*, 189–223.

Nietzsche, F. (1935/37). *La volonté de puissance I/II* (G. Bianquis, Trans.). & F. Würzbach (Ed.), Gallimard.

Nietzsche, F. (1968). Über Wahrheit und Lüge im Außermoralischen Sinn. In J. Habermas (Ed.), *Erkenntnistheoretische Schriften* (pp. 97–111). Suhrkamp.

Nietzsche, F. (1980a). In G. Colli & M. Montinari (Eds.), *Sämtliche Werke – Kritische Studienausgabe in 15 Bänden, Vol. 11: Nachgelassene Fragmente 1884–1885*. Deutscher Taschenbuch Verlag/de Gruyter.

Nietzsche, F. (1980b). In G. Colli & M. Montinari (Eds.), *Sämtliche Werke – Kritische Studienausgabe in 15 Bänden, Vol. 12: Nachgelassene Fragmente 1885–1887*. Deutscher Taschenbuch Verlag/de Gruyter.

Nietzsche, F. (1980c). In G. Colli & M. Montinari (Eds.), *Sämtliche Werke – Kritische Studienausgabe in 15 Bänden, Vol. 9: Nachgelassene Fragmente 1880–1882*. Deutscher Taschenbuch Verlag/de Gruyter.

Nietzsche, F. (2001). *The Gay Science* (J. Nauckhoff, Trans.). In B. Williams (Ed.), Cambridge University Press.

Schäfer, A., & Vogl, J. (2004). Feuer und Flamme: Über ein Ereignis des 19. Jahrhunderts. In H. Schmidgen, S. Dierig, & P. Geimer (Eds.), *Kultur im Experiment* (pp. 191–211). Kadmos.

Schmidgen, H. (2008). Fehlformen des Wissens. In G. Canguilhem (Ed.), *Die Herausbildung des Reflexbegriffs im 17. und 18. Jahrhundert* (pp. VII–LVIII). Fink.

Stiegler, B. (2000). De Canguilhem à Nietzsche: La normativité du vivant. In G. Le Blanc (Ed.), *Lectures de Canguilhem: Le normal et le pathologique* (pp. 85–101). Lyon.

Stiegler, B. (2001). *Nietzsche et la biologie*. PUF.

Talcott, S. (2019). *Georges Canguilhem and the Problem of Error*. Palgrave Macmillan.

Valéry, P. (1972). Introduction to the Method of Leonardo Da Vinci. (M. Cowley & J. R. Lawler, Trans.). In J. Mathews (Ed.), *Collected Works, Vol. 8. Leonardo, Poe, Mallarmé* (pp. 3–63). Princeton University Press.

Würzbach, F. (Ed.). (1940). *Das Vermächtnis Friedrich Nietzsches: Versuch einer neuen Auslegung allen Geschehens und einer Neuordnung aller Werte*. Pustet.

Henning Schmidgen is Professor of Media Studies at the Bauhaus University in Weimar, Germany. Among his books are *The Helmholtz Curves. Tracing Lost Time* (2014), *Bruno Latour in Pieces* (2014) and *Horn, or The Counterside of Media* (2022).

Neither Angel Nor Beast: Life and/Versus Mind in Canguilhem and Merleau-Ponty

Sebastjan Vörös

> *We suspect that, to do mathematics, it would suffice that we be angels. But to do biology, even with the aid of intelligence, we sometimes need to feel like beasts ourselves.*
>
> (Canguilhem 2008: xx)

Abstract The chapter addresses the problem of the relationship between life (vitality) and mind (thought) by drawing on the resources available in Canguilhem's and Merleau-Ponty's philosophies. It consists of six sections. In the first and second section, I outline the so-called 'mind-life problem' and two diametrically opposed responses to it: *life philosophy* (life subsumes mind) and *transcendentalism* (mind subsumes life). Against this background, I flesh out Canguilhem's 'slantwise' resolution, which argues that, while it is true that life feeds into mind, it is equally true that mind takes up and subl(im)ates life. In the third and fourth section, I focus on the first half of the proposed solution: I start by putting forward a non-reductionist account of life grounded on the idea of vital normativity, and then go on to show how this vital dynamism translates into human cognition via praktognosia (embodied and techn(olog)ical know-how). In the fifth section I tackle the second half of the solution: by delineating the idea of symbolic behavior and ex-centric positionality I try to indicate how mind, while grounded in life, is nonetheless able to transcend it. Finally, in the last section, I suggest that this Janus-faced dynamism

The research included in this chapter was funded by the Slovenian Research Agency (ARRS) under the research program "Philosophical Investigations" (P6-0252).

S. Vörös (✉)
Department of Philosophy, University of Ljubljana, Ljubljana, Slovenia
e-mail: sebastjan.voros@ff.uni-lj.si

G. Bianco et al. (eds.), *Canguilhem and Continental Philosophy of Biology*, History, Philosophy and Theory of the Life Sciences 31,
https://doi.org/10.1007/978-3-031-20529-3_9

between life and mind also holds true for the researcher investigating these topics, and hint at the broader philosophical implication of such a view for the practice of science and philosophy.

Keywords Canguilhem · Merleau-Ponty · Vitality · Mindedness · Transformativism · Technique · Eccentricity · Symbolism

The thematic thread of this chapter weaves around an issue taken up by Georges Canguilhem in many of his works, but perhaps most straightforwardly in his essay "Thought and Living" (TL). The problem in question might seem far removed from the topics one generally encounters in the Anglo-American philosophical context, but was actually a source of considerable controversy among the German and French intelligentsia in the first half of the twentieth century. With the reintroduction of the notion of 'life' into philosophical and scientific discourse[1] in an attempt to revitalize what to many seemed like ossified conceptual schemes, crippled by the pseudo-problems of their own making and divested from the fabric of everyday existence,[2] the question emerged as to how this rediscovered *vitality*, the domain of life and lived experience, relates to *mindedness*, the domain of thought and knowledge.[3]

As is often the case in such instances, one can, and many in fact did, espouse two diametrically opposed views on the topic: either life is but another *object of thought*, and therefore must be, if it is to be incorporated into the edifice of knowledge, ultimately derivable from the categories of reason; or, alternatively, thought is but another *expression of life*, and therefore must be, if it is to be incorporated into the totality of existence, ultimately derivable from the dynamics of life. The epistemic and existential stakes are high: in the first case, *thought engulfs life*, which means that life and lived experience are banished from the realm of cognition; in the second case, *life engulfs thought*, which means that thought and conceptual knowledge are subservient to the dictates of vitality.[4]

---

[1] I am, of course, referring here to so-called *philosophy of life*, or *Lebensphilosophie*, a philosophical movement – if, given its enormous heterogeneity, so it can be called – from the end of the nineteenth and the beginning of the twentieth century, which included such diverse authors as Henri Bergson, Wihelm Dilthey, Ludwig Klages, Oswald Spengler, and others (see e.g. Albert, 1995).

[2] This sentiment is expressed with remarkable vividness by Helmuth Plessner: "Every age finds its own redeeming word. The terminology of the eighteenth century culminated in the concept of reason; that of the nineteenth in the concept of progress; that of the current one [early twentieth century] in the concept of life. […] The only thing capable of enchanting was something irrefutable, to be grasped on this side of all ideologies, on this side of God and the state, of nature and history […] in short: life." (Plessner, 2019, 1–2)

[3] In what follows, I will use 'thought,' 'cognition,' and 'knowledge' as *roughly* synonymous. This is so not only because Canguilhem himself uses them in this way but also because, for our present purposes, what is crucial is that they can all be said to belong to the *domain of mindedness*.

[4] I provide a more thorough account of this distinction in the next section.

Soon, reconciliatory proposals started to emerge, pleading for a 'middle-way solution', but often doing little more than restating the problem without actually solving it. Canguilhem's take on the issue was to take a step back and approach it *slantwise*. That is, instead of plunging, head on, into a clear-but-insuperable conflict between life and knowledge, he first transformed it into a difficult-but-tractable tension between human being and its environment, and then, from this oblique standpoint, argued that life and thought are *two continuous, yet qualitatively distinct modes of being*: two structures of behaviour, of coping with the world.

In an attempt to elucidate the nature of this oblique standpoint and its ramifications, I will augment Canguilhem's reflections by enlisting conceptual resources from another French philosopher, Maurice Merleau-Ponty. Textbook classifications tend to place these two authors into separate and, I would argue, misleading categories – historical epistemology and phenomenology, respectively -, which has the unfortunate consequence that scholars from different backgrounds who work on one of the authors usually end up ignoring the other. However, I believe, and have argued elsewhere (Vörös, 2022), that their philosophical visions are much more congruent than is commonly accepted.[5] Another, and final point, I would like to make before plunging into the topic at hand is that, in my inquiry, I will be primarily thinking *with* Canguilhem and Merleau-Ponty, and not *about* them. Put differently, my investigation, although grounded in close readings of the two authors, is not so much an exegetical piece as it is an attempt to creatively use the two authors' ideas to tackle what I feel to be an intensely important philosophical question: the relation between life and mind. Whether I have been, in so doing, pushing them further than either would be willing to go – as Merleau-Ponty famously noted of his treatment of Husserl (PSM, 72) – is something I leave the reader to decide.

1 Mind-Life Problem: Between Transcendentalism and Life Philosophy

Human beings are both *living* (sentient) and *thinking* (sapient) beings: they not only breathe, sleep, and procreate, but also contemplate, deliberate, and predicate. However, how do these two aspects interrelate? Are my thoughts categorically different from, and/or constitutive of, my organic functions – or are they ultimately

[5] In fact, these similarities did not escape the attention of Canguilhem who, in his preface to the second edition of the *Essay on Some Problems Concerning the Normal and the Pathological* (published in 1950), points out that, during the inception of the *Essay* (first published in 1943), he would have profited by drawing on Merleau-Ponty's *The Structure of Behavior* (SB; first published in 1942). However, as it was brought to his attention when the manuscript had already been in print (NP, 29), he could give it but a passing nod of recognition. Yet to his mind, such an omission is not necessarily something to be regretted, since "a convergence whose fortuitous character better emphasizes the value of intellectual necessity to an acquiescence, even fully sincere, in the view of others" (NP, 29–30).

reducible to, and/or expressive of, these selfsame functions? Does the light of rational spontaneity, once kindled, dispel the shroud of vital impulsivity; or is the radiance of sapience merely a manifestation of the murky currents of sentience?

This, in a nutshell, is what could be called the *mind-life* or *mind-living-body problem*. While bearing a certain resemblance to the more prominent *mind-body problem*, the mind-life problem is both broader and deeper in scope. It is broader in that, if the former presupposes that life is not an independent phenomenon but merely an extension of matter and can therefore be excluded from discussion, the latter brings a new discussant – that of *living* or *lived body* – to the table, thus expanding the set of possible solutions. But even more importantly, it is *deeper* in the sense that, if determinism of causal mechanism (domain of 'matter') and spontaneity of categorical reasoning (domain of 'mind') are supplemented by precarious autonomy[6] of vital normativity (domain of 'life'), the conceptual landscape, in which the problem is couched and discussed, becomes much more fine-grained, and the infamous chasm between matter and life less foreboding.[7]

In this chapter, I will not attempt to provide a comprehensive cartography of possible responses to the mind-life problem, but will focus solely on the two positions used by Canguilhem to frame his inquiry:

> We accept far too easily that there exists a fundamental conflict between knowledge and life, such that their reciprocal aversion can lead only to the destruction of life by knowledge or to the derision of knowledge by life. We are then left with no choice except between a crystalline (i.e., transparent and inert) intellectualism and a foggy (at once active and muddled) mysticism. (TL, xvii)

In brief, Canguilhem distinguishes two antithetical views on the topic: on the one extreme, we find approaches that undermine life through knowledge; on the other, approaches that undermine knowledge through life. In the first case, mind is seen as (epistemologically, ethically, and/or existentially) *superior* to life: we are primarily *minded* lives; in the second case, mind is seen as (epistemologically, ethically, and/or existentially) *subservient* to life: we are primarily *living* minds.

Canguilhem's stark dichotomy between what he terms "intellectualism" and "mysticism" reflects a well-known antagonism in French thought between

[6] This is an implicit reference to Hans Jonas's idea that life is characterized by "needful freedom" or "hazardous independence": on the one hand, the organism constitutes an autonomous whole separate from its material surroundings; on the other hand, it has to engage in ongoing interactions with this selfsame milieu to maintain its autonomy. Thus, the living form "is never the same materially and yet it persists as its same self, by not remaining the same matter" (Jonas, 2001, 4, 76).

[7] A similar impetus can be found in recent attempts to transmute the mind-body problem into a *body-body* (Thompson, 2007, 235ff) or a *mind-body-body* problem (Hanna & Thompson, 2003), i.e., into a question about the interrelationship between the mind, the *physical* body (Ger. *Körper*), and the *living-lived* body (Ger. *Leib*) (the conceptual distinction between *Körper and Leib* was developed in the context of phenomenology and philosophical anthropology in the first half of the twentieth century, and has been recently taken up and developed further by so-called enactivist and embodied approaches in cognitive science). Here, too, the notion of life or vitality (here in the form of the animate vehicle of experience) is considered as a means to bridge and surpass the age-old philosophical dichotomies.

"philosophies of concept" (as exemplified by Léon Brunschvicg) and "philosophies of experience" (as exemplified by Henri Bergson) (see, e.g., Bianco, 2011; Foucault, 1985). Both of these approaches can be seen as two sides of a uniform reaction against the dualist and naturalist solutions to the mind-life problem. Their main grievance with these 'externalist' approaches – the approaches maintaining that we can philosophize about phenomena 'from the outside' – is that they stem from an unjustified supposition that the knower, although herself a living and cognizing being, can somehow extricate herself from the reality she lives in and reflects upon. In so doing, dualist and naturalist approaches fail to erect a coherent philosophy, since the 'emaciated reality' they work with is one premised on an ineradicable *subtraction*: the supposedly self-subsistent domain of being is self-subsistent only if we tacitly exclude from it the living-and-thinking being for whom this domain is given and in whose symbolic models it features.

Thus, the two approaches Canguilhem criticizes agree that life and mind do not designate simply something that 'is', but also, and essentially, something that 'it is like to be'. Put differently, before I can conceive of living and thinking things, I *am life*, I *think thoughts*; thus, life and mind cannot simply be taken as yet another thing, for they are *that for which there are things*. The reification of life and mind overlooks that 'things' are, fundamentally, not things-in-themselves, but rather things-*for-myself*, and that, for this 'my-self', life is primarily a life-lived, cognition a cognition-cognized. If vital and cognitive categories of my existence are taken on board, how can I pretend to be a mere passive onlooker for whom the self-subsistent reality is given through a series of causal affectations, be they of material or spiritual kind?

At this point, the two approaches part ways: they both agree that mind and life constitute *incommensurable* domains, but differ in which of the two they prioritize. The 'transcendentalist approaches' (Canguilhem's intellectualism) prioritize *mind*: the living, as well as the material, are *phenomena* – that which is given in experience – whereas the mind or consciousness is the *condition of possibility of phenomena* – that for which things are given in experience. The domain of life ultimately belongs to the domain of in-itself, which is dependent on the domain of mind or consciousness, the fundamental domain of for-itself. Mind is therefore not simply a *different type* of thing, but that which – as the transcendental center of constituting acts – enables *all types* of things, *vital* things included. 'What is' is 'what can be thought', so any ambiguity that pertains to the domain of the living needs to be exorcised in the name of the epistemological ideal of *clare et distincte*: the indeterminate, the vague – all that is but muddled thought that needs to be brought under the clarity of the concept.

In contrast, the 'life-philosophical approaches' (Canguilhem's mysticism) prioritize *life*: the vital, on this view, is the domain of pre-reflective and pre-linguistic 'lived experience', an ineffable and conceptually elusive foundation of mindedness and objectivity. As such, life *as lived* is neither in-itself nor for-itself, for it defies, and is the condition of, all categories; it is what gives rise to thought, and since the latter is in its service, our cognition can never grasp and elucidate it. Defying reflection, it is accessible solely to the intuitive seeing, to the sub- or trans-rational *Schau*:

all cognitive models are *ultimately* inadequate, ushering forth from, and paying their homage to, the ineradicable evanescence of life untamed. Reason may build its elaborate conceptual artifices, yet life-spirit bloweth where it listeth.

2 Reflexive Scientist: Circularity Between Life and Mind

Canguilhem's response to this dilemma is a variation of 'a plague on both of your houses', but one with a pronounced dialectical twist. For he wants to not only transcend but, so to speak, *subl(im)ate* the two antagonistic *view*points into a new *oblique stand*point, from which it will be seen that, while erroneous, the two extremes are nevertheless *motivated* errors, i.e., errors resting, as Merleau-Ponty puts it, "on an authentic phenomenon which philosophy has the function of making explicit" (SB, 216). In other words, the new vantage point must be able to retain the positive aspects of transcendentalist and life-philosophical approaches, particularly their insistence on the inclusion of the experiencer/cognizer (e.g., the scientist examining life and mind) in her overall account, while resisting the allure of (i) sliding into a self-enclosed dynamism, be it in the form of crystalline rationality (transcendentalist horn) or arcane vitality (life-philosophical horn), or (ii) reverting to a full-blown dualism or naturalism.

Canguilhem starts to carve out his position by subtly shifting the focus of the discussion: "Now, the conflict is not between thought and life in man, but *between man and the world* in the *human consciousness of life*" (TL, xvii: my emphases). What does he hope to achieve with this reformulation? One way of approaching the matter is as follows: When we *think* about life – when we set out to analyse, explain, measure, etc., it (ibid.) – what is our starting point? Inescapably, it is the *phenomenon* of life – the way in which life is *given to us in experience*. Now, life, in its most fundamental form, is given to us in terms of *active living wholes*, i.e., in terms of "totalities whose sense resides in their tendency to realize themselves as such in the course of their confrontation with their milieu" (TL, xix). Put differently, before we break life apart into its constituent elements and elucidate its mechanical scaffolding, it discloses itself to us in the form of *a living being actively coping with its environment*.

However, this is only the first half of the story. For this *consciousness of life* is itself permeated with currents of vitality, and is therefore also a *life of consciousness*: a scientist who is aware of life is simultaneously a scientist who lives this awareness. This has two important implications. To begin with, as a living being, the scientist thinking about life is, in-and-through her thinking activity, endeavouring, like all forms of life, to come to terms with the world she lives in. The life she studies – say, a given plant or an animal – is a problem for her, and is therefore something she must cope with. There is, then, a strong correspondence between the two poles of inquiry, between a scientist under-taking investigation and an organism taken-under investigation. In both cases, we are dealing with a living being coming to terms with its environment, and thus the scientist could be said to *live* the vital dynamics she is trying to *understand*.

Secondly, there seems to be a certain *tension* in this coming-to-terms of the scientist, for – as the matter stands in contemporary science – while she *lives* it, she *doesn't think* it, and while she *thinks* it, she *doesn't live* it. That is, if the cognitive mode of coming-to-terms – one that includes analysis, explanation, etc. – involves "a benefit for intelligence", it also comes "at the cost of enjoyment": "One enjoys not the laws of nature but nature itself, not numbers but qualities, not relations but beings" (TL, xvii). Thus, instead of seeing life and thought as two distinct self-enclosed and antithetical 'forces', they need to be construed as two *distinct modes of human existence*, two distinct ways of how human beings cope with their environment. So, if we want to understand the relationship between life and mind, we need to understand different ways in how humans relate to their world.

It could be said that, in their impassioned reaction against the 'externalist' – i.e., dualist and naturalist – approaches and their valorisation of the *creativity* of life and the *spontaneity* of reason, the transcendentalist and life-philosophical approaches end up 'throwing out the baby with the bathwater' in that they unjustifiably trivialize the intimate connection between the organism and its world. This is the "truth of empiricism" (in Canguilhem's words; NP, 105) or the "truth of naturalism" (in Merleau-Ponty's words; SB, 224), namely that the living/cognizing being is always both confronted with and embedded into, both resisted and beckoned by *the other*. Thus, when speaking of vitality and cognition, we are referring not to unbridled but *restrained* creativity, not to absolute but *precarious* spontaneity. Philosophy must make room for the ineradicable *element of passivity* (SB, 216) without turning it into a self-subsisting reality. In other words, it must never cease to be a "philosophy of intellectual adventure" (NP, 105), grounded in the fundamental dialogue between the living/cognizing being and its world. This, in a nutshell, is the 'authentic phenomenon' mentioned above, which the transcendentalist and life-philosophical approaches draw upon but fail to thematize.

Having shifted the focus of his discussion, Canguilhem expounds on his basic idea in a lengthy paragraph, which is worth quoting at length:

> Thought is nothing but a disentangling of man from the world that permits us to retreat from, to interrogate, and to doubt (to think is to weigh, etc.) in the face of obstacles that arise. In concrete terms, knowledge consists in the search for security via the reduction of obstacles; it consists in the construction of theories that proceed by assimilation. It is thus a general method for the direct or indirect resolution of tensions between man and milieu. Yet to define knowledge in this way is to find its meaning in its end, which is to allow man a new equilibrium with the world, a new form of organization of his life. It is not true that knowledge destroys life. Rather, knowledge undoes the experience of life, seeking to analyze its failures so as to abstract from it both a rationale for prudence (sapience, science, etc.) and, eventually, laws for success, in order to help man remake what life has made without him, in him, or outside of him. In consequence, it must be said that if thought and knowledge are inscribed within life so as to regulate it – as is the case with man – this very life cannot be the blind and stupid mechanical force that one likes to imagine it when one contrasts it to thought. Besides, if it were mechanical it could be neither blind nor stupid. Only a being that searches for light can be blind; only a being that claims to signify can be stupid. (TL, xvii)

If we follow Canguilhem's line of reasoning and consider thought not in isolation, but as embedded into a broader perspective of the human being's coping with the world, it can be said to serve a certain aim, namely that of "allow[ing] man a *new equilibrium* with the world, a *new form of organization* of his life" (ibid.; my emphasis). This has two important implications. On the one hand, the common conception of thought as self-sufficient, i.e., as independent of life, becomes tenuous. Just as one does not eat simply in order to eat but to satiate hunger (ibid., xvi), one does not know simply to know but to achieve a certain end, namely that of embodying a new *Lebensform*, a new equilibrium with the world: "*Vulgar*? Perhaps. *Blasphemous*? But why? Must we believe that, because certain men dedicate themselves to a life of knowledge, man can only really live in and through science?" (ibid., xvii) Thought, according to Canguilhem, is [a] the capacity to *break* with the directness of lived experience and extricate from it a rationale for prudent action, the foundation of science and philosophy, so as to [b] arrive at "laws for success", i.e., symbolic models, principles, rules, etc., that recapitulate what life has tacitly achieved without human interference (ibid.). However, this 'break' cannot be absolute, for while it is true that it may include, at its very extreme, a negation of life – say, in the form of ascetic self-abnegation -, the latter still remains a *life* of negation: even an ascetic breathes and sleeps, gets weary and sick.

On the other hand, recognizing that thought is *in the service of* life does not mean that it is *enslaved by* life. For if knowledge can be said to assist life it does so by engendering a qualitatively new form or structure (Ger. *Gestalt*) of life,[8] i.e., it *preserves* it by *transcending* it. In order to grasp itself, life has to distance itself from itself, and it is precisely this *hiatus,* this *gap*, that is provided by thought. This does not mean that thought negates or destroys life, but rather that it eliminates it as a self-standing dynamics while simultaneously conserving and integrating it into a new form or structure (SB, 207–8). Just as a visual patch acquires a completely different meaning if it is grasped within a larger perceptual *Gestalt* – what was previously a dark-gray blotch becomes an ear of a cat -, whereby its very qualities (the density and texture of its colours, etc.) become altered in the process, so thought

[8] Having both been influenced by Kurt Goldstein (1995) and Gestalt psychologists (Ash, 1995), Canguilhem and Merleau-Ponty often talk about the unity of life in terms of *Gestalten* (usually translated into English as 'structures' or 'forms'). The reason for doing so has to do with their conviction that the notion of structure/form allows them to steer the middle path between the extremes of mechanism and substantivist vitalism (cf. Wolfe, 2011). According to Merleau-Ponty, for instance, 'Gestalt' stands for a *dynamically structured whole*, such that: (i) it has "original properties with regard to those of the parts" (SB, 47), and therefore cannot be (*pace* mechanism) reduced to a sum of independent and causally interrelated elements (SB, 50); and yet, (ii) it is not (*pace* vitalism) an ontologically distinct 'substance' or 'principle,' as it is nothing over and above the network of its interdependent and dynamically co-constitutive parts (SB, 131). In *Gestalt*, what matters is not its 'elements' or 'constituents,' but the overall configuration of their relations, for it is only by taking place in such a configuration that a given 'aspect' of the whole can be called its 'part' (ibid.; see also Sheredos, 2017).

trans-forms or re-structures (Ger. *umgestalten*) life,[9] investing it with new meaning, without fully eluding its grasp (more on this in the last two sections).

What remains to be shown in the following sections is how thought can be considered as that in which life "becomes other without becoming another" (Bortoft, 2012, 71), how it can withdraw from life and become an *other*, without completely outrunning it and thus becoming *an-other*. To achieve this, we will pursue Canguilhem's suggestion that thought sublimates a dynamic that is already inherent in life, which means that mindedness cannot be completely foreign to life. This in turn means that, since mind is the bastion of *meaning* and *normativity*, the mechanicist-reductionist understanding of life as "blind" and "stupid" physico-chemical mechanism devoid of all meaning and normativity is spurious, for, as Canghuilem puts it, "[o]nly a being that searches for light can be blind; only a being that claims to signify can be stupid" (TL, xvii). That is, if life were merely a mechanical process, the chasm between life as that which belongs to the realm of causality and thought as that which belongs to the space of reasons would be unbridgeable. Instead, we must learn to recognize a "germ" or "dialectical ferment" of thought (PP, 101, 131) – a *vital normativity and signification* – in the very aliveness of life, and see how it gives birth to, and is then superseded by, the mind.

3 Dialectical Ferment: Vital Normativity

What, then, is this dialectical ferment of mindedness? In this section, I will briefly outline why, instead of construing life in mechanical terms, it should be seen as the domain of vital normativity and thus as carrying within itself a germ of mindedness. A good entryway into the topic is by way of Canguilhem's critique of mechanicist-reductionist accounts of the living. At its core, there lies a simple, yet far-reaching idea, namely that *life is a dynamic polarity*: "[T]he fundamental fact [of] life is [that it is] not indifferent to the conditions in which it is possible" (NP, 127). That is, the organism's behaviour, its engagement with its surroundings, is *never neutral* but always *polarized*, it involves "preference and exclusion", "propulsion and repulsion" (NP, 136). Put simply: it is because interactions with matter *matter* to the organism, that *living* is, fundamentally, a *normative activity* (NP, 123, 126, 228).

This normativity of vital behaviour expresses itself in the fact that the living being is not, as Merleau-Ponty puts it, akin to a keyboard (SB, 12–4), which passively registers, in a predetermined way, a set of external determinants, but actively participates in the process of selection and formation of the triggers to which it is sensitive. The organism does *not react* but instead *responds* to the environmental stimuli: it is not enough for a physico-chemical stimuli to be present; it also has to

[9] On this view, the transition from vitality to mentality is characterized not by addition (of some extra entity, faculty, etc.) but rather by *transformation*, by "a retaking and a 'new' structuration of the preceding [order]" (SB, 184). For a critical comparison of additivist and transformativist accounts of mind see Boyle, 2016.

be noticed and addressed by the organism. The exact nature of the organismal response will depend on the state in which the organism finds itself, and this state in turn will depend on the organism's organisation, behavioural aptitudes, developmental history, etc. (SB, 15), all of which are in the service of "a permanent and essential vital need" – the *need for self-maintenance/actualization*[10] (NP, 127).

Vital behaviour is thus an *irreducibly qualitative notion*: it is not a sum of mechanical reactions governed by a set of physico-chemical factors, but "a new articulation [...one] that cannot be entirely translated into analytical and quantitative terms" (NP, 80). On the one hand, this articulation pertains not to an assembly of anatomico-physiological parts, but to the living being *as a whole*, as a *center of normative activity*: it is the animal, and not its brain, endocrine glands or skeletal muscles, that eats, hunts and procreates. On the other hand, it is directed not at the physico-chemical surroundings, but at the organism's *environment* (Ger. *Umwelt*), the domain of structures that are significant for that particular organism:

> Physics is a science of fields, of milieus. But it has been discovered that, in order for there to be an environment, there must be a center. It is the position of a living being, its relation to the experience it lives in as a totality, that gives the milieu meaning as conditions of existence. Only a living being, infra-human, can coordinate a milieu. (AV, 70; cf. LM, 113–4)

The blindspot of the mechanicist-reductionist conceptual framework, according to Canguilhem, is its inability to make room for the idea that *quantitative continuity* does not preclude *qualitative discontinuity*, and that, correlatively, the normative character of the living, which expresses itself in its eudaimonic and functional states, has "no meaning on a scale where the biological object is reduced to colloidal equilibria and ionized solutions" (NP, 110). This is why, when arguing that health and disease[11] cannot be reduced to statistical terms, e.g., to the (non)divergence from statistical averages of physiological constants (blood pressure, sugar levels, etc.), Canguilhem compares normal and pathological states to euphonious and cacophonous states in music, stating that, although the laws of acoustics are not violated in cacophony, this doesn't imply that all sound combinations are pleasant (NP, 56). He accentuates this point by a very telling example:

> The laws of physics and chemistry do not vary according to health or disease. But to fail to admit that from a biological point of view, life differentiates between its states means condemning oneself to be even unable to distinguish food from excrement. Certainly a living

[10] This idea harkens back to Goldstein (1995, cf. esp. 162ff), and finds an intriguing echo in the autopoietic theory (Varela, 1979; Maturana & Varela, 1987; Thompson, 2007) whose central tenet is the idea that life is characterized by an ongoing recursive process of self-production.

[11] The reason why Canguilhem puts so much emphasis on issues related to health and disease is because, in his view, "[t]he distinction between the normal and the pathological holds for living beings alone" (MO, 90). A famous example he uses to substantiate this claim is "a massive and often neglected fact: life tolerates monstrosities" (MO, 90). Canguilhem elaborates: "There are no mineral monsters. There are no mechanical monsters. [...]. One could say that a rock is enormous, but not that a mountain is monstrous. [...] The monster is a living being with a negative value." (MM, 135)

> being's excrement can be food for another living being but not for him. What distinguishes food from excrement is not a physicochemical reality but a biological value. (NP, 220)[12]

We can elaborate on this last idea – the idea of food having a biological value – *via* the phenomenon of *pain* (NP, 96ff). Canguilhem argues that, instead of seeing pain as either (i) a "physicochemical reality" (NP, 220), which is devoid of value, or (ii) "logical signification" (PP, 7), which has "encyclopedic value" (NP, 98), we should rather see it as (iii) a "vital sensation" (NP, 98) imbued with "affective", "expressive" or "vital value" (NP, 136; P, 7). In other words, to say that pain is a 'sensation' is not to say that it is a mere *qualia* – an experiential 'oomph' bereft of all efficacy, as the term is commonly understood in the empiricist tradition – but that it is *that which makes sense* for the organism understood as a *concrete, living whole* (NP, 97). Pain, in short, pertains to "the sphere of concrete awareness": it is "a fact at the level of behaviour", of the organism as a whole, and not on "the plane of abstract science", of partitive mechanical processes (NP, 99).[13]

Thus, although pain does not necessarily provide accurate information about bodily topography or functioning – a person in pain could be, and often is, mistaken as to the location and origin of her pain (NP, 89) – it still has *significance for the organism*; and although this significance is not of an intellectual-conceptual nature, it stirs the organism in its *sensuo-motor-affective being*.

> The physiologist can indeed denounce the illusions of pain as the physicist does those of sight; this means that sensation is not knowledge and that its normal value is not a theoretical value, but this does not mean that it is normally without value. (NP, 98)

However, the same holds, *pace* the physicist from the last quote, for vision and all the other sensations properly understood. For instance, Merleau-Ponty notes that "[v]ision is already inhabited by a [… *sens*] which gives it a function in the spectacle of the world and in our existence", where the French word *sens* stands for both 'meaning' and 'direction' (cf. Smith's comment in a footnote at PP, 229). Thus, sensation in this broader sense is invested with an "immanent meaning" or a "non-thematic significance" (PP, 57, 320), which is related *not to notion*, but to *(e)motion*:

---

[12] This bears striking resemblance to work in the autopoietic-enactivist lineage. Take, for instance, the now famous example of a bacterium swimming in a sucrose gradient. While the sucrose molecule itself is characterized by a series of physico-chemical properties it "has significance or value as food […] only in the milieu that the organism itself brings into existence" (Thompson, 2007, 74). Put differently, the organism's normative activity brings forth what Varela calls the *surplus of signification*: "Remove the bacteria as a unit, and all correlations between gradients and hydrodynamic properties become environmental chemical laws, evident to us as observers but devoid of any special significance" (Varela, 1992, 79). What often gets overlooked in these accounts is that, with the onset of human mindedness, this surplus of *signification* gets trans-formed into the surplus of *negation* expressive of the ex-centric positionality of the human (see Section VI).

[13] See also the following passage in Merleau-Ponty: "This rich notion [of sensation] is still to be found in the Romantic usage, for example in Herder. It points to an experience in which we are given not 'dead' qualities, but active ones." (PP, 60)

it excites and orients the organism's attention, it solicits and constrains, demands or prohibits.[14]

However, isn't all of this – the talk of organism as vital normativity, of sensation as vital signification, etc. – but an anthropomorphic spasm brought about by what Henri Bergson called an "illusion of retroactivity", whereby one invests non-human beings with traits found exclusively in humans (NP, 129; cf. also 127; SB, 49)? In Canguilhem's view, the very opposite is true: foregrounding the normative character of life is not a matter of unjustifiably spilling humanness over 'everything organic', but a matter of seeing *human activity – even theoretical, scientific activity – as an extension* (Sect. 4) and *elaboration of vitality* (Sect. 5). In other words, it is a call to conceive of 'normative action', usually associated with the human faculty of producing normative judgements, against a larger and more fundamental background of life as "that which establishes norms" (NP, 127). After all, "how [could] normativity that is essential to consciousness [...] be explained if it did not in some way exist in embryo in life" (NP, 127)?

4 Not an Angel: On Praktognosia

How, then, *does* what has been said so far square with the realm of the human? In what way can vitality be said to feed into human cognition, scientific or otherwise? We have noted that a scientist, being a living being herself, is actively involved in coming to terms with her environment. As such, all her activities, scientific ones included, need to be situated against, and understood from, this general background, which is characteristic of all living beings.

For instance, Canguilhem notes that a physiologist, examining the biological constants and classifying them as normal or abnormal, goes beyond "the strict work of science". For, by considering life as polarized (normative) activity, he no longer observes it with a disinterested eye of the physicist, but rather with an engaged eye of a flesh-and-blood cognizer who is also permeated by vitality (NP, 222). Similarly, Merleau-Ponty writes how a psychologist, investigating human behaviour, could try and imitate a chemist or physicist in considering his own body and the bodies of others as "mechanical things with no inner life". However, in doing so, he is constantly drawn back into himself, for – unlike the said chemist or physicist – he is both the subject and the object of his investigation: "[This body] which he approached in a detached frame of mind was himself; he lived it while he thought about it [...] he *was* all that he was *talking about*." (PP, 109–110)". In sum:

> [T]here is nothing in science that has not first appeared in consciousness [lived experience]. [...] If, today, the physician's knowledge of disease can anticipate the sick man's experience

[14] See also: "Sense experience is that *vital communication* with the world which makes it present as a familiar setting to our life. It is to it that the perceived object and the perceived subject owe their thickness. It is the *intentional tissue* which the effort to know will try to take apart." (PP, 61; my emphases)

of it, it is because at one time this experience gave rise to, summoned up, that knowledge. (NP, 92–3)

This conclusion has at least two important implications. Firstly, and as indicated in the last paragraph, we see that *life* – even in human beings – *grounds, and calls forth, thought*. All our *cogitationes*, says Merleau-Ponty, are sustained "by our [vital] communication with the world as primary embodiment of rationality" (PP, xxiii). For the most part, however, we are not aware of this tacit background, because we *live through* it, we *are* it. It is often, as we have suggested above, life's drawbacks that bring it to light, and in doing so, transform it – or at least certain aspects of it – from the implicit horizon to the explicit object of thought.

Take, for instance, a memorable passage, in which Canguilhem underscores the noetic import of disease for life sciences:

Disease is the source of the speculative attention which life attaches to life by means of man. If health is life in the silence of the organs, then, strictly speaking, there is no science of health. Health is organic innocence. It must be lost, like all innocence, so that knowledge may be possible. Physiology is like all science, which, as Aristotle says, proceeds from wonder. But the truly vital wonder is the anguish caused by disease. (NP, 100–1)

Lived experience is the shadow of knowledge and, as Whitehead so readily puts it, "[t]here is no parting from your own shadow" (Whitehead, 1925, 20): the notion of vital wonder teaches us that it is in-and-through the ruptures in the normative activity of life that the life of active (reflective) normativity is born.

The second implication is that what I, following Merleau-Ponty, refer to as *praktognosia*[15] – embodied, technical, and, by extension, technological activity[16] – occupies a more central place in the epistemic edifice than is generally assumed. This implication is based on the recognition that vital normativity is not only an external accompaniment of reflection – a mere motivating factor that is genetically primary, yet epistemologically secondary – but rather that it *feeds into, and internally stimulates*, thought. That is, if thought is situated into the framework of life, and if scientists themselves are living – i.e., active, affective, embodied – beings, then the implicit, practical, and techn(olog)ical dimensions of knowing cannot be mere complements of thought, but are constitutive vehicles in-and-through which the latter manifests itself. That is to say, the *ideal of episteme,* of disengaged knowledge, needs to be tied to the *facticity of praktognosia* – the *logos* of our living and acting

[15] Merleau-Ponty himself borrows the term from the language theorist Abraham Grünbaum (PP, 162), and uses it to designate the implicit or tacit modes of somatic knowledge as expressed in our habitual, skilled engagements with the world.

[16] I will, in what follows, underlie the close interrelatedness between "technique" and "technology" by utilizing a perhaps somewhat cumbersome term "techn(olog)ical". The main purpose of the said neologism is to remind the reader how, in human behaviour, *techniques* (stylized ways of attending to, and engaging with, a specific problem domain) have a tendency to actualize themselves as, and to be recursively modified by, *technologies* (systematized implementation of various types of appropriated or constructed 'externalities' – tools, instruments, symbolic frameworks, etc. – for the more efficient, off-loaded implementation of a given technique or set of techniques).

bodies, and their numerous techn(olog)ical extensions, ranging from comportments, gestures, and habits to tools, machines and institutions.

The constitutive relevance of praktognosia for scientific knowledge is often downplayed, as practical-techn(olog)ical action tends to be (mis)construed as something derivative, as a *post-festum* practical application of scientific theories (NP, 99, 104–5). However, such a view errs in at least two respects. To begin with, it ignores the fact that our scientific theories often grow out of the techn(olog)ical, praktognosic tinkering, and not *vice versa*:

> To deny technology [*la technique*] a value all its own outside of the knowledge it succeeds in incorporating, is to render unintelligible the irregular way of the progress of knowledge [...]. If technology's rashness, unmindful of the obstacles to be encountered, did not constantly anticipate the prudence of codified knowledge, the number of scientific problems to resolve, which are surprises after having been setbacks, would be far fewer. (NP, 105)

Secondly, and relatedly, by putting the cart (science, theory) before the horse (technique, practice), the classical approach *severs mind from life*, transposing it – in theory if not in practice – into a pristine realm, in which a disembodied gaze hovers over the mechanical landscape. However, this noetic chasm can be bridged if we consider praktognosic tinkering not as a product of science, but as an extension and elaboration of "vital impulses at whose service it tries to place systematic knowledge" (NP, 130; cf. also MO, 93). The exploration of the embodied and techn(olog)ical modes of action characteristic of human beings thus not only affirms the relative creative independence of "arts and crafts" from theoretical knowledge, but also shows that, through praktognosia, human is continuous with life and that this continuity lays the vital ground against which the rupture wrought by thought (see next section) can manifest itself (MO, 197). Life not only motivates thought, which would be a rather trivial conclusion, but *permeates* it: its normative dynamics feeds into the dynamics of knowledge through the medium of embodied- techn(olog)ical action.

5 Nor a Beast: On Symbolic Behaviour

However, by entering the domain of praktognosia, we have already found ourselves at the chiasmic point in which the vital not only extends into, but is also transformed by, the mental. For what praktognosic, particularly techn(olog)ical, tinkering so vividly expresses is the *unique behavioural variability* of human beings:

> Man has succeeded in living in all climates; he is the only animal – with the possible exception of spiders – whose area of expansion equals the area of the earth. But above all he is the animal who, through technology, succeeds in varying even the ambience of his activity on the spot, thereby showing himself now as the only species capable of variation. (NP, 178)

While the environment of all living beings is, in an important sense, the result of their activity, techn(olog)ical activity allows for a *qualitatively novel recursivity of action*, whereby behavioural interventions aimed at the environment become themselves the aim of subsequent behavioural interventions, and thus, in a sense, constitutive dimensions of that very environment.

To stave off a potential objection that we have just entered a murky terrain of human exclusivity, let me point out that the same principle applies here as in the case of vitality: "The progressiveness of an event does not exclude the originality of an event." (NP, 87) That is to say, we must not, on account of recognizing the *continuity* of phenomena – the continuity between human technology and vital behaviour – ignore their *originality* (NP, 75), and overlook the fact that, just because there are intermediaries in the series under scrutiny, there are also *qualitative differences* at its extreme points (NP, 122).

It is against this background that we must read Merleau-Ponty's perhaps somewhat surprising statement:

> Mind is not a specific difference which would be added to vital or psychological being in order to constitute a man. Man is not a rational animal. The appearance of reason and mind does not leave intact a sphere of self-enclosed instincts in man. (SB, 181)

On this view, mindedness is not a mere addition to vitality, but a *wholesale transformation* or *re-structuration* of life – it is, as we have noted, a different mode of being, a different structure of behaviour. Does this mean that we have slipped back into Cartesian dualism? Is thought, once it emerges, some-*thing* over and above life? By no means. Merleau-Ponty is clear that what we are dealing with here is a "functional opposition", which cannot be transformed into a "substantial opposition" (ibid.). To understand this, however, we must allow for the fact that "there are *several ways for the body to be a body*" (PP, 143; my emphasis). But how is that possible and what does it entail?

We have seen that, in dealing with their surroundings, all living beings embody *a certain perspective* – a normative center from which they respond to external stimuli in a way that is meaningful to them, to their organization and their developmental history. It is from this perspective that they engage with their environment, and in doing so, alter it. With the onset of mindedness, however, this perspective, without being discarded, transmutes into the capacity to manifest not one, but *a multitude of perspectives*. As Canguilhem puts it:

> [T]he human reaction to provocation by the milieu is diversified. Man can give several different solutions to a single problem posed by the milieu. The milieu proposes, without imposing, a solution. To be sure, in a given state of civilization and culture, the possibilities are not unlimited. But the fact of considering as an obstacle something that may later be seen as a means to action ultimately derives from the idea [...] that man [...] builds himself out of his possibilities, his needs. [...I]t results from what he represents to himself as desirable, which is inseparable from the ensemble of values. (LM, 109)

This perspectival variability, found in humans, allows not only for a far greater number of possible engagements with, and thus modifications of, the environment, but also – and even more importantly – for *transforming the very conditions of possibility* of such engagements. Let us investigate this further.[17]

To begin with, I, as a *minded* being, am able to *alternate* my perspective on a thing; the latter, in turn, is not exhausted by any of its momentary aspects, but exhibits a certain *temporal continuity across its aspects*. A chimp, say, gets lost in its 'aspectival transformations': for instance, if it succeeds in using a tree branch as a stick, the tree branch can no longer be used as a stool. Put crudely, for a chimp, 'stick-as-branch' and 'stick-as-tool' are not two aspects of the same thing, but literally *two different things*: "It is a branch which becomes a stick […] the way a shake of a kaleidoscope makes a new pattern appear without my being able to recognize the old one in it." (PW, 119–20). In the case of humans, however, "the tree branch which has become a stick will remain precisely a tree-branch-which-has-become-a-stick, the same thing in two different functions" (SB, 175).

Secondly, and relatedly, the thing, in its newly acquired permanence, becomes a possible anchor for my alternating standpoint: I can lodge myself into its temporal thickness, and from there, so to speak, *turn the perspective onto myself*: I myself, or more precisely, my living body – the center out of which I have thus far unconsciously lived – can now become a yet another thing in the world of things. Note that this does not mean that I have to actively – say, *via* imagination – project myself into the thing, but that my minded mode of being encompasses, as the horizon of possibilities, feasible permutations of my standpoint.

Of course, none of this happens *in vacuo*. I am not a *solus ipse*, but am *immersed in the realm of intersubjectivity*: it is because I am, since my very childhood, enmeshed in a socio-cultural context, whose normative modes of dealing with things and people I have in-corporated (literally: integrated into my body schema) through learning, education, etc., that I know how to adopt, and meaningfully alternate, different perspectives – to see some-thing first as a mug, then as a paperweight, etc. -, and thereby gradually learn to see myself – by, as it were, transcending myself – as a dynamic node in this interrelated framework of varying meanings.

However, an in-depth account of this would take us too far afield. What is relevant for our discussion, is that mindedness developed against a background of culture and intersubjectivity (of what Canguilhem calls a "collective man"; LM Canguilhem, 2008, 109) brings with it a certain *multiplicity and mobility* of perspective, whereby I am able to plastically shift between different normative frameworks, different "ensembles of value" (ibid.), instead of being imprisoned in them, and thereby engender, on its utmost edges, a new normative dimension – that of *objectivity and truth*: "Man is a being who has the power of elevating to the status of objects the centers of resistance and reaction of his milieu […] among which animals live entranced." (SB, 176; quoting Scheler).

[17] The point of departure for my reflections will be Merleau-Ponty's critical engagement with Köhler's (1925) famous study of chimpanzee behaviour. In his analysis, Merleau-Ponty focuses primarily on those cases where chimpanzees failed to successfully solve the task they were presented with, and arrives to conclusions that are conspicuously close to those independently arrived at by Plessner (see esp. Plessner, 2019, Ch 6). For an exhaustive, and interpretatively brilliant, treatment of this topic, see Moss Brender, 2017, 142–7; see also Vörös, 2022.

But how can this malleability be achieved without our falling back upon a world-transcending Spirit? For a non-human living being what matters are *relations*. Its 'stimuli' are not individual physico-chemical factors, but meaningful configurations, i.e., specific relations between the physico-chemical factors that acquire their significance in relation to the organism in question: nutrient, predator, prey, etc. For a human being, on the other hand, what matters are *relations among relations*. I am able to establish relations in regard to the already established relations between my lived body and its meaningful configurations, and thus bring forth a *new type of relationality*, which invests human life with a qualitatively different signification.

However, and this is of utmost importance, this transition from relations to relations-between-relations still takes place in the realm of behaviour. Only now, the behaviour in question is a *second-order* or *symbolic* behaviour (SB, 104, 118 ff), in which *behaviour itself* becomes "the proper theme of activity" (SB, 103). A certain action I undertake – be it a gesture, drawing, or utterance – *expresses*, on a second (symbolic) level, a significance that is or could be *acted out* on the first (lived-through) level – and thereby *modifies* it. That is, a given 'first-order' significance (e.g., some biologically relevant meaning) becomes surrounded by the horizon of possible 'second-order' restructurations, which sap some of its existential urgency, as there are always other possible ways to respond to it or even to abstain from responding altogether. The realm of the symbolic is thus the *realm of the possible*: the capacity to alternate one's behavioral attitudes and thus change the significations one is confronted with.

This has three important implications. Firstly, when 'second-order' behaviours become sedimented, i.e., materialized in specific cultural objects such as hammers, sonatas, formulae, etc., they can become 'first-order' referents for subsequent 'second-order' behavioural manipulations not only here-and-now, but also – at least in principle – *everywhere-and-everywhen*. That is, I can, by means of a certain cultural object (e.g., a 2-D map), express a given meaning without having to enact it (e.g., actually traverse the depicted path); but I can then express the meaning of that expression – without, again, having to put it into practice – by means of another already-established or newly-minted cultural item (e.g., by writing a treatise on it or embedding it into a digital 3-D map). Thus, symbolic behaviour is *intrinsically recursive* in that each expressive means becomes a potential object of expression for subsequent expressive means.[18]

[18] The foremost techn(olog)ical vehicle of symbolic behaviour is, of course, *language* (see Vörös, 2021). The reasons for this are at least twofold. On the one hand, it is "able to settle into a sediment and constitute an acquisition for use in human relationships" (PP, 220): technologies of writing populate intersubjectively enacted domains of meaning with treatises, novels, and sagas, thereby instituting artistic, philosophical, etc., 'traditions', which lend themselves to the perusal of subsequent generations. On the other hand, and unlike some other cultural practices (e.g., painting and music), language can be infinitely recursive, which is why "it is possible to speak about speech whereas it is impossible to paint about painting" (PP, 221). Thus, linguistic meaning, once constituted, can serve not only to disclose new meanings, but also to use the latter to thematize the former – and so on, indefinitely. In fact, it could be said that it is *only with the onset of language* that thought, in its full sense, truly manifests itself: speech "does not translate ready-made thought, but accomplishes it" (PP, 207).

Secondly, with symbolic behaviour a *new center* appears, one that – unlike the normative center of the vital order – is *de-centered.*[19] The reason for its *ex-centricity* (lit.: out-of-centeredness) lies in its being *positioned askew* to its actual corporeal self and its milieu: the meaningful structures she inhabits at a given moment are surrounded by a halo of possible restructurations, alternate modes of acting-and-seeing. For a minded creature, every 'is-this' is surrounded by a surplus of 'could-be-other'. As such, a minded being is no longer centered in herself, but is always, even when enmeshed in everyday activities, (at least subliminally) *de-centered.* It is this excentric positionality that underscores the distinction between the living being inhabiting its milieu and the human being being open to the world:

> Animal behaviour aims at an animal setting (*Umwelt*) and centres of resistance (*Widerstand*). [...] Human behaviour opens upon a world (*Welt*) and upon an object (*Gegenstand*) beyond the tools which it makes for itself, and one may even treat one's own body as an object. [...] Human life 'understands' not only a certain definite environment, but an infinite number of possible environments, and it understands itself because it is thrown into a natural world. (PP, 380–1)

Finally, and relatedly, we have said that it is the failures, breakdowns, etc., that serve as the origin of vital wonder and thus as the cradle of thought; further, we have said that thought "undoes the experience of the world", allowing me to retreat from it and interrogate it "in the face of obstacles that arise" (TL, xvii). Thus, it could be said that what is a *contingency* in the realm of the living becomes a *principle* in the realm of thought: the ex-centric positionality of human mindedness has, as its condition of possibility, precisely that – a breakdown, failure, etc. – which constantly challenges each respective laying down of any conditions of possibility. More specifically, mindedness turns an occasionally occurring hiatus between myself and the world into a constitutive principle of the human mode of being: I am never fully centered in my current presence to the world but am always both before and behind it.

From this moment onwards, every *actual affirmation* is subtended by a *possible negation.* A cup is never exhaustively a cup – even though, in my everyday handlings it may seem as such -, for it could also be a weapon, a paperweight, or something else entirely; nor am I ever exhaustively a coffee lover, a professor, or – again – someone else entirely:

> [T]he human dialectic is ambiguous: it is first manifested by the social or cultural structures, the appearance of which it brings about and in which it imprisons itself. But its use-objects and its cultural objects would not be what they are if the activity which brings about their appearance did not also have as its meaning to reject them and to surpass them. (SB, 176)

It is against this horizon of possibilities, which surround each given actuality, that a qualitatively new determination emerges, namely that of *objectivity*. The thickness of each aspectival signification – of each partial biological meaning – must, so to speak, 'bleed out' in order for the thickness of the object to assert itself. It is only for a being whose *are's* are constitutively surrounded by *are-not's* or, more

[19] For a fascinating classical account of ex-centric positionality see Plessner, 2019, esp. Chap. 7.

generally, by what Plessner so aptly calls "a surplus of negativity" (Plessner, 2019, 251), that there can be *objects* – multidimensional things that *bite* into the cognizer (PW, 45). And it is only in the realm of objectivity – in the realm of meanings that are not only lived, but symbolically thematized – that one can be "open to truth" and "proper value of things", of what, and how, they are (not) (SB, 122).

6 Living Minds, Minded Lives: The Relation of Foundation

We are now better equipped to understand Canguilhem's oblique standpoint towards the subject of our inquiry. It is only if we position ourselves *aslant*, if we do not encase ourselves either (*pace* transcendentalism) in thought or (*pace* life philosophy) in life, but see both as unique ways of coming to terms with the resistances and beckonings of the world, that we are able to realize that *life is not completely foreign to mind,* for its vital murmurs are the silent nourishment of thought, and that *mind is not completely at home in life*, for while *in* life it is not *of* it. The life-philosopher is right in ascribing value and meaning to life, but errs in maintaining that this vital normativity exhausts the domain of mindedness; the transcendental philosopher is correct in insisting on the uniqueness of thought, but is wrong in uprooting it from its existential soil. It is only if we construe life and thought as two structures or forms of the same process, as two ways of an *ongoing dialogue* – of keeping distance while maintaining relatedness (Bortoft, 2012, 20) – between the organism and the world, that it is possible to consider thought as that in which life "becomes other without becoming another" (Ibid., 71).

We have seen that, life as a polarized, normative activity responds differentially to what environment presents it with, and is therefore a (trans)formation of forms, an ongoing (re)structuration of dynamic equilibria. This vital dynamism feeds into thought, it is its sustenance, its horizon of application. However, and as already mentioned, this is not to say that thought can simply be reduced to life.[20] Thought takes dynamics of life and, as it were, turns it against itself: it positions itself towards the fundamental positioning of the living, and thus realizes a new, *de*-centered center.[21] One is here reminded of a famous fragment from Blaise Pascal's *Pensées*, which reads: "Man is neither angel nor beast, and unhappily whoever wants to act the angel, acts the beast." (1995, § 557) Indeed, human beings are not angelic

[20] In Merleau-Ponty's words: "We are certainly not denying […] the originality of the order of knowledge *vis-à-vis* the [vital] order. We are trying only to loose the intentional web which ties them to one another, to rediscover the paths of the sublimation which preserves and transforms the perceived [lived] world into the spoken [thought] world." (PW, 123–4)

[21] To do justice to this two-way process, dynamic ways of thinking need to be developed that would allow us to recapture the circulatory relation between 'higher' cognitive superstructures and 'lower' vital substructures, without either absolutizing their separation or subsuming one under the other. Space constraints prevent me to pursue this matter further, but I would like to direct the reader's attention to one promising candidate for this role, namely the phenomenological notion of 'founding' or *Fundierung* (e.g., PP, 458; see also an illuminating and accessible account in: Matherne, 2018, esp. 783).

intellects fortuitously tied to their bodies, but *living minds* ineradicably enmeshed with temporality, worldliness, and sociality; and yet, their *minded lives* differ from those of other animals in that they are a manifestation of a radically new, and radically ex-centric, mode of being, which provides for a completely different structure of behaviour, completely different *dialoguing* with the world.

However, and crucially, the researcher investigating life-mind relation cannot extricate herself from the phenomena she is investigating, and must recognize in her capacity to position herself aslant the manifestation of the ex-centricity of the minded body she is scrutinizing. That is, it is only by enacting the ex-centric positionality which allows human beings to position themselves askew and thus *think what they embody*, that we, as investigators of the relationship in question, are able to grasp its dialectical nature.

This, in turn, reflects back on the very nature of all our philosophical and scientific endeavours, as they too, cannot be extricated from the outlined dynamic between vitality and mentality. As such, the ideal of pristine, crystalline rationality must give way to a more grounded understanding, but one that, although admitting the contribution of vitality, does not slip into the extreme of ineffable irrationality. In other words, we must move towards what Canguilhem calls "reasonable rationalism", a rationalism which "recognize(s) its limits" by "incorporat(ing) the conditions of its practice" and is keenly, even painfully aware that, in order to do philosophy and science, we cannot withdraw into the realm of angels but have to, even when engaged in rational reflection, sometimes "feel like beasts ourselves" (TL, xx).

References

Works by Canguilhem and Merleau-Ponty (Abbreviations):

AV Canguilhem, Georges. (2008). Aspects of Vitalism. In *Knowledge of Life* (S. Geroulanos & D. Ginsburg, Trans.) (pp. 59–74). Fordham University Press.

LM Canguilhem, Georges. (2008). The Living and its Milieu. In *Knowledge of Life* (S. Geroulanos & D. Ginsburg, Trans.) (pp. 98–120). Fordham University Press.

MM Canguilhem, Georges. (2008). Monstrosity and the Monstrous. In *Knowledge of Life* (S. Geroulanos & D. Ginsburg, Trans.) (pp. 134–146). Fordham University Press.

MO Canguilhem, Georges. (2008). Machine and Organism. In *Knowledge of Life* (S. Geroulanos & D. Ginsburg, Trans.) (pp. 75–97). Fordham University Press.

NP Canguilhem, Georges. (1991). *The Normal and the Pathological* (C.R. Fawcett, Trans.). Zone Books.

PP Merleau-Ponty, Maurice. (2002). *Phenomenology of Perception* (C. Smith, Trans.). Routledge Classics.

PSM Merleau-Ponty, Maurice. (1964). Phenomenology and the Sciences of Man. In *The Primacy of Perception* (W. Cobb, Trans.) (pp. 43–95). Northwestern University Press.

PW Merleau-Ponty, Maurice. (1973). *Prose of the World* (J. O'Neill, Trans.). Northwestern University Press.

SB Merleau-Ponty, Maurice. (1963). *The Structure of Behavior* (A.L. Fisher, Trans.). Duquesne University Press.

TL Canguilhem, Georges. (2008). Thought and The Living. In *Knowledge of Life* (S. Geroulanos & D. Ginsburg, Trans.) (pp. xvii–xx). Fordham University Press.

Other sources:

Albert, K. (1995). *Lebensphilosophie*. Verlag Karl Alber.
Ash, M. G. (1995). *Gestalt Psychology in German Culture, 1890–1967: Holism and the Quest for Objectivity*. Cambridge University Press.
Bianco, G. (2011). Experience vs. Concept? The Role of Bergson in Twentieth-Century French Philosophy. *The European Legacy, 16*(7), 855–872.
Bortoft, H. (2012). *Taking Appearance Seriously: The Dynamic Way of Seeing in Goethe and European Thought*. Floris Books.
Boyle, M. (2016). Additive Theories of Rationality: A Critique. *European Journal of Philosophy, 24*(3), 527–555.
Brender, N. M. (2017). On the Nature of Space: Getting From Motricity to Reflection and Back Again. In *Perception and its Development in Merleau-Ponty's Phenomenology*. University of Toronto Press.
Foucault, M. (1985). Life: Experience and Science. In J. Faubion (Ed.), *Aesthetics Method, and Epistemology. Essential Works of Foucault 1954–1984* (Vol. 2, pp. 465–478). The New Press.
Goldstein, K. (1995). *The Organism*. Zone Books.
Hanna, R., & Thompson, E. (2003). The Mind-Body-Body Problem. *Theoria et Historia Scientiarum, 7*(1), 24–44.
Jonas, H. (2001). *The Phenomenon of Life: Toward a Philosophical Biology*. Northwestern University Press.
Köhler, W. (1925). *The Mentality of Apes* (E. Winter, Trans.). Kegan Paul.
Matherne, S. (2018). Merleau-Ponty on Abstract Thought in Mathematics and Natural Science. *European Journal of Philosophy, 26*(2), 780–797.
Maturana, H., & Varela, F. J. (1987). *Tree of Knowledge: The Biological Roots of Human Understanding* (R. Paolucci, Trans.). Boston & London.
Pascal, B. (1995). *Pensées, and Other Writings* (H. Levi, Trans.). Oxford University Press.
Plessner, H. (2019). *Levels of Organic Life and the Human: An Introduction to Philosophical Anthropology* (M. Hyatt, Trans.). Fordham University Press.
Sheredos, B. (2017). Merleau-Ponty's Immanent Critique of Gestalt Theory. *Human Studies, 40*(2), 191–215.
Thompson, E. (2007). *Biology, Phenomenology, and the Sciences of Mind*. Harvard University Press.
Varela, F. J. (1979). *Principles of Biological Autonomy*. Kluwer.
Varela, F. J. (1992). Autopoiesis and a Biology of Intentionality. In B. McMullin (Ed.), *Proceedings of the Workshop "Autopoiesis and Perception"* (pp. 4–14). Dublin City University.
Vörös, S. (2021). Minding the Body: From Corporeal Mind to Minded Corporeality. *Phainomena, 30*(118–119), 5–33.
Vörös, S. (2022). Is There Not a Truth of Vitalism? Vital Normativity in Canguilhem and Merleau-Ponty. In C. Donohue & C. T. Wolfe (Eds.), *Vitalism and its Legacies in Twentieth-Century Life Sciences and Philosophy* (pp. 153–172). Springer.
Whitehead, A. N. (1925). *Science and the Modern World*. The Free Press.
Wolfe, C. (2011). From Substantival to Functional Vitalism and Beyond: Animas, Organisms and Attitudes. *Eidos, 14*, 212–235.

Sebastjan Vörös is Assistant Professor of Philosophy at the University of Ljubljana in Slovenia. His research interests span philosophy and history of science (esp. cognitive and life sciences), epistemology, and phenomenology. He is the author of *Podobe neupodobljivega* (*The Images of the Unimaginable*), in which he addresses mystical experiences from neuroscientific, phenomenological, and epistemological perspectives, and several articles on enaction, embodiment, and (neuro) phenomenology. He is currently writing a monograph on the life and work of Francisco Varela.

Canguilhem and the Promise of the Flesh

Charles T. Wolfe

Abstract The living body appears like an endlessly renewable reservoir of authenticity, hope, and taboo. But, for the sake of conceptual clarity, we are often been told that the (mere) *body* should be distinguished from the *flesh*. That is, it's undeniable that I have a body; that I notice yours; that we worry about their birth and death and upkeep. But the flesh is a more transcendentalized, loaded concept – not least given its frequently religious background (incarnation: the Word made Flesh). It is the body 'kicked upstairs', 'bumped up' one ontological level. Flesh is like a mantra, an obsessive leitmotif. Is the difference just one of abstraction? Indeed, crucial to the narrative of phenomenology (most obviously in Merleau-Ponty but really, throughout, including in enactivism), to the story of ancestor worship and identity it tells itself and its acolytes around the campfire, is a basic distinction between the merely physical body and the flesh as something requiring 'mineness', namely, an understanding of it as uniquely 'my own', a feeling of 'what it is like to be embodied'. This goes back to the Husserlian distinction between *Körper*, 'body' in the sense of one body among others in a vast mechanistic universe of bodies, and *Leib*, 'flesh' in the sense of a subjectivity which is the locus of experience. In this essay I reflect on this vision of the body's authenticity and its costs, and contrast it with insights derived from Georges Canguilhem, whose critique of mechanism/mechanicism is not done in the name of a wholescale organicism and/or an unproblematized *éloge* of embodiment and privacy.

Keywords Flesh · Embodiment · Subjectivity · Canguilhem

C. T. Wolfe (✉)
Département de Philosophie & ERRAPHIS, Université de Toulouse Jean-Jaurès, Toulouse, France

G. Bianco et al. (eds.), *Canguilhem and Continental Philosophy of Biology*, History, Philosophy and Theory of the Life Sciences 31,
https://doi.org/10.1007/978-3-031-20529-3_10

1 Introduction

Should Canguilhem be seen as a philosopher of the flesh and what would this mean? By this question I mean the following. If one side of Canguilhem's work or intellectual persona seems to be that of the 'historical epistemologist', not least following Foucault's influential portrayal of his mentor as an *épistémologue* whose patient demystification and contextualization of scientific problems put him miles away – indeed, on the opposite side of the table – from the fervid phenomenologists of subjectivity and consciousness,[1] then another side of the philosopher from Castelnaudary, which is more commonly studied in France, is that of the philosopher of 'life', of 'vital normativity', of patient-centred medicine, overall almost a kind of humanism. By the latter I mean simply a perspective which engages with the scientific study of human life primarily in a defensive or protective mode – an attitude on full display e.g. in Canguilhem's two essays on the behavioral sciences (Canguilhem, 1958, 1980, 1992).

It is not noticed often enough that there is something of a tension between these two sides or facets of Canguilhem: after all, how could the philosopher of life and embodiment (a term not used by Canguilhem but which seems fairly easily applicable, cf. Wolfe, 2015), or dare one say, the vitalist philosopher (Etxeberria & Wolfe, 2018) also accept that these concepts are historically located constructs? When Canguilhem speaks as a vitalist, he is not afraid to speak in ontologically affirmative terms, e.g. of "the recognition of the originality of the vital fact [*le fait vital*]" ("Le normal et le pathologique," in Canguilhem, 1965, 156; Canguilhem, 2008a, 122). I will not explore this particular tension in this essay, but will focus instead on the second dimension of his thought – that focusing on the personal and, perhaps, uniquely human aspects of vitality and embodiment, which we could term 'flesh'. Rather than reflect in comprehensive fashion on each of the texts where Canguilhem addresses the issue, which ends up as a vast enterprise given that concepts like health and sickness themselves can be said to be concepts of the flesh, I will seek to locate Canguilhem within some of the positions, aporias and overdeterminations of this concept. One key aspect will be the idea that the body possesses its own 'authenticity'; that there is a 'truth' of the body – both of which are somehow inaccessible to science. As we shall see, Canguilhem neither dismisses nor endorses such ideas, but he does provide some careful critique thereof.

[1] Foucault, 1985, 1989. Foucault does, admittedly, call attention clearly to the distinctive focus on Life and the life sciences in Canguilhem's historical epistemology, including the special role he grants vitalism (Foucault, 1985, 11–12).

2 Phenomenology's Romance of the Flesh

The phenomenological tradition, despite its divergences on other topics, seems to speak with one voice when it confidently asserts that the body, by which it always means the "lived" body, is not merely an object in space, an object studied and constituted by physics, a kind of externality. For Merleau-Ponty, "the mind does not use the body, but fulfills itself through it while at the same time transferring the body outside of physical space;" this is restated in rather 'pre-Kantian' terms by the prominent enactivist theorist Evan Thompson: "Life is not physical in the standard materialist sense of purely external structure and function. Life realizes a kind of interiority, the interiority of selfhood and sense-making. We accordingly need an expanded notion of the physical to account for the organism or living being."[2] The body thus understood is always *my body*, what Merleau-Ponty called "le corps propre": "my existence as subjectivity is one with my existence as a body" (Merleau-Ponty, 1945, 467 / Merleau-Ponty, 1962, 475).[3] Now, it seems undeniable that I have a body; that I notice yours; that we worry about their birth and death and upkeep. But why speak then about the flesh, which seems, notably to its neo-Platonic and Christian heritage, a more transcendentalized, loaded concept – manifest notably in the concept of incarnation (the Word made Flesh), and visible in Merleau-Ponty's recourse to strong Catholic imagery like transubstantiation and communion (for

[2] Merleau-Ponty, 1942, 225 / 1963, 208–209 (trans. modified); Thompson, 2007, 238. Phenomenologically inspired work on proprioception continues to sound this theme of 'my own body', 'my experience', etc., notably when it employs the Husserlian notion of kinesthesis, i.e., the way the body relates to the external world. All external motions which we perceive are first of all related to kinesthetic sensations, Husserl says when discussing the constitution of space. Our body already displays "originary intentionality" in how it relates to the world.

[3] Of course Merleau-Ponty claims, in his concept of "the flesh of the world," to articulate a level more primary than both objective corporeity (embodiment?) and subjective corporeity (embodiment?): "Flesh, not mind, constitutes the visibility of the self encountered through reflexivity. Flesh, not mind, gives form to the visible-invisible chiasm in which the self acquires meaning for itself through reflexive scrutiny" (Ashbaugh, 1978, 220). And I make no claims about the entirety of Merleau-Ponty's thought (e.g. the possibility that he came back on the strength of this highly subjectivist, transcendentalized vision of the body in late writings and lectures such as those on Nature, as G. Gandolfi has noted (Canguilhem workshop discussions, 2020–2021)). For more charitable readings see the work (in progress) of Thomas Ebke and of Sebastjan Vörös. But both in Merleau-Ponty and in later phenomenologists of the body like Michel Henry, we are faced with a "transcendentalization of life" (Barbaras, 2008, 9). On the broader question of the 'theological turn' in phenomenology, see Janicaud, 1991/2001. As for Husserl himself, to be fair, what I might term the conceptual overdetermination of the lived body in the phenomenology of embodiment with Merleau-Ponty and beyond (which, as I show below, Canguilhem is more than sanguine about) is not entirely to be laid at Husserl's door; *Leib* itself in German has no particular religious connotations (cf. *Leibarzt*, the personal physician), any more than the word 'flesh' does in English. But this is not the place to adjudicate the question that slithered, Ourobouros-like, through French phenomenology (as per Janicaud), namely, whether or not divinity is really 'bracketed' by Husserl or not (for a reading which emphasizes the theological affinities of Husserl's analysis of flesh and embodiment, see Depraz, 1993).

which Deleuze and Guattari ridiculed him).[4] The flesh is the body 'kicked upstairs', 'bumped up' one ontological level. Flesh is like a mantra, an obsessive leitmotif.

Is the difference just one of abstraction? Indeed, crucial to the narrative of phenomenology (most obviously in Merleau-Ponty but really, throughout, including in enactivism), to the story of ancestor worship and identity it tells itself and its acolytes around the campfire, is a basic distinction between the merely physical body and the flesh as something requiring 'mineness', namely, an understanding of it as uniquely 'my own', a feeling of 'what it is like to be embodied'. This way of talking goes back to the Husserlian distinction between *Körper*, 'body' in the sense of one body among others in a vast mechanistic universe of bodies, and *Leib*, 'flesh' in the sense of a subjectivity which is the locus of experience, a 'bodily or fleshly I' (*leiblicher Ichlichkeit*; Ferencz-Flatz, 2014). For Merleau-Ponty, "Flesh is not matter, nor mind, nor substance. In order to designate it we need the old and new term element, in the same sense as this term was used to speak of water, air, earth and fire, i.e. in the sense of a general thing – a sort of embodied principle … Flesh is in this sense an element of Being."[5] As much as one might smile at the unself-conscious respiritualization of the body underway in this kind of thinking, it is also, as we know from our studies, a reaction against the classic whipping-boy, Cartesian dualism.

Philosophers of all kinds and creeds have cut their teeth on the critique of Cartesian dualism. Differently put, to laugh at the absurdity of Cartesian dualism – how could we survive for one second if mind and body were two separate substances? How could they communicate? – has been a kind of standard rite of passage or entry condition into the club of reasonable philosophers. Now, one of the reactions against the stupidity of this dualism was to privilege 'embodiment', with core claims including 'I am not "in my body" like a body in space', the appeal to first-person experience, and to a foundational subjectivity, all of which constitute a phenomenology of embodiment. Indeed, the "lived body" we encounter in contemporary embodiment discourse is the body in pain, or in a state of enjoyment; in a reflexive, indeed intimate relation to

[4] Merleau-Ponty, 1945, 245–246 / Merleau-Ponty, 1962, 246; Deleuze & Guattari, 1991 / Deleuze & Guattari 1994, 168–169 / 176. That Merleau-Ponty 'mystifies' or 'transcendentalizes' the flesh more in writings like the *Phenomenology of Perception* and beyond, and much less in earlier writings like *The Structure of Behavior* should be noted, for the sake of fairness and precision (I thank A. Métraux for this point). The earlier work, as can be seen from the materials on which it draws, is much closer to Goldstein and Canguilhem; it can still be charged with 'biochauvinism' but just like Canguilhem in that way. Its way of emphasizing how the living body does not "fit" in reductionist schemes is less ontologized, e.g. when Merleau-Ponty describes bodily activity as non-reducible to "a blind mechanism, a mosaic of causally independent sequences" (Merleau-Ponty, 1942, 30/1964, 30) or how "the living body does not organize time and space indifferently" (Merleau-Ponty, 1942, 122/1964, 112). Building on Merleau-Ponty, the phenomenology of embodiment will tirelessly repeat that the living body or the organism is 'not simply its psycho-chemical reality', and appeal to the 'flesh' as a category to be understood on the basis of the self-constitution of the human body (Ashbaugh, 1978).

[5] Merleau-Ponty, 1964, 182, cit. and discussed in Negri, 2008, 118 (why a political project like Negri's should need to appeal to a phenomenology of the flesh is a puzzling question that cannot be addressed in the context of the present essay).

itself – quite different from the more generic body in space. The most flesh-fixated of the embodiment theorists – the least naturalistic, in contemporary parlance – maintain that the lived body (which really is *the* body for embodiment discourse) exists at least in part "outside of physical space," as Merleau-Ponty would have it.

Canguilhem, as I will indicate below, does not at all opt for this kind of antinaturalism. But he shares the hostility to Descartes (itself rather generic for several generations of philosophers, Continental and other, but *this* hostility to the 'disembodied' Descartes is more specific both to phenomenologists of the 'flesh' and to Canguilhem).[6] My question here is not whether Canguilhem should be counted as a phenomenologist or not (a difficult question to answer in any sharp sense, although my instincts point in one direction), but rather, given that both the embodied-phenomenological tradition and Canguilhem share a 'rich' or 'thick' concept of the flesh, what is that concept and is it really so univocal?

3 Canguilhem on Embodied Experience

Rather than dwelling on consciousness, perception, and intentionality, Canguilhem's way of occupying an adjacent conceptual space to that of embodied phenomenology comes out notably in the way he sometimes speaks the language of *experience*, including when discussing the nature of sickness or the status of the patient's discourse in medicine. The appeal to experience can come with or without direct appeal to the *subjectivity* of experience. In the more indirect form, we can read statements like "the life of a living being … only recognises the categories of health and illness on the level of experience, which is first of all an *épreuve* in the affective sense of the term – not on the level of science" (Canguilhem, 1972, 131). And even in this indirect form, Canguilhem is stressing that the life of a creature is first and foremost affective experience rather than the dimensions restated and analyzed by science. In a more direct, almost blunt tone, Canguilhem also states that "In short, it is impossible to cancel out the subjectivity of the patient's lived experience in the objectivity of medical knowledge."[7] And the overall project of *The Normal and the Pathological* is to push back against a certain biological concept of objectivity (quantitative and impersonal) in favor of a more medical approach which recognizes, again, the subjective dimension both of the experience of health and illness, and of our capacity to posit norms ("vital normativity").

This experiential emphasis is, recall, an important part of the concept of 'flesh' as opposed to the merely spatial concept of 'body'. In what can be taken as a sympathetic commentary on Maine de Biran, Canguilhem writes that as humans, we cannot be treated as "intelligence served by organs" (i.e., the proverbial sailor in the

---

[6] On Canguilhem's relation to Descartes see Guillin, 2008; for a newer, more 'embodied' perspective on Descartes see Hutchins et al., 2016.

[7] Canguilhem, "Puissance et limites de la rationalité en médecine" (1978), in Canguilhem, 1994, 409 (this essay was added to the last edition of Canguilhem's collection).

ship of the body); instead, we are "a living organization served by an intelligence"; thus "the soul is necessarily embodied and there is consequently no psychology without biology."[8] Life is necessarily experiential and the soul is necessarily embodied: perhaps this is what Canguilhem meant by the rather intriguing expression he used in some lectures and essays of the mid-to late 1940s, never really defined, namely the "ontological originality" of life. For instance, he described the project of a "biological philosophy" as a philosophy which "draws its inspiration and material for thought from … a vision of the vital phenomenon *understood in its originality.*"[9] In a 1947 essay (Canguilhem, 1947) he positions this "biological philosophy" as closer to a Romantic philosophy of life but also to Bergson and Nietzsche, and opposed to the Cartesian denial of properties of life – of life's "ontological originality": "Life, in the Cartesian system, is not granted any ontological originality."[10]

One may quite legitimately be puzzled by the nature and scope of this "biological philosophy" in relation to the experiential emphasis detailed above. That is, are they different aspects of one and the same anti-mechanistic, weakly vitalistic vision? Or – which seems at least to be the more obvious reading – do they actually pull in quite different directions, given that the "subjectivity of the patient" is a strongly anthropological concept, whereas the "originality of the vital phenomenon" seems to be a concept about *Life itself*? Indeed, Canguilhem in some later essays, uses the term 'anthropology' quite explicitly: "I think that human biology and medicine are, and always have been, necessary parts of an 'anthropology'."[11] This anthropology is also what he is aiming at in his magnum opus (1943, 1966), *The Normal and the Pathological*, since in each case Canguilhem insists on the inextricable intrication of biology and normativity in human life. In contrast, in some other writings (and not only from the late 1940s) he indicates quite clearly that the real issue, the real challenge, the real explanatory ground is certainly not 'the human' but 'the vital': "I do not ascribe a human content to vital norms," he writes (Canguilhem, 1972, 77; 1989, 127). As regards the former, he is also sometimes quite careful – bordering on the critical – when it comes to excessive anthropocentrism and/or subjectivism, which makes him seem further removed from the phenomenological orbit.

4 Canguilhem and the *garde-fou* of Scientific Truth

As I have noted above, Canguilhem did at times call for a "biological philosophy," something closer to a Romantic philosophy of life but also to Bergson and Nietzsche, and opposed to the Cartesian denial of properties of life – of life's "ontological

[8] Canguilhem, 1958, 19; 2019, 760; 1980, 44 (trans. modified).

[9] Lecture course of 1946–1947 on "Philosophie et biologie," cit. in Limoges, 2018, 27, emphasis mine.

[10] Canguilhem, 1947, 324; on this text cf. Wolfe, Forthcoming.

[11] "Le normal et le pathologique," in Canguilhem, 1965, 169; 2008a, 133 (translation revised).

originality." Yet unlike most of these thinkers, as well as the phenomenologists of corporeity, Canguilhem does not rely on a romantic subjectivity, in the sense that Jean-Marie Schaeffer, for example, describes it: "It's only in phenomenology that the problem of corporeity situates itself in the frame of an approach that continues to accept as an axiom the epistemic privilege of consciousness's investigation of itself."[12] When Canguilhem seeks to understand the living body, vital norms, health, the organism, he does not do so in the name of the 'self' or of 'interiority'. Granted, he sometimes invokes "the subjectivity of the lived experience of the patient" as irreducible to medical objectivity. But the subjectivity in question is, to be honest, never disembodied, never reduced to a pure ego contemplating the reality of the flesh like a sailor on a ship, according to the Cartesian formula (where it is in fact the pilot of the ship).[13]

In his essay on health, Canguilhem also speaks of the "truth" of the living body[14]: "The truth of my body – its very constitution or its authentic existence *(authenticité d'existence)* – is not an idea open to representation."[15] In the same text he refers to a rather difficult passage in Merleau-Ponty in which the living body is described as "inaccessible to others, accessible only to its titular holder."[16] Recall that in the embodied-phenomenological vision of the flesh, my body is not in space, but that which animates space. In that sense it is not accessible to science. And, importantly, Canguilhem is wary of a vision of body and health that would be too personal, too 'mystical' one might say, and that would thus distance him definitively from the scientific universe: "The recognition of health as truth of the body in an ontological sense not only can but must admit the presence, at the limit and as a guard-rail (*garde-fou*), properly speaking, of truth in the logical sense, i.e., of science."[17]

The above sentence is crucial for my purposes, because it shows, as Canguilhem does elsewhere more broadly and with a less specific appeal to science itself, his overall naturalism. My experience of my body, my sense of embodiment are real

---

12 Schaeffer, 2007, 118. For interesting reflections on Canguilhem as phenomenologist (in relation notably to Erwin Straus) see Gérard, 2010; while I do not follow her line of interpretation she raises worthwhile questions.

13 The image that the (immaterial) soul is in the (material) body like a sailor in a ship is something that Aristotle considers (*De Anima* II, i, 413a5) and that Descartes in the Sixth Meditation rejects, without mentioning Aristotle, and sounding rather like a phenomenologist: "Nature . . . teaches me, by these sensations of pain, hunger, thirst and so on, that I am not merely present in my body as a sailor is present in a ship, but that I am very closely joined and, as it were, intermingled with it, so that I and the body form a unit" (AT IX, 64). A similar passage is found in the *Discourse on method*, part VI, AT VI 59).

14 The living body for Canguilhem seems to mean the human body, although he rejects anthropocentrism explicitly and most of his key concepts, like vital normativity, can apply equally well to a human being, an armadillo or, as Canguilhem would say in *The Normal and the Pathological*, an amoeba.

15 Canguilhem, 2002, 63; 2008b, trans. modified, 475; generally the translations in Canguilhem, 2008a and b are quite unreliable, I have revised them.

16 Canguilhem, 2002, 65; 2008b, 476.

17 Canguilhem, 2002, 68; 2008b, 477.

and not to be neglected (least of all in the doctor-patient relationship),[18] but they are not hypostatized or transcendentalized into 'the flesh' as other than the body. Truth "in a logical sense," i.e. science, as Canguilhem puts it above, serves as a limiting condition (*garde-fou*) of otherwise infinitely extensible and ineffable claims about "the truth of my body." Granted, that Canguilhem is 'pro-science' and refuses to endorse an anti-naturalist mysticism of the flesh does not mean he is the sort of accelerationist who thinks that "the human can eventually be discarded as mere drag to an abstract planetary intelligence rapidly constructing itself from the bricolaged fragments of former civilisations."[19] It's rather that his approach to the body is not 'biophobic', in the way that a lot of embodiment discourse is, e.g. when it opposes the embodied body to "the body-as-organism of biology" (Dale, 2001, 11).

5 Conclusion

I have tried here to establish distinctions within a series of positions, which do not overlap, and in which the 'lived' body or 'Life is presented as special, and generally opposed to the rest of physical nature. Canguilhem does not oppose modern biology, and does not seek in any way to "reintroduce the category of subject in biology," unlike not only Merleau-Ponty but also Varela.[20] He develops his argument from the properties of existing biological entities – sometimes cells, sometimes monsters, or environments, but more often people, whether considered as agents or patients. As he says in the introduction to *The Normal and the Pathological*, he is not so presumptuous as to declare that he could renew medicine by incorporating a metaphysics into it (Canguilhem, 1972, 8; 1989, 34). I did not seek here to set out a Canguilhemian philosophy of embodiment, the way some have sought to articulate a Canguilhemian philosophy of medicine. Such a project would undoubtedly have to draw on Kurt Goldstein's thought (Wolfe, 2015; Moya Diez, 2018), in the direction of a 'patient-centred' philosophy of medicine, but also a more metaphysically ambitious project in which the person is a creator of vital norms.

Is it possible to understand embodiment otherwise than as a kind of ontological surplus (a kingdom within a kingdom, as Canguilhem would have put it, citing Spinoza),[21] an epistemic privilege, or a private state, inaccessible to others as Canguilhem put it (this inaccessibility being able to be integrated into an ontologi-

[18] On medical authority in this context see Canguilhem, 2002, 64 / 2008b, 475; 1973, 13 and Lefève, 2014.

[19] Thus the authors of the Accelerationist Manifesto (Williams & Srnicek, 2013) criticize other, more utopian versions of their idea.

[20] Weber & Varela, 2002, 117. See Etxeberria & Wolfe, 2018 for more on this comparison.

[21] "In sum, the classical vitalist grants that living beings belong to a physical environment, yet asserts that they are an exception to physical laws. This is the inexcusable philosophical mistake, in my view. *There can be no kingdom within a kingdom* [*empire dans un empire*], or else there is no kingdom at all" (Canguilhem, 1965, 95; 2008a, 70, emphasis mine).

cal or epistemic reading, as one chooses)? Let us imagine that we remove these cards from the game one by one; what is left? Minimally, of course, there remains a kind of biologism, within which the 'subject' can be described by its cerebral processes, its immune system, its genetic signature, etc. But the possibility of a less cryptodualist understanding of flesh and embodiment, is not limited to this – just as the possibility of a living subject-body, often rethought and reconfigured as an agent, in the feminist sense of 'agency', is not reduced to an egology.

References

Ashbaugh, A. F. (1978). The Philosophy of Flesh and the Flesh of Philosophy. *Research in Phenomenology, 8*(1), 217–223.

Barbaras, R. (2008). *Introduction à une phénoménologie de la vie*. Vrin.

Canguilhem, G. (1947). Note sur la situation faite en France à la philosophie biologique. *Revue de métaphysique et de morale, 52*(3/4), 322–332. Reprinted in Canguilhem 2015, 307–320.

Canguilhem, G. (1958). Qu'est-ce que la psychologie? *Revue de Métaphysique et de Morale, 63*(1), 12–25. Reprinted in Canguilhem 1994, 365–381 and Canguilhem 2019, 747–772.

Canguilhem, G. (1965). *La connaissance de la vie* (Rev. ed.), Vrin. (First published 1952)

Canguilhem, G. (1972). *Le Normal et le pathologique* (3d rev. ed)., PUF. (First published 1943)

Canguilhem, G. (1973). L'autorité du médecin (autoritarisme et dogmatisme), entretien avec P. Roy et H. Péquignot, *Médecine de l'homme, Revue du Centre catholique des médecins français* 55: 20–25; in Canguilhem 2018, 607–617.

Canguilhem, G. (1980). What is Psychology? (translation of Canguilhem 1958), (H. Davies, Trans.). *Ideology and Consciousness* 7: 37–50.

Canguilhem, G. (1992). Le cerveau et la pensée (1980). In E. Balibar, D. Lecourt, et al. (Eds.), *Canguilhem, philosophe, historien des sciences* (pp. 11–33). Albin Michel. Reprinted in Canguilhem 2018, 895–932.

Canguilhem, G. (1989). *The Normal and the Pathological*, R.S. Cohen (Ed.), (C.R. Fawcett, Trans.). Zone Books.

Canguilhem, G. (1994). *Études d'histoire et de philosophie des sciences*, 7e édition augmentée. Vrin.

Canguilhem, G. (2002). La santé, concept vulgaire et question philosophique (1988). In Canguilhem (Ed.), *Écrits sur la médecine* (pp. 49–68). Éditions du Seuil. Reprinted in Canguilhem 2018, 1143–1154.

Canguilhem, G. (2008a). *Knowledge of Life*. P. Marrati & T. Meyers (Eds.), (S. Geroulanos & D. Ginsburg, Trans.). Fordham University Press.

Canguilhem, G. (2008b). Health: Crude Concept and Philosophical Question (translation of Canguilhem 1988/2002, by T. Meyers and S. Geroulanos). *Public Culture, 20*(3), 467–477.

Canguilhem, G. (2018). *Œuvres complètes* vol. 5: *Histoire des sciences, épistémologie, commémorations: 1966–1995*, C. Limoges (Ed.), J. Vrin.

Canguilhem, G. (2019). *Œuvres complètes* vol. 3: *Écrits d'histoire des sciences et d'épistémologie*, C. Limoges (Ed.), Vrin.

Canguilhem, G. (2015). *Œuvres complètes* vol. 4: *Résistance, philosophie biologique et histoire des sciences 1940–1965*, C. Limoges (Ed.), Vrin.

Dale, K. (2001). *Anatomising Embodiment and Organisation Theory*. Palgrave.

Deleuze, G., & Guattari, F. (1991). *Qu'est-ce que la philosophie?* Éditions de Minuit.

Deleuze, G., & Guattari, F. (1994). *What Is Philosophy?* (H. Tomlinson & G. Burchell, Trans.). Columbia University Press.

Depraz, N. (1993). L'incarnation phénoménologique. Un problème théologique ? *Tidjschrift voor filosofie, 55*(3), 496–517.
Descartes, R. (1964–1974). *Œuvres*, In C. Adam & P. Tannery (Eds.), 11 vols. Vrin (cited as AT followed by volume and page number).
Etxeberria, A. & Wolfe, C. T. (2018). Canguilhem and the Logic of Life. *Transversal: International Journal for the Historiography of Science*, 4, special issue on Canguilhem (2018): 47–63, online at https://www.historiographyofscience.org/index.php/transversal/article/view/69; A revised version appears. In C. Donohue & C.T. Wolfe (Eds.), 2022. *Vitalism in 20th-century Life Science and Philosophy*. Springer.
Ferencz-Flatz, C. (2014). Husserls Begriff der Kinästhese und seine Entwicklung. *Husserl Studies, 30*, 21–45.
Foucault, M. (1985). La vie: l'expérience et la science. *Revue de métaphysique et de morale*, 90/1, *Canguilhem*, 3–14.
Foucault, M. (1989). Introduction. In G. Canguilhem, *The Normal and the Pathological* (C. R. Fawcett, Trans.), (pp. 7–24). Zone Books.
Gérard, M. (2010). Canguilhem, Erwin Straus et la phénoménologie: La question de l'organisme vivant. *Bulletin d'analyse phénoménologique, 6*(2), 118–145. http://popups.ulg.ac.be/bap.htm
Guillin, V. (2008). Les études cartésiennes de G. Canguilhem. *Cahiers philosophiques, 114*(2), 65–84.
Hutchins, B., Wolfe, C. T., & Eriksen, C. B. (2016). The Embodied Descartes: Contemporary Readings of *L'Homme*. In D. Antoine-Mahut & S. Gaukroger (Eds.), *Descartes' Treatise on Man and its Reception* (pp. 287–304). Springer.
Janicaud, D. (1991). *Le tournant théologique de la phénoménologie française*. Editions de l'Éclat.
Janicaud, D., et al. (2001). *Phenomenology and "the Theological Turn:" The French Debate*. Fordham University Press.
Lefève, C. (2014). De la philosophie de la médecine de Georges Canguilhem à la philosophie du soin médical. *Revue de métaphysique et de morale, 82*, 197–221.
Limoges, C. (2018). La confirmation de l'historien des sciences et la mise à l'épreuve de sa philosophie biologique: Georges Canguilhem 1966–1995. In G. Canguilhem, *Œuvres complètes, vol. 5: Histoire des sciences, épistémologie, commémorations: 1966–1995*, ed. C. Limoges (pp. 7–57). Vrin.
Merleau-Ponty, M. (1942). *La structure du comportement*. PUF.
Merleau-Ponty, M. (1945). *Phénoménologie de la perception*. Gallimard.
Merleau-Ponty, M. (1962). *Phenomenology of Perception* (C. Smith, Trans.). Routledge Kegan Paul.
Merleau-Ponty, M. (1963). *The Structure of Behavior* (A.L. Fisher, Trans.). Beacon.
Merleau-Ponty, M. (1964). *Le visible et l'invisible*. Gallimard.
Moya Diez, I. (2018). Canguilhem avec Goldstein: De la normativité de la vie à la normativité de la connaissance. *Revue d'histoire des sciences, 71*(2), 179–204.
Negri, A. (2008). Towards an Ontological Definition of The Multitude (2002). In A. Negri, *Reflections on Empire* (Ed. & Trans. E. Emery). Polity.
Penoncelli, M., Wolfe, C., & Wong, A. (2020). Is the *Normal and The Pathological* Vitalist? In P.-O. Méthot (Ed.), *Vital Norms. Canguilhem's The Normal and The Pathological in the Twenty-First Century* (pp. 219–250). Editions Hermann.
Schaeffer, J.-M. (2007). *La fin de l'exception humaine*. Gallimard-NRF.
Thompson, E. (2007). *Mind in Life: Biology, Phenomenology, and the Sciences of Mind*. Harvard University Press.
Weber, A., & Varela, F. J. (2002). Life after Kant: Natural Purposes and the Autopoietic Foundations of Biological Individuality. *Phenomenology and the Cognitive Sciences, 1*, 97–125.
Williams, A, & Srnicek, N. (2013). Accelerate! Manifesto for an Accelerationist Politics. http://criticallegalthinking.com/2013/05/14/accelerate-manifesto-for-an-accelerationist-politics/ Reprinted in *#Accelerate The Accelerationist Reader*, R. Mackay & A. Avanessian (Eds.), (pp. 349–362). MIT Press/Urbanomic.

Wolfe, C. T. (2015). Was Canguilhem a Biochauvinist? Goldstein, Canguilhem and the Project of 'Biophilosophy'. In D. Meacham (Ed.), *Medicine and Society, New Continental Perspectives* (pp. 197–212). Springer.

Wolfe, C. T. (Forthcoming). A Note on the Situation of Biological Philosophy. Special Issue of *Revue internationale de philosophie* on Canguilhem.

Charles T. Wolfe is a Professor in the Department of Philosophy, Université de Toulouse-2 Jean-Jaurès. He works primarily in history and philosophy of the early modern life sciences, with a particular interest in materialism and vitalism. He is the author of *Materialism: A Historico-Philosophical Introduction* (2016), *La philosophie de la biologie avant la biologie: une histoire du vitalisme* (2019) and *Lire le matérialisme* (2020), and recently coedited (w. D. Jalobeanu) the *Encyclopedia of Early Modern Philosophy and the Sciences*. He is co-editor of the book series 'History, Philosophy and Theory of the Life Sciences' (Springer). Papers available at [https://univ-tlse2.academia.edu/CharlesWolfe]

Part III
Beyond Canguilhem

What Is Biological Normativity?

Paul-Antoine Miquel

Abstract This essay focuses on two philosophical assumptions.

According to the first one, biological normativity is not an irreducible property of the living, but rather the living is the historical result of its normative activity. There is therefore a ***logic of life*** at work in every living organism that makes it a subject and an agent. It is not the fact that it is already a subject that explains the presence of this logic. It is therefore not impossible to naturalise biological normativity, even if this concept proposed by Georges Canguilhem makes us bifurcate from a world of facts to a world of values.

According to the second, we need to extend Varela's concept of operational closure in order to naturalise biological normativity. We propose a new way of writing it that takes into account the fact that ***architectural constraints*** (**ϕ_1 ϕ_2**) are always at stake in a biological system. By such constraints, we can predict the presence of specific ***propulsive and repulsive devices*** in every organism, by which its organisation can be constantly rebuilt, and through which biological disruption can also be amplified.

Keywords Constraint · Operational closure · Normativity · Pathology

1 Introduction

This paper proposes a three-step analysis of the concept of biological normativity, as developed by the French philosopher Georges Canguilhem (1966).

The first step is that a biological system is also a living organism. It is not simply determined by physical laws. A living organism follows its own rules, that are also biological ones. Since a living obeys to its rules, they are specific (1968, 119).

P.-A. Miquel (✉)
Department of Philosophy, Université de Toulouse Jean Jaurès, Toulouse, France
e-mail: paul-antoine.miquel@univ-tlse2.fr

G. Bianco et al. (eds.), *Canguilhem and Continental Philosophy of Biology*, History, Philosophy and Theory of the Life Sciences 31,
https://doi.org/10.1007/978-3-031-20529-3_11

The second step, is that these rules are internal ones. An inclined plane is an external constraint. But the temperature of the body is an internal physiological constant. The living human organism has not 37° of temperature. It makes its temperature. A biological norm is lived and a physiological constant is nothing but a habit. It cannot be simply represented as a property of an organism.

The third step is to understand that an organism depends on a "polarized dynamic activity of norms" (1966, 77, 79, 118, 137, 151). Following Canguilhem, to be normative, is to be in good health. But health is not the absence of pathology. Health is the ability to resist and insert new biological norms in order to overcome illness. To be an agent, means to be able to resist to pathology, and not to avoid them.

This three-step analysis is intended to highlight the close link between the concept of "normativity" and what Canguilhem calls "the dynamic polarity of the living". The aim is to show that this link is of interest in contemporary theoretical biology, since it allows us to explore an extension of the concept of operational closure, as it has been developed in the recent literature (Mossio & Moreno, 2010; Mossio & Montévil, 2015; Kauffman, 2019; Montévil & Mossio, 2020).

1.1 A Physiological Constant Is Not an Invariant

A living organism is an individual. But what is an individual? Following Canguilhem, an individual is not simply placed in a "cosmic milieu" of "physical and chemical constants" (1968, 131). It has its own milieu, or "Umwelt". As commented by Canguilhem from Von Uexküll (1956), the tick is not a machine that responds to excitation. It grows on the warm of mammals, and it reacts to noticed "perceptive" signs, by producing "actantial" ones, that are nothing but biological operations. Von Uexküll's conclusion is that the relation between perceptive and actantial signs is ***closed*** on a "functional circle" (24) characterizing biosemiotics as such.

However, Canguilhem disagrees with this last proposal (1980, 144), because he understands the organism as fundamentally heterogeneous. On one hand, the environment is nothing but a part of the living organism. But on the other one it is more. The role played by the functional biosemiotics circle is already an extension of the role played by the internal milieu, so that internal milieu and functional biosemiotics circle are already *entangled.* The organism cannot be defined only by its internal milieu, as asserted by Claude Bernard.

It we go further in this way, the liver cell is already living in an internal milieu, as shown by Claude Bernard. But it is at the same time a milieu for its infra-cellular elements (1980, 144), meaning that the limit of individuality is not the cell: a living organism does not loop on itself without depending at the same time on other levels of organization. The living organism shapes its own "immanent" norms (1966, 144, 186), but in order to do this, it has to deal with heterogeneity: "from the biological point of view it is necessary to understand that between an organism and the environment there is the same relationship as between the parts and the whole inside the organism itself. The individuality of the living does not stop at its ectodermal borders, not more than it begins with the cell" (1980, 144).

As a consequence, the physiological closure is never *fully* accomplished. As a second consequence it can be accomplished by various ways and functions. It explains the polyfunctionality of every organ in a mammal, and also the redundancy of each functional element. As Canguilhem says, we have everything in excess. The rule is not parsimony. It is prodigality – a lesson coming from Bergson (1907).

The second problem, now is: why are immanent norms not simply physiological constants, like in Claude Bernard's approach? Reading Claude Bernard carefully, we realize that the concept of "internal milieu" is much more complex than it seems to be at first sight. It is true that in a week sense, every organism has an internal milieu, but in a strong sense it is proper to superior vertebrates, as long as the temperature of their circulatory fluid remains constant. It means more precisely that every change in the cosmic world is compensated by a physiological "calorific regulation" (1878, 117) in the internal one, so that the temperature remains constant. However we must carefully distinguish external condition and internal ones. The presence of water, sun and oxygen are the physical environmental conditions that will "determine" (1878, 61) the growth of a plant, but the elements composing both sympathetic and cerebrospinal system, which regulate the temperature in a mammal are internal physiological conditions. Now, if someone cuts the rabbit's spinal cord at the seventh cervical vertebra, its temperature will drop sharply, but if someone cuts the great sympathetic nerve of a horse, it immediately gets a fever (1878, 118). Thus the internal milieu is nothing but the fact the a biological system is maintained in the same internal conditions represented here, by the sympathetic nerve, and the spinal cord. In other words, physiological organization is closed on its own "maintenance" expressed by the fact that the temperature does not change (1878, 117). This concept of internal milieu has a very big heuristic power, since it can explain the difference between a "thermodynamical machine" and a "living organism" (153) which can "repair" itself and "evolve".[1] A thermodynamic balance is made by comparing what goes in and what comes out of a thermal machine. But an organism is not simply consuming sugar. It ***consumes itself*** (1878, 180), since maintenance means also a ***permanent compensation*** between catabolic ("forces de destruction") and anabolic chemical forces ("forces de création", "synthèse chimique" (1878, 180). Thus, the liver is consuming sugar, but Claude Bernard assumed that the organism consumes the glycogen produced by the liver. The famous experiment of the "washed liver" related in all his books (1865, 1878) showed that he was wright. It was a great scientific discovery. Finally, it is true that "the fixity of the internal milieu is the condition of a free and independent life" (1878, 113), but this fixity passes by the compensation between creation and vital destruction.

If however Canguilhem does not completely restore the subtlety of Claude Bernard's reasoning, it is undeniable that the latter conceives the conditions which allow the internal environment to remain stable as constants. Finally, biological

[1] "Evolution" means "development," in Claude Bernard's language.

organization itself is only considered as a "complex mechanism" (1878, 35). There are thus special conditions that define the living, but they are not distinguished in nature from the general laws of physics and chemistry. And therein lies the major dispute between the nineteenth century physiologist and the twentieth century philosopher. According to Canguilhem, on the contrary, a biological norm is *lived*. Since each life is experienced from the point of view of the living in a non-objective way, since each norm is in a sense "instituted" (1966, 77) by the living itself, it seems that Canguilhem remains stuck on a vitalist position. What is immanent is lived, and what is lived cannot be fully explained by science.

He has been criticized for this epistemological position: for him, life is not simply a complex property that cannot easily be reduced to simpler ones. Life can only be felt from the point of view of the living. This does not preclude the existence of life sciences, but it obviously limits enormously the possibility of scientific explanation in biology. Let us go further, this characterization of life is not for him an anthropomorphism. On the contrary, it allows a better understanding of the specificity of human consciousness which is already in germ in every living organism (1966, 77). However, this epistemological position is not coupled with a more metaphysical speculation, as with Bergson (1907). No vital impulse, no spiritual principle at work in matter are postulated by Canguilhem. In this sense, he has learnt from Claude Bernard. Nevertheless, one may wonder to what extent such a philosophical position does not lead us once again to set limits to scientific investigation, considering the frontier between what is explained by science and what is experienced by the living as an impassable gap. This objection is relevant, but it does not negate the critical scope of the French philosopher's reflection, which shows to what extent the traditional concepts and methods of natural science do not fit well with the specificities of a biological system. Let us now ask ourselves: what could justify the hypothesis that the living makes its own norm and that physiological constants are *habits* rather than invariants in the traditional sense of the term?

The first fundamental argument is that variation does not play the same role in a physical-chemical system and in a living organism. In a physical-chemical system, variation is the deviation from the mean, in a living organism it marks the limit between anomaly and abnormality. Anomaly is in fact already a specificity of the living. Each organism is irregular by nature, and not by accident, ***in the sense precisely of an irregularity which does not verify the average, but forces us to recompose it permanently***, so that this concept ends up losing much of its relevance. There is thus a major conflict between the probability and statistical approach to biology and medicine and Canguilhem's hypotheses. He does not forbid the use of such formal techniques, as Claude Bernard did in his time. However, following the French philosopher, the naive approach of the statistician can in no way constitute a miracle solution in biology.

Anomaly is indeed a radical irregularity, not in the sense, for example, that there are monsters in addition to normal living beings, but rather in the sense that every living organism is in some way already a monster. Canguilhem (1980) takes this lesson from the work of Isidore Geoffroy Saint-Hilaire. It is important to add that an anomaly does not ipso facto become a pathology. A human can very well spend his life with only one kidney or with the heart on the right without even noticing it. An irregularity, as Canguilhem says, is "a variation on a specific theme" (1966, 84) which "prevents two beings from being able to substitute one for the other". In a physical system, any object is generic, in the sense that whatever the differences in state between one object and another, these do not affect the nature of the equation that defines its behaviour (Longo & Montévil, 2014). These differences can therefore be described in an invariant framework, which is also called a phase space. Any change, for example the motion of a planet, also appears as a trajectory in this space that is governed by an equation.

On the contrary, as far as living beings are concerned, "nothing is random, but everything happens in the form of events" (1966, 131). What characterises the living is *contingency*. "Contingency" (1966, 131) means that every biological object is also *specific*, rather than generic. Variation occurs in relation with a specific theme, and it is not governed by an equation. Why, then, should we say that every living being is specific? First of all because an organism is an agent. In a physical system, any transformation is also explained by the structure of the system. This is even the case when this system is not deterministic and when the formalism of statistics and probabilities is introduced. On the contrary, the identity of a biological system is the result of its activity, because it is an "autopoetic" agent ("autopoétique" Canguilhem, 1980; discussed by Etxeberria & Wolfe, 2018 and in this volume). Thus, the organisation of such system is a "specific" one, and it depends on its history as a *thematic*, rather than as an invariant structure. It is an immanent norm. It is not a law of nature.

Let us take two examples to illustrate this point of view, starting with diabetes. First of all, there is not one diabetes, there are many. And the basic symptom of diabetes is not necessarily glycosuria. There can be hyperglycaemia without glycosuria. If we wanted to define in a causal way how diabetes is triggered, we would be faced with a difficulty that has not yet been completely resolved. No contemporary biologist would assert that the problem of glycemia can be solved in terms of positive and negative regulatory loops. Which regulations are of interest? The ones that come from the pancreas (insulin, glucagon)? But there are also those that come from the pituitary gland, as Canguilhem already noted. There is also the activity of the cells of the gut and muscle tissues (Soto et al., 2020). Finally, how can we understand by the use of these regulatory loops that the hypoglycaemia of certain African populations is in no way a synonym for disease? This is a typical example of an anomaly that is not easily explained by attempts at causal analysis of the disease. Finally how to draw quantitatively the limit between anomaly and pathology, since they can both be considered as standard deviations from the mean? Since every disease is highly context-dependent and also the result of an individual history, the answer to this question is far from clear.

But the genetic example of the peppered moth reported by Teissier (1938) and re-examined by Canguilhem is perhaps the most striking. This example emphasizes the special status of hereditary variations, directly related to the Darwinian principle of descent with modifications (Darwin, 1859). As Ronald Fisher himself (1930) wrote in the Preface to his most important book: "natural selection is not evolution". The deep meaning of this formula was obvious: the quantitative and statistical analysis of the action of natural selection on a genetic pool could only work if the set of possible hereditary variations was fixed at the outset. In order to calculate statistical frequency and mathematical probability, the universe of possibilities for the system under analysis must be a priori posed. This is why Fisher defended a "particular theory of inheritance" (1930, 11) analogous to kinetic theory of gases in physics and opposed to the Darwinian "blending theory". Genes have to be equivalent to atoms or particles that interact only by elastic collisions. Such a requirement goes directly against the Darwinian principle, according to which the universe of possibilities is *redefined* at each step of evolution, by the fact that ***new possible hereditary variations*** can constantly arise. Unlike in all physical theories, the theoretical principle of descent with modifications is also a *historical* principle, not a *conservative* one.

It could be argued (for instance, Dennett, 1995) that the universe of possibilities is precisely fixed by the set of all possible genetic mutations that can occur in a biological system, because the space of possible DNA sequences is also a set of finite sequences of the four symbols A, T, G, C. But first of all, this set is extremely vast, even for the genome of the simplest of all living beings. Second, if we call a gene a basic unit of DNA sequences that encodes the synthesis of a protein, the description of the transcription and the translation of the gene, is certainly not already embedded in the gene itself, because of alternative splicing, non-coding DNA, epigenetics effects, proteome dynamics, etc.... Thus, variations of inheritance would certainly not be reduced to nucleotides variations. Third, the relationship between the gene and the gene product is subject to a multiplicity of ***interpretations*** that is never directly dictated by the grammar of the genome, rather like a musical phrase played by a pianist or violinist (Montévil, 2018). Four, and a least the most important point: a protein synthesized is certainly not equivalent to a phenotype expressed.

The deep meaning of all this analysis is the fundamental assumption that what comes first in biology is variation. Variation precedes the rule, it cannot be subsumed under it. On the contrary, at each historical and evolutionary step, the rule is likely to be recomposed by variation, for good or for ill. No biological norm is therefore fixed in advance. It depends each time on the way the organism will react, not only to changes in the environment, but to the variations that make it up at all its levels of organisation, without it being necessary to focus solely on the genetic level. The anomaly characterized in this way is also what specifies the plasticity of a living organism. Plasticity is not simply synonymous with robustness. One could say that a robust system maintains its functional structure over time, despite changes in the environment. Such a definition would have suited Claude Bernard perfectly. But a plastic system is a system whose functional structure is never fully given. It is constantly recomposed according to the nature of the internal and external variations that define the biological organisation at all its levels. This is why this functional structure is likely to be modified over time, in a way that is linked to the history of a system, i.e. in an unpredictable way. What a biological system does is not defined by its structure, rather its structure is the *recursive result* of what it does, because this system is a process, rather than an object.

1.2 Physiology and Pathology

An organism is an agent. But an agent is not a subjective entity in the classical meaning of the term (Descartes, Kant, Husserl), precisely because external and internal variations precede the rules. Biological organization is intrinsically specific and contingent. This can be easily shown by the fact that biological organisation is never simply characterized at the physiological level. Biological organisation always depends on an intimate relationship between physiology and pathology. It is the major contribution of Canguilhem to have emphasised this point, which reverses the logic of Claude Bernard's reasoning. Following the latter, the physiological structure of a biological system defines its individuality, completely characterised by a certain number of constants. Illness is then no more than a quantitatively measurable deviation that will gradually become a deviation from the statistical average, during the twentieth century. In order to accept this conclusion, we must be able to determine by a theoretical reasoning what this physiological structure could be. Following Canguilhem this is precisely impossible, because pathology is part of physiology, so that they cannot be separated one from the other. Therefore, illness is by no means simply an anomaly, for it opens up the new possibility of an organism destroying itself. At each step of its construction, it runs the radical risk of its destruction, precisely because no organism is indifferent to the variations it encounters within and without itself. ***Pathology is negativity understood as otherness.*** It is neither a simple opposite, nor a contradictory: "the pathological is not the absence of a biological norm, it is *another* norm but one that is comparatively pushed back by life" (1966, 91).

It means first that ***the absence of pathology is not equivalent to health***, since the logic of life doesn't obey to the third excluded principle:

$$\neg\neg A \neq A. \tag{1}$$

Health is the ability to resist and insert new biological norms in order to overcome illness, meaning that there is no health without pathology. That is why, to be normal is not to be normative. This point is central: being a healthy organism is the result of this active process by which the living being learns to overcome pathology, and by which at any moment it can fall into disease. We are not in good health. We become healthy or ill. Agency is the recursive result of this polarised dynamic, it is not the faculty of a transcendental subject. And since agency is a result, personal history matters, memory matters, learning matters.

Second, health and pathology are too different ways of life, "deux allures de vie" (1966, 51). The logic of life is not an Aristotelian one, since Canguilhem learnt from Bergson's " two orders theory" (1907, chap. 3) that the negation is precisely never a "nothing". ***Thus, pathology is not nothing.*** Pathology is passivity understood as a way of living that can lead to death. It is not a deviation from the average, it is another pace of life, and in conclusion pathology is also instituted and experienced by the living, as an immanent norm. If we follow this way of thinking, pathology is already organised, like ageing. It is not simply the result of a wear and tear mechanism. ***Pathology is disruption as a biological norm, and not entropy as a mere physical property.*** We don't get sick from the flu because our bodies wear out like a machine, and the cancerous tumour is disorderly only from an anthropomorphic point of view. From another point of view it is wonderfully organised to resist ageing and destruction.

Finally, a living being is never identical to itself. Rather, it is always both more and less than itself, because pathology is related to normativity by an internal link. This is why the French concept of "normativity" has nothing to do with the concept of adaptation. Normativity is the capacity of every living organism to establish new norms, insofar as it can also fall into pathology at any moment. Each time a living being prefers vitality to passivity and each time it can fall back into passivity. As a consequence, its organization is never maintained, it is constantly rebuilt, because it has to deal permanently with what is ***other than vitality. In*** this logic of alterity, vitality is otherness, and pathology is what is other the otherness. ***By*** this logic of alterity, biological organisation is irreducibly contingent: it can ***not to be,*** because an organism depends on its own "polarized dynamic activity of rules".

A living organism is therefore not indifferent to the risk of no longer existing, in the sense that, at each stage of its life, it ***must*** resist to its annihilation. It ***prefers*** to continue to live rather than to disappear. This preference is thus instituted by the organism, and not simply determined by internal or external conditions. It makes its norm, over-looping itself. And so it also makes its own normativity. Of course this value position is "spontaneous" (1966, 186), rather than reflexive. But the normative power to institute and to modify biological norms is an "immanent" one (1966, 186). Any norm, from this point of view, is at the same time a goal, rather than a cause, since it comes from this immanent power. Canguilhem's organicism does not reject the fact that living beings can learn and set goals for themselves. Simply, the fact of setting goals is not a faculty. It can be understood as the result of a process of overcoming pathology, rather than as a final cause. The logic of life is not an Aristotelian one.

So let us spell it out once again to avoid any misunderstanding: normativity does not mean that a living being obeys a biological norm, in the same way that a planet obeys the laws of mechanics. First of all, variation takes precedence over the rule and variation is a source of heterogeneity. The norm is the result of a reaction of the biological system to the variations it contains. This is how it becomes an agent, by recomposing what it was, and not by remaining identical to itself.

Now the author adds that this reaction is 'instituted' (1966, 77), as he repeats many times. It results from a "dynamic polarity", so that the identity of the living being is built on a radical duality. The figure of what it is not, is part of what it is. To say that the reaction is instituted, results less from the arbitrariness of choice (terms that are never used) than from the fact that any biological norm is also defined by its transgression. The fact of not obeying is part of its definition. And this possible openness to transgression is the result of an appreciation, a judgement. Every norm position is also a value judgment. And now we understand why: it comes from the agent. Even if in one sense the agent is the result of the system's activity, in another sense the system is the result of the agent's appreciation. Thus, there is no normativity without this normative circularity. The French term 'immanence' is therefore nothing more than a translation of this radical circularity, the properties of which we shall examine more closely.

One last word to conclude this second part, one could obviously think that if normativity is thus a matter of individual appreciation of the living, biological organisation cannot be explained by science. First of all, this is not what Canguilhem thinks, despite certain ambiguous formulas. For him, the living does not violate the laws of physics, and his vitalism, as we have seen, is more epistemological than ontological. There is nothing to prevent one from believing that life is a natural phenomenon. We must hasten to add, however, that if it is a natural phenomenon, it cannot be explained directly from the laws of physics, even if it does not violate them, nor is it susceptible of being reduced to a classic causal explanation, because of this dimension of circularity that we mentioned. In conclusion, we can now ask ourselves whether Canguilhem's rationalist vitalism is still relevant today, or whether it is simply a figure from a bygone era.

1.3 Propulsive and Repulsive Constants

This analysis of the dynamic polarity at work in biological normativity comes up against a direct observation: even admitting that physiological constants are habits, at the same time they do exist. One may therefore wonder whether all this philosophical reasoning is not largely speculative. After all, when we fall ill, the fever may rise, but the organism returns to its normal temperature if we recover. And if we do not recover, if for example the blood sugar level remains high, new and more serious disasters may occur. For Canguilhem's philosophical position to be interesting, it would be necessary to be able to respond forcefully to this objection, by showing that there are indeed ***specific devices*** in a living organism that attest to what we have called with him its normativity.

Let's come back first to the crucial point: following Canguilhem, the identity of an organism is its normativity.

First, normativity means some kind of ***creative virtuous circle*** through which under heterogeneous internal and external conditions, its identity is also the result of its permanent re-composition. However, since normativity is already related to the dynamic polarity, biological organisation can ***switch*** permanently from the emergence of creative virtuous circles to the emergence of ***disruptive vicious circles***. As a matter of consequence, its identity is never the same, since it is all the time threatened by the very same process of re-organisation that build it (including of course the relation with its environment). Thus biological identity is always *a* singular one. However, would it be possible to find specific devices in the biological organisation of life that attest, through their functioning, that the emergence of biological norms can be instituted and amplified by a living organism? In other words, are there specific biological devices that show the limits of the concepts of physiological constant and homeostasis inherited from Claude Bernard?

Canguilhem mentions two of them: ***immune recognition*** and ***anaphylactic shock*** (1966, 138). His interpretation of the functioning of the immune system is centred on the concept of normativity, and makes it a system that can learn, because it keeps antigen presentations in memory, but also because certain cells of the immune system become memory cells. Memory T cells can activate a response very quickly when pathogens are presented. This is where the concept of the "propulsive constant" comes from. New constants emerge so that the system stabilises, in a logic of transformation by non-identical variations that is not a simple logic of maintenance. At the same time, what is striking for Canguilhem is that the immune system can make catastrophic mistakes. A catastrophic "error" occurs when a first error engages "repulsive constants". The first error is stored in memory, and gives rise to a process of amplification of the immune response by the mediation of immunoglobulins which transforms it into a catastrophe. Finally an organism can learn by the same process through which it can make catastrophic errors.

If Canguilhem is right, such positive and negative amplification effects, are instituted by the logic of life at stake in every organism. They can hardly be explained by an understanding of the immune system in terms of clonal selection, since the result of its activity is not simply governed by the external presence of antigens, because it depends at the same time on propagative and repulsive *second order* constants, by which it learns, and through which it makes mistakes. Second, if we follow this logic, the self cannot be characterized as such before acting, since it is also the result of its polarized activity. This also explains why it is absurd to try to distinguish the self from the non-self by identifying its specific constituents. Finally, since what defines the biological individual is its potential for re-composition, it can of course be defined at several levels of organisation. Such a challenging vision is fully compatible with the fact that lymphocytes could also react to endogenic patterns, and tolerate foreign entities, like symbiotic bacteria, even if immune interactions with them would occur (Pradeu, 2012).

1.4 An Extended Definition of Closure?

Canguilhem's philosophy is somewhat similar to Spinoza's. Its tragedy is not the audacity of its author, but rather that it is too far ahead of its time, a time when the dominant wind in science was blowing towards reductionism and the advent of molecular biology. Let us summarise what we have just seen. We started with a simple idea: living beings make their own rules. They do not simply obey the laws of physics and chemistry. Yet this idea does not necessarily go against the scientific approach, quite the contrary. To make both compatibles, ***we must simply admit as a first crucial assumption, that the structure of a biological system is also the recursive result of its activity***. It derives its identity from its operativity. Francisco Varela (1989), for example, defined an "autopoietic" biological system in precisely these terms. We will now show in what sense the concept of normativity can be thought of as an extension of the concept of autopoiesis.

Following Varela, autopoiesis means: (1) the existence of a define class of networks of components which "participate recursively to the same network of components which produced them", so that the same organization is continuously rebuilt, in interaction with an environment. (2) This organisation is rebuilt, "as a unity in the space in which the components exists" (1989). Through these two properties, but especially the first one, biological organisation has to be understood as a biological ***norm***, since maintenance is nothing for such an autopoietic system, but the recursive result of its ***own*** operations, in interaction with an environment. In other words, biological organisation is the ***maintenance*** of its own activity, and as a consequence, the system is autonomous. Rephrased in a more symbolic language by a fixed point equation:

$$\mathbf{B} = \phi\mathbf{B} \tag{2}$$

In interaction with its milieu, the B system is closed on its maintenance (ϕ). That is why biological organization is not simply a cause; it is also a ***norm***, as an ***anticipation*** of the future, because depending on this biological normative rule, and not only from a physical attractor. However, the emergence of such temporal properties, like the anticipation of the future, are not coming from a supranatural condition. They are just emerging from the recursive self-referential structure of such complex systems. A biological system is not an object. It is an agent. Agency is not a mysterious property, it appears simply as the result of a recursive process, under certain chemical and physical conditions.

As a second important point, autopoiesis is not a physical property that can be compared with the cycle of water. The cycle of water is not autonomous, it is just the mere attractor of a physico-chemical dynamics under precise boundary conditions (Mossio & Moreno, 2010). But in order to understand the difference, Varela's characterization of autopoiesis has to be rephrased. A biological system is not simply defined by its physical structures, but also by its biological functions. Functions are not simply matter, they are what matters. We will add that functions, whatever the level could be (tissue, cell, molecules) are the memory of a biological system. The memory of a system is what matters *in order to* preserve its self-maintenance. Thus, such a system is also the result of a temporal looping by which self-maintenance can be characterized recursively as *an aim* depending on a *set of constraints* that is represented as *closed on itself* (catalysis, allosteric regulation, canalization, heredity; Mossio & Montévil, 2015; Kauffman, 2019). Constraints are what matters in order to.

As a third point, biological constraints cannot be considered equivalent to physical laws. Constraints are conditions that allow a certain local transformation invariance to be maintained (Mossio & Montévil, 2015). For example, insulin and glucagon can be considered as constraints that negatively and positively regulate blood sugar levels through the reversible transformation of glucose into glycogen.

But as biological norms, constraints are always contingent. They can be replaced in the course of phylogeny or ontogeny by other constraints.

However, for Canguilhem, biological organisation is not only maintenance, but also normativity, because the fact that an organism makes its own norm is linked to the hypothesis of a dynamic polarity of the living. Rephrased in the language of contemporary theoretical biology (Miquel & Hwang, 2016), a living appears to be closed, not directly on its maintenance, but on two new second order architectural constraints, *from which* its maintenance depends.

$$B = \phi_1 \phi_2 B \tag{3}$$

If we refer to the work of Canguilhem, these two architectural constraints (ϕ_1 ϕ_2) are nothing but the symbolic translation of the concepts of heterogeneity and normativity.

Following Canguilhem, a biological system is heterogeneous (ϕ_1), because it admits within itself through specific devices, not only the presence of random variations, but also of "*anomalies*" (1966, 81). We can interpret it as *non-identical iterations* able to modify the biological standards themselves (Soto et al., 2016). As we know today such biological variations occur at all levels of organisation (West-Eberhard, 2003): genetic and epigenetic variations, instability in morphogenetic processes, cell proliferation, etc. However we must add that these iterations are not identical, because of the weight of the individual history of an organism, and of the collective history of the species, or even of the family of species to which it belongs, in so far that every organisms always derive from a common ancestor. They are therefore certainly not only the effect of chance, since the past influences what happens in the present. As Canguilhem himself says: "everything happens in the form of events" (1968, 131).

As a second crucial point, a biological system is heterogeneous because pathological norms are a potential part of its definition, since they cannot only push it to its disruption, but even to the *amplification* of its disruption. Such disruptive pathological norms do exist in biological systems, and they are now beginning to be studied seriously, not only at the ecological level, but also at the level of the physiology of organisms. Firstly, it should be noted that constraints that play a functional role during development can, on the contrary, induce desynchronisation and dysfunction during ageing. This is indeed the case if we take the example of the role played by certain molecules that make up the extracellular matrix. The pioneering and little-known work of Ladislas Robert in France showed, for example, that fibronectin and elastin could be degraded into peptides under the combined action of calcium and lipids. But this is not a simple process of degradation and wear. Robert et al. (1989) showed that these peptides were recognised by molecular receptors inducing an amplification of the degradation effects in the cell by what they called ***molecular vicious circles***. The receptor plays in this example exactly the role of a repulsive device that induces a self-destruction phenomenon not really programmed by genes. Its activation involves an uncoupling of calcium flows and a release of free radicals which will contribute to the new degradation of elastin into peptides. It illustrates perfectly the fact that aging is a biological norm, and certainly not only a mere physical property.

The second architectural constraint (**ϕ_2**) can be related to "normativity" understood as an overcoming of pathology. First of all, the living is not simply individuated, like in Simondon philosophy (1964). The individuation process is the result of this articulation between "normativity" and "dynamic polarity". It means first that propulsive diachronic constraints[2] are instituted, as biological devices that permit the resistance to disruption, by the emergence of new functions. Second through such propulsive constraints a dynamic of functional changes can emerge in a short period of time. Such propulsive norms have already been highlighted in molecular biology. A paradigmatic example is the role of mutator genes in the SOS box (Radman, 1999). When E. coli is stressed by radiation, chromosomal breaks appear blocking the replication mechanism. Mutator genes are activated and are at the origin of targeted and non-targeted mutations in the bacterial genome. In a short time interval, these mutations give rise to new biological functions allowing the bacterium to adapt and better resist the stress. There is obviously a close link here between what Radman calls "adaptability" in his articles (1999), and the concept of "normativity" proposed by Canguilhem, through the *amplification* of mutations by the action of mutator genes, that are activated by the stress, and that synthetize repair polymerase enzymes.

A new vision emerges through these two lines of coherence. Life is no longer defined simply by its biological constants, but rather by the fact that there are always propulsive and repulsive devices among them. Such an idea is not incompatible with the definition of life as autopoiesis, since a biological system can always be said to produce the conditions that allow it to continue to exist in interaction with the environment. But through these two lines of coherence, it constitutes an ***extension*** of it.

When we take a closer look at the conditions of persistence of a living, we realise that they are not limited to the existence of a set of physiological constraints operationally closed on itself. Indeed, it must be added that under the action of repulsive and propulsive architectural constraints, it loops on itself in such a way that it has the potential to continuously transform its organisation over time.

And it is this potential that allows us to redefine in another language what the French philosopher called biological normativity. At any moment of its life, everything can collapse, and at any moment of its life, a living can reinvent itself. Clearly, a living organism is not a machine like any other, since it is a machine capable of continually reinventing its arrangements. We can still speak of a system, no doubt. But can we still talk about a machine? Maybe yes, but in a very new acceptance of the word.

[2] In a recent very insightful article Montévil and Mossio (2020) call "χ" such diachronic constraint intrinsically present in the diagram of constraints, reflecting the importance of history in biological organization. However in one of its characterizations, "χ" can also be a tool of propagation involving the continuous emergence of new biological constraints during a certain period of time. We think that "χ" can also be a tool of disruption.

References

Bergson, H. (1907). *L'évolution créatrice*. Alcan.
Bernard, C. (1865). *Introduction à l'étude de la médecine expérimentale*. PUF.
Bernard, C. (1878). *Leçons sur les phénomènes de la vie communs aux animaux et aux végétaux*. J.B. Baillère et fils.
Canguilhem, G. (1966). *Le normal et le pathologique*. PUF.
Canguilhem, G. (1968). *Etudes d'histoire et de philosophie des sciences*. Vrin.
Canguilhem, G. (1980). *La connaissance de la vie*. Vrin.
Darwin, C. (1859). *On the Origin of Species by Means of Natural Selection*. John Murray.
Dennett, D. (1995). *Darwin's Dangerous Idea. Evolution and the Meanings of Life*. Simon & Schuster.
Etxeberria, A., & Wolfe, C. (2018). Canguilhem and the Logic of Life. *Transversal: International Journal for the Historiography of Science, 4*. https://doi.org/10.24117/2526-2270.2018.i4.06
Fisher, R. (1930). *The Genetical Theory of Natural Selection*. Oxford University Press.
Kauffman, S. (2019). *A World Beyond Physics*. Oxford University Press.
Longo, G., & Montévil, M. (2014). *Perspectives on Organisms: Biological Time, Symmetries, Singularities*. Springer.
Miquel, P. A., & Hwang, S. Y. (2016). From Physical to Biological Individuation. *Progress in Biophysics and Molecular Biology, 122*(1), 51–57.
Montévil, M. (2018). Possibility Spaces and the Notion of Novelty: From Music to Biology. *Synthese, 196*(11), 4555–4581.
Montévil, M., & Mossio, M. (2020). The Identity of Organisms in Scientific Practice: Integrating Historical and Relational Conceptions. *Frontiers in Physiology, 11*, 611.
Mossio, M., & Montévil, M. (2015). Biological Organisation as Closure of Constraints. *Journal of Theoretical Biology, 2015*(372), 179–191.
Mossio, M., & Moreno, A. (2010). Organisational Closure in Biological Organisms. *History and Philosophy of Life Sciences, 32*, 269–288.
Pradeu, T. (2012). *The Limits of the Self: Immunology and Biological Identity*. Oxford Scholarship.
Radman, M. (1999). Enzymes of Evolutionary Change. *Nature, 401*, 866–869.
Radman, M., et al. (1999). Evolution of Evolvability. *Annals of the New York Academy of Sciences, 870*, 146–155.
Robert, L., Fülop, T., et al. (1989). Elastonectin and the Elastin Receptor. *Pathologie et Biologie, 37*, 736–741.
Simondon, G. (1964). *L'individu et sa genèse physico-biologique*. PUF.
Soto, A. M., Longo, G., Montevil, M., & Sonnenschein, C. (2016). The Biological Default State of Cell Proliferation With Variation and Motility, A Fundamental Principle for a Theory of Organisms. *Progress in Biophysics and Molecular Biology, 122*(1), 16–23.
Soto, A. M., Mossio, M., & Bich, L. (2020). Glycemia Regulation: From Feedback Loops to Organizational Closure. *Frontiers in Physiology, 11*, 69.
Teissier, G. (1938). Une controverse sur l'évolution. *Revue de l'Encyclopédie française*, N°3.
Varela, F. (1989). *Autonomie et connaissance*. Seuil.
Von Uexküll, J. (1956). *Streifzüge durch die Umwelten von Tieren und Menschen. Ein Bilderbuch unsichtbarer Welten/Bedeutungslehre*. Rowohlt.
West-Eberhard, M. J. (2003). *Developmental Plasticity and Evolution*. Oxford University Press.

Paul-Antoine Miquel is Professor of Contemporary Philosophy at the Jean Jaurès University in Toulouse. His research focuses on French philosophy and the philosophy of biology. He is the author of several works on Bergson, Simondon, Canguilhem, and he is the editor of the works of Bergson published by Garnier Flammarion.

Self-Organizing Life: Michel Serres and the Problem of Meaning

Massimiliano Simons

> *« L'important était de signaler que la nouvelle biologie, en cherchant l'Inde, avait trouvé l'Amérique ». (Morin, 1973, 28)*
>
> *"Maybe that Brillouin, the information physicist, is more of a philosopher than Jean-Paul Sartre." (Latour, 1987, 84)*

Abstract Within continental philosophy of biology the work of Michel Serres has not received a lot of attention. Nonetheless, this chapter wants to argue that Serres was part of a group of thinkers – together with Jacques Monod and Henri Atlan – that started to think about biology in terms of second-order cybernetics and information theory. Therefore, this chapter aims to do four things. First of all, it maps the relation between Serres and Canguilhem, one that was mediated by authors such as Louis Althusser or Jacques Monod. Secondly, it fleshes out Serres's own 'biophilosophy'. I label this alternative tradition as a 'biophilosophy without a subject'. Finally, this chapter explores the consequences of this alternative biophilosophy through a brief examination of two authors whose work lies in the line of this tradition: René Girard and Bruno Latour. Though at first sight different, they both draw inspiration from this biophilosophy to develop a framework that, paradoxically, 'jumps over' the subject. Hence, the reason why biology is neither a prominent theme in Girard's nor in Latour's work. This is not because of a lack of biophilosophy, but because of an implicit one: a biophilosophy without a subject.

Keywords Michel Serres · Georges Canguilhem · Jacques Monod · Henri Atlan · René Girard · Bruno Latour

---

M. Simons (✉)
Department of Philosophy, Maastricht University, Maastricht, The Netherlands
e-mail: massimiliano.simons@maastrichtuniversity.nl

G. Bianco et al. (eds.), *Canguilhem and Continental Philosophy of Biology*, History, Philosophy and Theory of the Life Sciences 31,
https://doi.org/10.1007/978-3-031-20529-3_12

1 Introduction

Philosophy of biology in French philosophy has been mainly linked with the work of Georges Canguilhem (Méthot, 2020; Méthot, Chap. 2, this volume), or authors linked to his work, such as Henri Bergson, Gilbert Simondon or Jean Gayon. This chapter wants to focus on one of Canguilhem's students, who is often forgotten: Michel Serres. A student of Canguilhem whose work was not primarily focused on biology, Serres nevertheless wrote several texts on how his more general ideas would shed new light on biological matters.

Specifically, this chapter will do four things. First of all, it maps the relation between Serres and Canguilhem, one that was mediated by authors such as Louis Althusser or Jacques Monod (Sect. 12.2). Secondly, it fleshes out Serres's own 'biophilosophy' (Sect. 12.3). Whereas a philosophy of biology applies certain philosophical insights to the field of biology, a biophilosophy uses insights from the life sciences to reconceptualize one's philosophical worldview. My claim is that Serres endorses a different biophilosophy than Canguilhem. He builds this alternative on insights from information theory, inspired by Léon Brillouin, Jacques Monod and Henri Atlan. I will label this alternative tradition as a 'biophilosophy without a subject'. Finally, this chapter explores the consequences of this alternative biophilosophy through a brief examination of two authors whose work lies in the line of this tradition: René Girard and Bruno Latour (Sect. 12.4). Though at first sight different, they both draw inspiration from this biophilosophy without a subject to develop a framework that, paradoxically, 'jumps over' the subject. Hence, the reason why biology is neither a prominent theme in Girard's nor Latour's work. This is not because of a lack of biophilosophy, but because of an implicit one: a biophilosophy without a subject.

2 Serres and Canguilhem

Michel Serres (1930–2019) was a French philosopher, originally trained in mathematics. After finishing his *agrégation* in philosophy in 1955, he received his doctorate in philosophy 1968 at the Sorbonne. With Jean Hyppolite and Georges Canguilhem as supervisors, his main thesis was *Le Système de Leibniz et ses modèles mathématiques* (1968) and his minor thesis *Epistémologie de l'interférence*.[1]

As his dissertations indicate, Serres's main occupation was not biology, but formal questions concerning structure and information. Nonetheless, Serres saw biology as one of the fields in which this novel approach took shape. But he identified this new biophilosophy not with Canguilhem, but with the new molecular biology,

[1] See Serres (1972).

embodied by Jacques Monod and François Jacob.[2] For their contributions to molecular biology Monod and Jacob, together with Lwoff, would indeed get their Nobel Prize in 1965 and in 1970 they would publish their philosophical treatises: Monod's *Le hasard et la nécessité* (1970) and Jacob's *La logique du vivant* (1970). Serres would write extensive and enthusiastic reviews of these books, expanding on his own biophilosophy as well.[3]

Serres invokes molecular biology against Canguilhem. In a later interview with Latour, Serres would recall how "biochemists understood rapidly that their own revolution would come, after information theory, from the questions posed in Schrödinger's *What Is Life?* and in France from Monod and Jacob's discoveries. Now, that was certainly not what epistemology was teaching about biology."[4] To which Latour replied that French epistemologists only concerned themselves with "cells and the reflex arc", allusions to Canguilhem.[5] Though Serres adds that these are "perfectly respectable things" which we should "at least preserve in our memory" but that are, nonetheless, "things which, at the time, became abruptly outdated. Once again the epistemologists didn't follow."[6] In another interview, Serres recounts his failure to arrange a meeting between Canguilhem and Monod:

> I even tried to introduce Monod to Canguilhem, who was after all the philosopher of the life sciences. Except that the paradigm he supported dated from the physiology of the 1940s. He had no idea what biochemistry could entail, let alone the genetic code, nor that one would soon consider deducing the totality of a living being from the DNA algorithm! He was in the past and Monod in the future. I tend to think he made me pay for this paradigm break. It must be said that such an epistemological bifurcation was difficult to swallow for a man who had dominated the discipline for so long. Anyway, he didn't want to meet Monod after all.[7]

In contrast, Serres saw his own earlier encounter with Monod as a greater success:

> A gentleman came to me at the end of one of my classes to say, 'This is what brings me here. I took a lot of philosophy courses to find a philosopher that I would take as an advisor, because I wrote a little book that I wanted to have reread. I sighed but hey, I accepted. The man gave me his manuscript titled *Le Hasard et le nécessité*! It was Jacques Monod.[8]

Serres adds that he and Monod, from that day on, "became very good friends. He introduced me to a small circle that met at his home, a sort of club where we met some nice people: René Thom, François Jacob, Marco Schützenberger and some others."[9]

---

[2] See Debru et al. (2012), Erdur (2018).

[3] See Serres (1974a).

[4] (Serres & Latour, 1995, 12).

[5] (Canguilhem, 1952, 1955).

[6] (Serres & Latour 1995, 12).

[7] (Serres, 2014, 50).

[8] (Serres, 2014, 49).

[9] (Serres, 2014, 49).

However, such a picture is too one-sided. For Canguilhem did engage with molecular biology.[10] For instance, he wrote a review of Jacob's *La logique du vivant,*[11] and in his archives one can find notes on Jacob's and Monod's books.[12] Canguilhem even invited Lwoff, Jacob and Monod for a seminar on 17 June 1971, where he presented them with a set of questions focusing on the epistemological and philosophical dimensions of their work.

Many commentators have argued that there was a certain antagonism between Canguilhem and molecular biology.[13] In his notes on Monod, Canguilhem was very critical and concluded that Monod's conceptual apparatus did not add anything that was not already found in Kant.[14] Similarly, Talcott argues that "[t]he stunning success of post-war molecular biology presented a major challenge to Canguilhem"[15] and Morange concludes that "his philosophy of life, inspired by a form of holism, opposed itself to that of the molecular biologists" and that "Canguilhem was not a keen and insightful observer of developments in molecular biology".[16]

On the other hand, Loison (2018) claims that if you look at the texts Canguilhem published on molecular biology, the latter actually enthusiastically saw molecular biology as a revival of Aristotelianism: "There is in the living a *logos*, inscribed, preserved and transmitted".[17] According to Loison, this enthusiasm faded after 1970, when Canguilhem read Jacob's *La logique du vivant*, where the latter took distance from a too-easy equation between DNA and language. We can thus conclude that, contrary to what Serres suggests, Canguilhem did not ignore molecular biology. Why then did Serres feel the need to draw such a dichotomy? I will explore at least three reasons: a biographical, a political and a philosophical one.

2.1 Serres as an Isolated Philosopher

The biographical reason refers to Serres's personal break with Canguilhem. Whereas in his doctoral thesis Serres still praises Canguilhem, the doctoral defense itself turned out to be a breaking point. Serres later recounted how something went wrong, though he refused to go into the specifics. "Let's say there was a tragic moment in my personal and academic history that I don't like to talk about. Until then,

[10] See Morange (2000), Talcott (2014), Loison (2018), Erdur (2018).

[11] (Canguilhem, 1971).

[12] (Loison, 2018, 277).

[13] E.g. Morange (2000) and Talcott (2014).

[14] (Loison, 2018, 278).

[15] (Talcott, 2014, 263).

[16] (Morange, 2000, 85, 91).

[17] (Canguilhem, 1966, 221).

Canguilhem had taken me under his wing. I was his favorite student of sorts. He got mad at me that day."[18]

Though hard to reconstruct, there are indications that Canguilhem found Serres's self-presentation rather arrogant. Serres did not sufficiently acknowledge his debt to French epistemology, in particular to Gaston Bachelard. Serres indeed published a critical text on Bachelard in 1970, distancing himself from French epistemology.[19] Evidence for Canguilhem's disapproval is found in *Idéologie et rationalité dans l'histoire des sciences de la vie* (1977), where Serres is mentioned as one of the two recent challenges to French epistemology. The first one comes from Dominique Lecourt (1969), also a student of Canguilhem and follower of Althusser (see below). The second challenge was Serres: "Another young epistemologist, Michel Serres, raises a different objection. The history of science, he says, does not exist".[20] Canguilhem refers to Serres (1974b), in which the latter argued that a general history of science has not been written so far, since it first required a critical history of classifications.[21] Canguilhem is skeptical, since for him the project already exists as the 'regional rationalism' in Bachelard's *Rationalisme appliqué* (1949) Canguilhem concludes "that Bachelardian epistemology confronted this problem well before anyone had thought of accusing historians of ignoring it."[22]

In his turn, Serres felt isolated from philosophers in France. Instead of getting a position in philosophy, Serres ended up in a history of science department: "I found myself banned from philosophy" and "had to teach outside of my profession. I used to have five hundred people in my philosophy class, and at one time I only had a handful in history of science".[23] In a similar vein, Serres reported how "Claude Lévi-Strauss and François Jacob wanted me to join the Collège de France because they liked what I was doing", an effort which failed because "the philosophers opposed it".[24] In the end, Serres and Canguilhem only met again in 1995, just a few months before Canguilhem's death: "He just asked me at the end if I had had a lot of PhD students in my life. I replied: You know, sir, that I never had one since I was not in my discipline. I was teary-eyed ... and so was he."[25]

---

[18] (Serres, 2014, 49).

[19] (Simons, 2019).

[20] (Canguilhem, 1988, 18).

[21] This project that Serres (1974b) invokes is mainly inspired by his reading of Auguste Comte and the problem of classification: if we classify all the sciences, where must be place this act of classification itself? Is it itself a scientific act, thus implying another science to be added to the classification, an act that requires another act of classification, etc. Or it is something outside of science, rather inspired by politics or culture? (see Simons, 2022, Chap. 1).

[22] (Canguilhem, 1988, 18).

[23] (Serres, 2014, 51).

[24] (Serres, 2014, 54).

[25] (Serres, 2014, 51).

2.2 Serres and Althusserianism

The second, political reason for Serres's strong dichotomy has to do with the political context in France in the 1960s. The main authority at the ENS when Serres was studying was Althusserianism. Serres was unconformable with how, what he saw as exciting new scientific developments, such as molecular biology, were dismissed as ideological by the Althusserians. Moreover, they did so, following Althusser, by invoking the authority of Bachelard, arguing for an 'epistemological break' between science and ideology (see Simons, 2018).

Biology was often one of the main objects of these debates, where Althusserians neatly tried to separate the scientific from the ideological. One example is the infamous Lysenko affair which raged at that time, and dismissed genetics as a 'bourgeois' science, in opposition to Lysenkoism which was a 'proletarian science'. In the Soviet-Union opponents were dismissed, imprisoned or even executed, while Lysenkoism became the official state-endorsed biology. Communist parties in other countries soon endorsed this new party line. As Serres reported, the consequences were often horrible, even in philosophy departments:

> I also remember a guy in my class, biologist or zoologist – well, a brilliant guy – who committed suicide after a well-watered dinner during which one of the guests, who was both a professor at the Sorbonne and a member of the Communist Party central committee, had explained to him at length that the 'proletarian biology' of Michurin and Lysenko – which he taught, however – was in fact a fraud from a scientific point of view. This is the atmosphere of the *Ecole Normale* at that time, with the blessing of Althusser. (Serres, 2014, 38)

Neither Althusser nor Canguilhem explicitly endorsed Lysenkoism, but left room to criticize and problematize it. For instance, they encouraged Lecourt to write a study on the social history of the whole episode.[26] Lecourt later recounted:

> Canguilhem, like Jacques Monod who slammed the door with a crash, distanced himself from the Communist Party he had rubbed shoulders with during the Resistance. It was the occasion of a real break-up about which he happened to speak to me angrily twenty years later. Like Monod, he could not accept this intellectual swindle which wrongly directed all the attention towards the Lamarkian notion of inheritance of acquired characters. As a science historian, Canguilhem saw this as a real regression.[27]

But while Lecourt aligns Canguilhem and Monod, Serres again has the tendency to separate them. Though Canguilhem and Althusser might have distanced themselves from Lysenko, in the eyes of Serres it was too little, too late. Serres saw in Monod and Jacob a more clear and swift response to Lysenko. Already in 1948 Monod wrote an article in the newspaper *Combat*: "La Victoire de Lyssenko n'a aucun caractère scientifique". Jacob would similarly oppose Lysenkoism, later claiming that his decision to focus on genetics was a product of this opposition.[28]

---

[26] (Lecourt, 1976).

[27] (Lecourt, 2016, 138–139)

[28] (e.g. Jacob, 1981, 36; 1987, 234).

A second clash between Monod and Althusser happened in the 1960s. In 1967, Monod was elected to the faculty of the Collège de France. In his inaugural lecture he gave a first rough sketch of what would later become *Le hasard et la nécessité* (1970). That same year, Althusser was organizing a lecture series for a 'course in philosophy for scientists'. Though not originally planned, Althusser took advantage of this lecture series to comment on Monod.

As recent scholarship has made clear, Althusser was actually very positive about Monod.[29] Althusser described Monod's lecture as "an exceptional document, of an unparalleled scientific quality and intellectual honesty"[30] and saw in Monod an ally to rethink dialectical materialism. But the problem was Monod's subsequent step, where Monod extrapolated from biology the world of ideas, language, history and ethics. Monod ends his plea with the claim that 'language created man', something that Althusser saw as "idealist".[31] Althusser thus invoked the Bachelardian epistemological break against Monod, by showing where the scientist leaves science and enters ideology. In the case of Monod, this was his illegitimate move to "arbitrarily impose upon another science which possesses a real object, different from that of the first, the materialist content of the first science".[32] Soon other Althusserians would repeat this dismissal of Monod.[33]

In his own review of Monod's *Le Hasard et la nécessité,* Serres responded to this accusation. According to Serres, there are only two ways to meaningfully define idealism. Either idealism entails a mathematical idealism, in the Platonic sense, or an idealism of the subject, where "the world is nothing but my representation".[34] Serres dismisses the second type of idealism: "A century and a half of critique has shown, I believe definitely, that it was nothing but a mythology."[35] Since a focus on the subject or on representation is absent in Monod's work, he is not part of this subjective idealism. Hence, his idealism must be of the first Platonic type, where ideas have an objective existence, outside of the subject. In Monod's case these ideas can be materially situated: "Monod knows *where* its invariant form is: it is written on the DNA tape. Finally, genetics was one of the first sciences to relativize, once and for all, the activity of the individual subject."[36]

The opposition created by Serres between Canguilhem and Monod thus had biographical and political reasons. However, there is also a third, philosophical reason for this opposition. Rather than contextual, this reason has to do with the content of molecular biology and the effect it should have on philosophy: For Serres,

---

29 (Turchetto, 2009; Tirard, 2012).

30 (Althusser, 1990, 145).

31 (Althusser, 1990, 150).

32 (Althusser, 1990, 151).

33 (Pêcheux & Fichant, 1969).

34 (Serres, 1974a, 55).

35 (Serres, 1974a, 55).

36 (Serres, 1974a, 56).

Canguilhem did not really take the revolutionary character of molecular biology into account.

3 Another Biophilosophy Is Possible

Serres never developed an extensive philosophy of biology. Biology is rather an exemplary domain of a larger reconceptualization of philosophy and science. In his early writings, Serres captures this revolution under the banner of a 'new new scientific spirit' (*le nouveau nouvel ésprit scientifique*), alluding to Bachelard's *Le nouvel ésprit scientifique* (1934).[37] The new wave of sciences Serres had in mind were (second-order) cybernetics and information theory, and their effects on subfields in physics, chemistry and biology (ranging from chaos theory to molecular biology). For Serres, this new new scientific spirit had four characteristics: a new ontology, a new place for epistemology, a new classification of the sciences, and new role for the philosopher.

3.1 The New New Scientific Spirit

First of all, the new new scientific spirit implies a radically different ontology. Starting from information theory, Serres suggests that in these new sciences all worldly processes are interpreted as exchanges of information. Originally, information theory concerned engineering problems. In the case of Claude Shannon, for instance, it was about telephone communication: how to think about the fidelity of messages transported by telephone lines? Shannon (1948) breaks up this problem in three subproblems. First of all there is the *technical problem* about how accurate a message can be transmitted. Secondly, there is the *semantic problem*, which deals with the question whether the transmitted message conveys the desired meaning. Finally, there is the *effectiveness problem*, dealing with how effective the received message affects conduct.

Information theorists like Shannon mainly understood information in relation to noise, i.e. unwanted and undesired disturbances of the message one is trying to convey. Communication is understood as a struggle against noise. Hence the importance of the medium that guarantees the fidelity of the message. This was typically captured through the terms such as *robustness* and *redundancy*: the manner in which a system was still able to convey the message, even if there were disturbances in the medium, for instance by making sure that there are multiple ways to do it. This central point can also be found in Serres's early work, in the form of the 'excluded

[37] (Simons, 2019).

third' (*le tiers exclu*): "To hold a dialogue is to suppose a third man and to seek to exclude him; a successful communication is the exclusion of the third man".[38]

But within this informational framework, information exchange is not restricted to human communication, but refers to all material and biological interactions. The picture that follows from this is a world interpreted as one giant network of communication:

> There is a constant and continuous dialogue between things which form the historical fabric of events and laws, among whom my intervention is exceptional [...] The general informational language is the fundamental and continuous relation between objects. Even before their deciphering, the certainty that it exists induces the certainty that the external world exists, in the mode of a communicating network, of which all the networks I know and could constitute are singular, exceptional cases, approximating to imitate the real world.[39]

For this claim, Serres draws on the French physicist Léon Brillouin (1889–1969). Brillouin wrote on the new science of information theory, works that deeply inspired Serres.[40] For Serres, Brillouin is, therefore, also exemplary for the second characteristic: the new place for epistemology.

French epistemologists, such as Bachelard, had argued that the norms of science were produced intrinsic to the scientific practices. The task of the epistemologist was not so much to impose certain external norms on scientific practices, but to make the implicit norms explicit. Serres goes one step further: not only are scientific practices 'auto-normative', but they also produce their own explicit epistemology. This is what Serres finds in the work of Brillouin:

> the theory of information has constituted a philosophy of physics, intrinsic to the discipline itself. It is remarkable, for instance, that Brillouin has chosen as the title of his last work: *Science and Information Theory.* One finds in this work, indeed, a complete, descriptive, quantified, normative and founding epistemology, expressed in the language of physics itself, of the notion and practice of experimentation, scientific laws, precision and approximate knowledge, the limits of knowledge (what can I know?), in short all the classical topics; and all the 'modern' ones: a theory of codes, language, writing and translation. Philosophers need neither look for nor write a handbook of the epistemology of experimental knowledge: it is found here.[41]

To understand these claims, let us have a look at Brillouin's work. Brillouin mainly defines information as "a function of the ratio of the number of possible answers before and after".[42] When we say that information increases, this means that the number of possible answers to the initial question decrease. For example, if we are wondering where our friend is and we enter his room and we see a hot cup of tea, we say that this is more informative than if we would enter a room without the hot tea. Why? Because the hot tea narrows down the number of possible scenarios: the tea indicates that someone is likely in the neighborhood.

---

38 (Serres, 1969, 41).

39 (Serres, 1972, 110).

40 (Brillouin, 1956).

41 (Serres, 1974a, 45).

42 (Brillouin, 1956, x).

This example, moreover, also highlights the link between information and entropy. In defining information in this manner, Brillouin aimed for a "generalization of the second principle of thermodynamics"[43]:

> Every physical system is incompletely defined. We only know the values of some macroscopic variables and we are unable to specify the exact positions and velocities of all the molecules contained in a system. We have only scanty, partial information on the system, and most of the information on the detailed structure is missing. Entropy measures the lack of information; it gives us the total amount of missing information on the ultramicroscopic structure of the system.[44]

Brillouin defined information as negative entropy, or what he would call: negentropy.[45] In this way information theory can provide a solution to an infamous thought experiment proposed by J. C. Maxwell in 1867, now known as Maxwell's demon. According to Maxwell the second law of thermodynamics was seemingly violated when you imagine a demon controlling a trapdoor between two chambers filled with gas. If we imagine that the demon has the capacity to open the door when fast particles approach while keeping it shut in the case of slower ones, the demon seems to be capable to decrease entropy in a closed system, thus violating the second law. Brillouin's solution is to apply information theory to the actions of the demon, which have their own informational cost (and thus show that the system is not really closed). The demon uses energy to manipulate the door, but more importantly, also produce information about the whereabouts of the approaching particles. This information itself can only be gained by spending energy.

It is here that the new new scientific spirit becomes self-reflexive. Brillouin applies this framework not only to physical systems, but also to the practice of experimentation: doing an experiment is itself a matter of creating information by spending energy outside of the experimental set-up. In other words, information theory offers us a framework to understand scientific research as an negentropic activity. In this sense, Serres can concludes that information theory carries its own epistemology: "What is experimentation in general, if not an informational as well as an energy balance of the laboratory?"[46]

But, according to Serres, this applies to the molecular biology of Monod as well, whose work embodies the "new new biological spirit".[47] Monod's work similarly carries its own epistemology in itself:

> No critic has yet noticed that biochemists, and Monod especially, have a 'natural philosophy' intrinsic to their scientific activity. [...] Monod, it is true, sometimes refers to the great names of the academic pantheon, such as Descartes, Kant, Hegel; but the efficient operators of his work are not the tools forged in and by this tradition. These are new tools, dating from around this century, and which you will find in [Norbert] Wiener, [Percy] Bridgmann, [Erwin] Schrödinger and [Léon] Brillouin [...]. And the philosophy of physics is informa-

[43] (Brillouin, 1956, xii).

[44] (Brillouin, 1956, xii).

[45] (Brillouin, 1956, vii).

[46] (Serres, 1977a, 287).

[47] (Serres, 1972, 60).

> tion theory. And so, when a biochemist announces that he is writing a natural philosophy, it clearly means that he is applying the theory of information (the natural philosophy of natural philosophy) to his own discipline.[48]

Not only are biological organisms understood as information processing machines – exchanging information through the genetic code and other chemical interactions – Monod's *Le hasard et la nécessité* (1970) expands this framework to the realm of ideas and the social. "For a biologist it is tempting," Monod says,

> to draw a parallel between the evolution of ideas and that of the biosphere. For while the abstract kingdom stands at a yet greater distance above the biosphere than the latter does above the nonliving universe, ideas have retained some of the properties of organisms. Like them, they tend to perpetuate their structure and to breed; they too can fuse, recombine, segregate their content; indeed they too can evolve, and in this evolution selection must surely play an important role.[49]

It was against this continuity Althusser so strongly objected. But for Serres this continuity highlights how Monod embodies the new new scientific spirit: the insights of molecular biology can also be applied to the constitution of molecular biology as a scientific practice. Just as Brillouin's information theory, Monod's molecular biology contributes to a general framework that understands the world – from atom to society – in informational terms. This generalizing tendency brings us to the third characteristic of the new new scientific spirit: a new classification of the sciences.

The model Serres proposes for the new new scientific spirit, echoing Comte, is that of a new Encyclopedia. The different sciences are no longer classified according to a traditional hierarchy with physics at the bottom and sociology at the top. Instead, the new new scientific spirit follows the model of horizontal translation: all fields are on the same level, but relate through a set of internal cross-references and cross-fertilizations: "the new new spirit is developing into a philosophy of transport: intersection, intervention, interception."[50] It is not so much that the physical information theory is 'applied' to biology, but both are characterized by a set of isomorphic structures that can mutually inspire one another. New insights from physics can inspire biology, but also vice versa: developments in molecular biology can also inspire isomorphic frameworks to look at physical phenomena. But if ideas from the physical realm can without problem jump to the level of social practices, such as experimentation, what then is the place of biology? For this, we turn to a final source of Serres, Henri Atlan.

48 (Serres, 1972, 43–44).

49 (Monod, 1972, 165).

50 (Serres, 1972, 10).

3.2 Henri Atlan and the Problem of Meaning

In *L'organisation biologique et la théorie de l'information* (1972) and *Entre le cristal et la fumée* (1979), the French biologist Henri Atlan (°1931) developed a similar project as Serres. The central question of his work is the question of biological self-organization: how does biological organization emerge? Atlan uses information theory to answer this question. But an element missing in the first generation of information theorists is the *generation* of order and information. As we saw, Shannon (1948) was mainly concerned with the perseverance of information, but how does new information and organization arise in the first place?

For this Atlan invokes the work of Heinz von Foerster and his principle of *order from noise*. Whereas for Shannon (1948) noise was seen as negative, Von Foerster (1960) argued that noise plays a positive role in the generation of new levels of order. Von Foerster gives the example of a set of magnetized cubes (e.g. three sides positive, three sides negative). If you put them in a box and then shake them, they seemingly 'self-organize' in geometrical figures when you open the box again. The noise of the shaking results in the creation of order. Thus self-organizing systems create and uphold their order, not just by excluding noise, but also by productively incorporating parts of its noisy environment: "the system is in close contact with an environment, *which possesses available energy and order,* and with which our system is in a state of perpetual interaction, such that it somehow manages to 'live' on the expenses of this environment."[51] According to Atlan, this offers us a model to understand self-organization:

> Within the framework of this theory, self-organization can be described as a dynamic process by which random perturbations or noise acting on the channels of communication in an organized system are able to produce, not only disfunction and disorganization, but also a change in organization to a state with more complexity and less redundancy.[52]

Though a step in the right direction, Von Foerster's principle does not suffice to understand self-organizing biological systems. The example of Von Foerster only works if one simulates ignorance about magnetic forces. If you take these forces into account, the produced order is not surprising, but the expected product of magnetic forces. It is closer to crystal formation than to a living system: there is growth, but it tends to follow the same and predictable repetitive structure. Therefore Atlan proposes his own principle, namely that of *complexity from noise.*

Though information theory deals with communication, the early information theorists stressed a clear restriction: information theory ignores the human value of information and thus the question of meaning. The theory does not differentiate between a 100-letter sentence from Shakespeare or an equally improbable sentence of 100 randomly selected letters. "In other words, we define 'information' as distinct from 'knowledge'," Brillouin acknowledges, "for which we have no

[51] (Von Foerster, 1960, 33).

[52] (Atlan, 1987, 564).

numerical measure. We make no distinction between useful and useless information, and we choose to ignore completely the value of the information."[53] As Atlan summarizes, information theorists

> can do very well without having to bother with understanding and formalizing how meaning is created: both the meaning of information transmitted in a channel and that of a computer program are assumed to exist but the mathematical theories which deal with these problems do not have to take this into account. In information theory, the fact that messages have meaning is obvious, but remains implicit since this theory treats problems of coding and transmission efficiency without having to consider the actual meaning of the messages to be coded and transmitted.[54]

Interestingly, Atlan and Serres try to turn this restriction into an advantage. They will argue that it shows that we can develop a convincing analysis of meaningful information without the need for the notion of the subject. In the words of Atlan, "[w]e proceed in a negative way, by taking advantage of what is usually considered a flaw in information theory – namely the fact that Shannon's probabilistic information theory does not make any distinction between meaningful and meaningless information."[55]

Atlan achieves this by focusing on the relativity of the observer's position, and thus the acknowledgment of a multilayered, hierarchical model of reality, where noise and information are never absolutely given, but relative to the layer in which the observer situates itself:

> This observer, external to the system, is in fact, in a hierarchical system, a higher (and encompassing) level of organization compared to the elementary systems that constitute it; it is the organ in relation to the cell, the organism in relation to the organ, etc. It is in relation to this that the effects of noise on a channel within the system can, under certain conditions, be positive.[56]

In other words, that there is no clear distinction between information and noise is not a problem, because such a distinction only exists relative to the layer from which one is making that distinction. What is noise for one level can be meaningful information for another, higher level. "The meaning of the message, in contrast, is never intrinsic to the message; the meaning is the relationship of the message to some reference point outside of the information borne by the message. Something or somebody has to 'read' the message. Meaning is referential and contingent."[57] In *Le Parasite* (1980), Serres gives the example of a telephone call at a banquet:

> At the feast everyone is talking. At the door of the room there is a ringing noise, the telephone. Communication cuts conversation, the noise interrupting the messages. As soon as I start to talk with this new interlocutor, the sounds of the banquet become noise for the new

[53] (Brillouin, 1956, 9).
[54] (Atlan, 1987, 563).
[55] (Atlan, 1987, 564).
[56] (Atlan, 1979, 70).
[57] (Atlan & Cohen, 1998, 713).

> 'us.' The system has shifted. If I approach the table, the noise slowly becomes conversation.[58]

Whereas the meaningful conversation becomes noise once one shifts to another level, it regains its meaning once one shifts back to the level of the conversation. On his turn, Atlan uses the example of noise in cells: whereas badly formed proteins, with non-proper enzymatic properties, can be seen as 'noise' for the metabolism of the cell, if one switches to the level of the organ or the organism, this noise can be a source for adaptation to new environments:

> From the point of view of the organ or physiological apparatus, this same noise has the effects of creating variety and heterogeneity among cells, which allows them to more adaptability. Therefore, up to a certain point. and providing the redundancy of the cell is large enough so that these false proteins are not going to impair the cell function, the same effects of the noise on the channel within the cell which are viewed as detrimental by the cell itself can be viewed as beneficial by the organ.[59]

Self-organization thus requires two conditions: on the one hand "enough initial redundancy" to be "used as a reservoir, or potential for self-organization;" and on the other hand a certain "inertia, i.e. its reliability or resilience must suffice to keep small perturbations from immediately destroying it."[60] In that sense, a living system is not to be reduced to a repetitive crystal as in Von Foerster's magnetized cubes, but neither is it completely in flux. A certain degree of repetitiveness and redundancy is required to maintain a stability and resilience against noise. Hence Atlan's central metaphor: a living system is located "between two extremes: a perfectly symmetrical repeating order of which the crystals are the most classic physical models, and an infinitely complex variety and unpredictable in its details, like that of the evanescent forms of smoke."[61]

4 A Biophilosophy Without a Subject

The work of Michel Serres can thus be seen as part of another philosophical project, one inspired by information theory, which thinks about biological organisms in a radically different way. References to the role of consciousness or experience are absent. Instead, the problem of meaning is translated in an informational framework, where information and noise are defined in relation to the framework of the observer. These observers, however, are not interrogated from a hermeneutical or phenomenological perspective – how does the information or meaning presents itself to the subject? – but in an operational way: what kind of effects does this information have on the observer, and how is this shown in a change in behaviour?

[58] (Serres, 2007, 66).
[59] (Atlan, 1977, 179).
[60] (Atlan, 1987, 565).
[61] (Atlan, 1979, 5).

We see this in Atlan, who explicitly equate the two: "We suggest to define the observed meaning of information as its observed consequence on the receiver. In other words, we suggest to unify the levels B and C of [Warren] Weaver (semantics and efficiency), although we know that in our linguistic experience it is not so."[62] The end result is a framework that enables one to analyze the different levels of reality (physical, biological, social), without the need to invoke questions concerning subjectivity. In that sense, the biological level does not possess any kind of qualitative uniqueness, but is but one level that situates itself relatively, but in an isomorph fashion between the others.

In a similar vein, Serres concludes that there is no genuine distinction anymore between subject and object, since all must be interpreted as layers of emitters and interpreters of information:

> I know who is the final observer, the receiver at the end of the chain: the one, precisely, who emits language. But I don't know who is the first transmitter on the other end. It's a black box indefinitely. A box of boxes, and so on. I can thus go as far as I want, to the cells, to the molecules, provided, of course, that I change the object observed. All I know, but of that I'm sure, is that they are all structured by the information-background, random-program, or entropy-negentropy pair.[63]

And therefore, according to Serres, "[n]othing distinguishes me ontologically from a crystal, a plant, this animal and the order of the world: we drift together towards the noise and the black background of the universe, and our various complexions of system up the entropic river in the direction of the solar source, itself derived from it."[64] All physical, biological and social phenomena can be analyzed in an isomorph way through this framework.

Hence my final claim: this framework has had an effect on a next generation of thinkers who, inspired by this informational framework, started to analyze social phenomena. Hermeneutical or phenomenological questions concerning subjectivity and meaning are absent in their work, as are specific reflections on biology and consciousness. Instead, we are faced with a paradoxical biophilosophy without a subject that jumps over subjectivity and started to study social phenomena through an informational lens.

Atlan already suggests that, though originating from reflections on biology, it is possible "to extend some of these considerations to other systems and other forms of organization, human in particular" and thus "educate us about the logical possibilities of organization in general."[65] Atlan refers to the work of Edgar Morin, Serres and even Canguilhem. In the case of Canguilhem, Atlan refers to his encyclopedic lemma about 'la vie', where Canguilhem argues that Atlan's complexity-from-noise principle can provide a cybernetic understanding of the death thrive of Sigmund

---

[62] (Atlan, 1977, 180).

[63] (Serres, 1977a, 270).

[64] (Serres, 1977a, 271).

[65] (Atlan, 1979, 6).

Freud.[66] More broadly, Atlan, Morin and Serres were member of the Group of Ten (*Groupe des Dix*), a group of French thinkers who, between 1968 and 1976, tried to apply this new framework to societal and political issues.[67] Created by the economist Robert Buron, the neurobiologist Henri Laborit, the sociologist Edgar Morin and the Jacques Robin, the group initially consisted of ten members but soon enough others joined, including Atlan and Serres.

The Group of Ten aimed to explore to what extent social and political phenomena, ranging from elections to economic cycles, could be interpreted as following the above framework of self-organization. The Group of Ten failed to have policy impact, but it did result in a number of influential publications, such as Rosnay's *Le macroscope* (1975), Passet's *L'économie et le vivant* (1979), but also Morin's *La méthode* (1977) and Atlan's *Entre le cristal et la fumée* (1979). Serres himself shifted away in the 1980s from strictly epistemological questions concerning the applicability of this framework to all kinds of phenomena. In his later work, and inspired by his idea that science produces its own epistemology, he conceived of a new role of the philosopher, namely the development of an 'anthropology of science', i.e. a political analysis of the violence that result from science, exemplified by the ecological crisis.[68] Two scholars that have been more effective in their analysis of social phenomena are René Girard and Bruno Latour.

4.1 *The Scapegoat as Self-Organization*

If we turn to the work of René Girard (1923–2015), we enter a different world. Girard was a French anthropologist and literary critic, famous for his book *La violence et le sacré* (1972). Girard offers a rereading of the history of religion, arguing for the omnipresence of the phenomenon of the scapegoat: a figure on which all the wrongdoings of a society are projected and which is outcast, while at the same time sacralized. The most obvious example would be Jesus, who was scapegoated, but simultaneously soon seen as holy (though the example of Jesus is actually more complex).

Girard's work offers a mechanism that explains the historical omnipresence of this scapegoat phenomenon. This has to do, according to Girard, with the nature of human desire. Central to this desire, is not so much a common appetite for food and shelter, but a 'metaphysical desire' that follows a mimetic structure: you desire something, not so due to its intrinsic properties, but because someone else desires that object. You want it because someone else wants it. "We must understand that desire itself is essentially mimetic, directed toward an object desired by the model."[69]

---

[66] (Canguilhem, 1973).

[67] (Chamak, 1997).

[68] (Simons, 2022, Chap. 5).

[69] (Girard, 2017, 146).

Whereas such desires are limitless, the object of desire is not. Only one can possess it. Hence it typically results in conflict. This violence is also structured in a mimetic way: I desire to hurt you because you wanted to hurt me. "Only violence can put an end to violence, and that is why violence is self-propagating. Everyone wants to strike the last blow, and reprisal can thus follow reprisal without any true conclusion ever being reached."[70] The result is, paradoxically, that the initial object of desire is soon forgotten and the individuals become doubles of one another, mirroring each other's violence while having long forgotten why they started to fight in the first place. This mimetic violence is moreover contagious: "if two persons are fighting over the same object, then this object seems more valuable to bystanders."[71] Soon enough, violence spreads throughout the whole collective.

Nonetheless, we live in semi-stable societies where violence is not omnipresent. Here the scapegoat mechanism enters the picture: mimetic violence is halted because mimetic spirals tend to reach points where the blame is projected on one scapegoat, namely an individual or group that, often by accident, is put into the spotlight:

> The killing of the scapegoat ends the crisis, since the transference against it is unanimous. That is the importance of the scapegoat mechanism: it channels the collective violence against one arbitrarily chosen member of the community, and this victim becomes the common enemy of the entire community, which is reconciled as a result.[72]

This also give us a reason why the scapegoat is often divinized, since it is the pacifier of societal violence, and explains the structure and functions of numerous rituals and taboos. They are attempts to ban or contain forms of dedifferentiation (such as blood, twins, death) that risk to provoke mimetic violence.

This is in a nutshell Girard's general framework. Though seemingly unrelated, there is in fact a strong connection with the informational framework we fleshed out in this chapter. This is first of all shows in the biographical connections: Serres and Girard were very good friends, meeting each other in Buffalo in 1975, later becoming colleagues in Stanford. In 2005 Serres would also sponsor Girard's election to the *Académie Française*. Serres was fascinated by Girard and found in the latter a similar framework as the one he was exploring in his own oeuvre. For the same reason also Henri Atlan had an interest in Girard:

> Girard's answer, at least in its logical form, is comparable to the answer biological theory brings to the same problem, in so far as both assign an important part to randomness or chance. Starting out from an undifferentiated state made repetitive through the labour of mimesis, very slight perturbations in the way in which these repetitions take place lead to a process of differences and of differentiations.[73]

Serres similarly endorses this isomorphism between Girard's scapegoat's mechanism and physical and biological processes of self-organization, even comparing it

---

[70] (Girard, 2017, 26).

[71] (Girard, 2008, 64).

[72] (Girard, 2008, 65).

[73] (Atlan, 1988, 194).

to a physical law: "I have long assumed that violence obeys, in groups, constants similar to those of energy. Just as mechanics and thermodynamics base their exact truths on a stable amount of force in the Universe, does politics rest on a permanent volume of violence in communities?"[74]

A third example is the work of Jean-Pierre Dupuy (°1941), another scholar fascinated with the question of self-organisation. Dupuy befriended both von Foerster and Atlan, and regularly refers to Serres as well. He met Girard in the United States, and together they organized a set of interdisciplinary conferences, such as the 1981 symposium on 'Disorder and Order' (with speakers such as Atlan, von Foerster, but also Ilya Prigogine, Isabelle Stengers, Cornelius Castoriadis and Francisco Varela). It was Dupuy who, according to Girard, "made me aware of the relationship between 'chaos theory' and the mimetic theory."[75] Dupuy indeed describes Girard's theory as "a special case of the logic of self-organisation":

> The autonomous system, in Girard, is violence, acquisitive mimesis, the closed circle that binds men to their rivals. He is the real subject of the story. Men, as individuals, are simple subsystems, producers of noise and fury that will only be transformed into meaning at the higher level, that of autonomous violence.[76]

Similar to information theory, we find in Girard's theory a form of reflexivity: though the violent participants do not see this, we are capable to realize the meaning of the scapegoat mechanism. This itself can be explained by the theory, i.e. it has to do with the fact that we situate ourselves on a higher level: "Only the external observer – modernity and its science – is capable of placing himself at the logical level higher than that of the social-historical, and of seeing the transmutation of noise into organization, of chance into meaning."[77] This reflexivity, however, did not start with science. According to Girard Christianity is the religion that uncovers the meaning of the scapegoat mechanism. Hence, the reason why the Jesus example is more complex: it follows the scapegoat mechanism, but in an explicit manner, unmasking its violent role in history. Dupuy and Serres endorse this interpretation of Christianity, while Atlan (1988) argues against it, claiming that this reflexivity is already present in Judaism.

Girard's theory is thus an instance of this broader biophilosophy without a subject. In the kind of Girardian analyses one does not find traditional questions of subjectivity and interpretation. Instead, phenomena are explained in a formal matter, by showing how the same spontaneous process of self-organization occurs, in this case through mimetic violence and the scapegoat mechanism. No particular place is left for biological phenomena, who are just seen as one isomorphic level between the others.

---

[74] (Serres, 2019, 304).

[75] (Girard, 2008, 41).

[76] (Dupuy, 1982, 276–277).

[77] (Dupuy, 1982, 276–277).

4.2 Science as Order Out of Noise

A similar story holds for Bruno Latour (1947–2022). Latour's early work was an ethnographical study of scientific laboratories, often interpreted as a form of social constructivism, claiming to explain the progress of scientific debates through social factors. Though Latour was indeed partly inspired by the Strong Programme of the Edinburgh School, embodied by David Bloor (1976), his work was more substantially framed through the informational framework. We find evidence for this in one of Latour's first articles, where he frames his sociology of science in informational terminology. Sociologists have shown, according to Latour, that science is not "less disorderly, less noisy, less fictional than the rest of history".[78] Hence, the task of the sociologist is to find a framework to map how science nonetheless transitions from this disorder to the order of a well-argued scientific statement. Latour is clear about where he finds this framework: "The only attempt has not been made by sociologists of science, but by isolated scientists dealing with information, or with turbulent phenomena."[79] He refers to Brillouin (1964), Atlan (1972), Prigogine and Stengers (1979) and Serres (1977b):

> In the old framework, disorder, turbulence, agitation, circumstances, were to be *eliminated* for a world of order, logics and rationality to appear and be maintained. In the new framework, order is nothing but local circumstances obtained from, maintained by, dissolved from time to time in disorder; if you eliminate the opportunism, the context, the fiction building, the agitation, the reconstruction, the rationalization you get *nothing* at all; if you introduce them you understand how the scientific facts, discoveries and theories emerge and are maintained.[80]

We find a similar story in the book that made his career: *Laboratory Life* (1979), written with the sociologist Steve Woolgar. Again the sociologist is faced with a "disordered array of observations with which scientists struggle to produce order".[81] The problem is how a chaotic network of actors in a laboratory stabilizes into a scientific fact. Once more they draw on information theory: scientific statements become meaningful by becoming informational, i.e. statistically improbable: "If a large number [of alternative interpretations] can easily be thought of, the original statement will be taken as meaningless and hardly distinguishable from others. If the others seem much less likely than the original statement, the latter will stand out and be taken as a meaningful contribution."[82] For this, they explicitly make us of the concept of "*noise* (or more exactly, the ratio of signal to noise)",[83] drawing on

[78] (Latour, 1981, 69).

[79] (Latour, 1981, 70).

[80] (Latour, 1981, 70).

[81] (Latour & Woolgar, 1986, 36).

[82] (Latour & Woolgar, 1986, 240–241).

[83] (Latour & Woolgar, 1986, 239).

"Michel Serres, who, in turn, had been greatly influenced by authors such as Brillouin and Boltzmann and by new developments in biology".[84]

Laboratory Life thus concerns the question of how one possible scientific statement about the world gets the upper hand over its rivals. This happens not so much through a social power struggle, but through an informational struggle: scientists mobilize data, experiments, arguments, other scientists, scientific articles, etc., in order to create an inequality between their own statement and the others. They thus constantly struggle to create a difference between information and noise within the scientific literature. "In principle, the number of alternative readings of [a] particular utterance is very large. The number which will be accepted as plausible by an informed audience, however, will be constrained by the particular context which is brought to bear upon the reading of the utterance."[85]

To argue for this perspective, Latour and Woolgar invoke several of the sources we already encounter. For instance, they draw a parallel with Monod's *Le hasard et la necessité*. Also in Monod's case it is a question of how a specific order (living organisms) follows from a state of disorder. Similarly how in biology there is no need to presuppose a given blueprint, in science there is no ordered reality that has to be represented. "Reality is constructed out of disorder, without the use of any preexisting representation of life".[86] A similar parallel is drawn with Brillouin's interpretation of Maxwell's demon:

> Maxwell's devil provides a useful metaphor for laboratory activity because it shows both that order is *created* and that this order in no way preexists the devil's manipulations. Scientific reality is a pocket of order, created out of disorder by seizing on any signal which fits what has already been enclosed and by enclosing it, albeit *at a cost*.[87]

The result is an application of this informational model on a whole set of societal phenomena. Initially restricted to the history of science, it was soon expanded to other domains, forming Actor-Network Theory (ANT). ANT was in fact initially known as the 'sociology of translation', where 'translation' was a notion explicitly borrowed from Serres.[88] This is not only the case for Latour, but also of Michel Callon.[89] Translation refers to the differentiation work of scientists: they have to translate a divergent set of phenomena and actors (e.g. a soil sample from Brazil; an experiment done in France; the opinion of a rival scholar in Japan) in the same register (e.g. a paper, a graph, an argument). The result is a statement, backed up by a network, that makes it improbable: not just any statement, but an unlikely and informative one.

Similar to Girard, the result is a perspective on social phenomena that draws inspiration from information theory, and glosses over the realm of biology and

---

[84] (Latour & Woolgar, 1986, 251).
[85] (Latour & Woolgar, 1986, 35).
[86] (Latour & Woolgar, 1986, 250).
[87] (Latour & Woolgar, 1986, 246).
[88] (Serres, 1974a).
[89] (Callon, 1981, 219n16).

subjectivity. Latour and Woolgar in fact provocatively propose a moratorium on these subjective explanations:

> Perhaps the best way to express our position is by proposing a ten-year moratorium on cognitive explanations of science. If our French epistemologist colleagues are sufficiently confident in the paramount importance of cognitive phenomena for understanding science, they will accept the challenge. We hereby promise that if anything remains to be explained at the end of this period, we too will turn to the mind![90]

The result is an analysis of science that has no need to speak about the subject, its experience, its thinking, or its interpretations. Again, a framework that maps the formal logic of how scientific practices self-organize into meaningful statements suffices. Nonetheless, as in the case of Girard, a level of self-reflexivity remains, this time inspired by David Bloor's original Strong Programme: "the realisation that observers of scientific activity are engaged in methods which are essentially similar to those of the practioners which they study."[91] In the same way that scientists have to translate the initial disorder into the order of a scientific statement, "[t]he observer's task is to transform notes […] into an ordered account".[92] We again have a reflexivity in the form of an isomorphism between the different levels: the work of the sociologists is formally structured in a similar way to that of the scientists themselves.

5 Conclusion

In this chapter I looked at how the work of Michel Serres embodied an alternative biophilosophy, diverging from themes associated with Serres's supervisor Georges Canguilhem. Serres accused Canguilhem of not systematically picking up the molecular revolution which was going on at that time. Though we saw that Canguilhem did engage with molecular biology, Serres had several reasons to argue for this break. First of all there were biographical and political reasons: a personal fallout with Canguilhem and a political struggle with Althuserianism.

But besides that, we also saw how Serres invoked an alternative biophilosophy, which drew inspiration from information theory. Inspired by Brillouin and Atlan, Serres sketched a biophilosophy in which the living organism, and the human mind, plays no exceptional role anymore. Rather, we end up with a more general informational framework focused on self-organization, a process that is seen at work at all levels of reality in an isomorphic way.

The result is a biophilosophy without a subject, a framework that has had impact far beyond philosophy of biology. It has had a profound influence on a number of authors, of which we have only briefly explored two: Girard and Latour. But one

[90] (Latour & Woolgar, 1986, 280).
[91] (Latour & Woolgar, 1986, 30).
[92] (Latour & Woolgar, 1986, 37).

could explore how a similar way of thinking is at work in a number of other authors. This is not only the case for authors who explicitly focused on biology, such as Gilbert Simondon, Raymond Ruyer, Edgar Morin or Francisco Varela, but also those where a biophilosophy is present in its absence, such as Cornelius Castoriadis, Michel Deguy, Gilles Deleuze, Judith Schlanger or Isabelle Stengers. For instance, one can refer to Deleuze and Guattari's infamous *L'anti-Œdipe* (1972) and *Mille plateaux* (1980), where they effortlessly shift from reflections on physics to social ontology, framing it in a common terminology of machines, rhizomes and lines of flight. In this framework biological topics are present, but there is no need to invoke the notion of the subject or experience in any traditional sense. In a similar vein, Castoriadis in his reading of historical change in *Les carrefours du labyrinthe* (1978), draws on complexity theory and the concept of emergence to conceptualize the rise of social institutions. Once again, it is a framework that 'jumps over' subjectivity, from the physical to the social sphere. In that sense, one could argue that continental philosophy of biology has a broader, yet invisible history than is often presupposed.

References

Althusser, L. (1990). *Philosophy and the Spontaneous Philosophy of the Scientists and Other Essays* (B. Brewster et al., Trans.). Verso.

Atlan, H. (1972). *L'organisation biologique et la théorie de l'information*. Seuil.

Atlan, H. (1977). Sources of Information in Biological Systems. *IFAC Proceedings, 10*(12), 177–184.

Atlan, H. (1979). *Entre le cristal et la fumée: essai sur l'organisation du vivant*. Seuil.

Atlan, H. (1987). Self-Creation of Meaning. *Physica Scripta, 36*(3), 563–576.

Atlan, H. (1988). Founding Violence and Divine Referent. In P. Dumouchel (Ed.), *Violence and Truth: On the Work of René Girard* (pp. 192–208). Athlone.

Atlan, H., & Cohen, I. (1998). Immune Information, Self-Organization and Meaning. *International Immunology, 10*(6), 711–717.

Bachelard, G. (1934). *Le nouvel esprit scientifique*. Alcan.

Bachelard, G. (1949). *Le rationalisme appliqué*. PUF.

Bloor, D. (1976). *Knowledge and Social Imagery*. Routledge and Kegan Paul.

Brillouin, L. (1956). *Science and Information Theory*. Academic.

Brillouin, L. (1964). *Scientific Uncertainty and Information*. Academic.

Callon, M. (1981). Struggles and Negotiations to Define What is Problematic and What is Not: The Socio-Logic of Translation. In K. Knorr, R. Krohn, & R. Whitley (Eds.), *The Social Process of Scientific Investigation, Sociology of the Sciences, a Yearbook* (pp. 197–219). Reidel.

Canguilhem, G. (1952). *La connaissance de la vie*. Librairie Hachette.

Canguilhem, G. (1955). *La formation du concept de réflexe aux XVIIe et XVIIIe siècles*. PUF.

Canguilhem, G. (1966). Le concept et la vie. *Revue Philosophique de Louvain, 82*, 193–223.

Canguilhem, G. (1971). Logique du vivant et histoire de la biologie. *Sciences: Revue de la civilisation scientifique, 71*, 20–25.

Canguilhem, G. (1973). Vie. In C. Grégory (Ed.). *Encyclopedia universalis* (Vol. 16, pp. 764–769). Encyclopedia Universalis.

Canguilhem, G. (1988). *Ideology and Rationality in the History of the Life Sciences* (A. Goldhammer, Trans.). MIT Press.

Castoriadis, C. (1978). *Les carrefours du labyrinthe*. Seuil.
Chamak, B. (1997). *Le Groupe des Dix, ou les avatars des rapports entre science et politique*. Éditions du Rocher.
Debru, C., Morange, M., & Worms, F. (Eds.). (2012). *Une nouvelle connaissance du vivant: François Jacob, André Lwoff & Jacques Monod*. Éditions de la Rue d'Ulm.
Deleuze, G., & Guattari, F. (1972). *Capitalisme et schizophrénie. 1: L'Anti-Œdipe*. Ed. de Minuit.
Deleuze, G., & Guattari, F. (1980). *Capitalisme et schizophrénie. 2: Mille plateaux*. Ed. de Minuit.
Dupuy, J. (1982). Mimésis et morphogénèse. In M. Deguy & J. Dupuy (Eds.), *René Girard et le problème du mal* (pp. 225–278). Grasset.
Erdur, O. (2018). *Die epistemologischen Jahre. Philosophie und Biologie in Frankreich, 1960–1980*. Chronos Verlag.
Girard, R. (2008). *Evolution and Conversion: Dialogues on the Origins of Culture*. Continuum.
Girard, R. (2017). *Violence and the Sacred* (P. Gregory, Trans.). Bloomsbury.
Jacob, F. (1981). *Le jeu des possibles: Essai sur la diversité du vivant*. Fayard.
Jacob, F. (1987). *La statue intérieure*. Seuil.
Latour, B. (1981). Is it Possible to Reconstruct the Research Process? Sociology of a Brain Peptide. In K. Knorr, R. Korhn, & R. Whitley (Eds.), *The Social Process of Scientific Investigation, Sociology of the Sciences, a Yearbook* (pp. 53–77). Reidel.
Latour, B. (1987). *Science in Action: How to Follow Scientists and Engineers Through Society*. Harvard University Press.
Latour, B., & Woolgar, S. (1986). *Laboratory Life: The Construction of Scientific Facts*. Princeton University Press.
Lecourt, D. (1969). *L'épistémologie historique de Gaston Bachelard*. Vrin.
Lecourt, D. (1972). *Pour une critique de l'épistémologie (Bachelard, Canguilhem, Foucault)*. Maspero.
Lecourt, D. (1976). *Lyssenko: histoire réelle d'une science prolétarienne*. Maspero.
Lecourt, D. (2016). Des penseurs des Lumières à la science contemporaine. In A. Wald Lasowski (Ed.), *Althusser et nous* (pp. 115–146). PUF.
Loison, L. (2018). Un enthousiasme paradoxal ? Georges Canguilhem et la biologie moléculaire (1966–1973). *Revue d'histoire des sciences, 71*(2), 271–300.
Méthot, P.-O. (Ed.) (2020). *Vital Norms: Canguilhem's the Normal and the Pathological in the Twenty-First Century*. Hermann.
Monod, J. (1972). *Chance and Necessity: An Essay on the Natural Philosophy of Modern Biology* (A. Wainhouse, Trans.). Collins.
Morange, M. (2000). Georges Canguilhem et la biologie du XXe siècle. *Revue d'histoire des sciences, 53*(1), 83–106.
Morin, E. (1973). *Le paradigme perdu: la nature humaine*. Seuil.
Pêcheux, M., & Fichant, M. (1969). *Sur l'histoire des sciences*. François Maspero.
Prigogine, I., & Stengers, I. (1979). *La nouvelle alliance: Métamorphose de la science*. Gallimard.
Serres, M. (1969). *Hermès I, La communication*. Éditions de Minuit.
Serres, M. (1970) [2019]. Reformation and the Seven Sins. *Parrhesia, 31*, 33–47.
Serres, M. (1972). *Hermès II, L'interférence*. Éditions de Minuit.
Serres, M. (1974a). *Hermès III, La traduction*. Éditions de Minuit.
Serres, M. (1974b). Les Sciences. In J. Le Goff & P. Nora (Eds.), *Faire de l'histoire* (Vol. 2, pp. 203–228). Gallimard.
Serres, M. (1977a). *Hermès IV, La distribution*. Éditions de Minuit.
Serres, M. (1977b). *La naissance de la physique dans le texte de Lucrèce: fleuves et turbulences*. Ed. de Minuit.
Serres, M. (2007). *The Parasite* (L. Schehr, Trans.). The John Hopkins University Press.
Serres, M. (2014). *Pantopie ou le monde de Michel Serres, de Hermès à Petite Poucette*. Le Pommier.
Serres, M. (2019). *Hominescence* (R. Burks, Trans.). Bloomsbury.

Serres, M., & Latour, B. (1995). *Conversations on Science, Culture and Time* (R. Lapidus, Trans.). Michigan University Press.

Shannon, C. (1948). *A Mathematical Theory of Communication*. University of Illinois Press.

Simons, M. (2018). The Janus Head of Bachelard's Phenomenotechnique: From Purification to Proliferation and Back. *European Journal for Philosophy of Science, 8*(3), 689–707.

Simons, M. (2019). Surrationalism after Bachelard: Michel Serres and *le nouveau nouvel esprit scientifique*. *Parrhesia: A Journal of Critical Philosophy, 31*, 60–84.

Simons, M. (2022). *Michel Serres and French Philosophy of Science*. Bloomsbury.

Talcott, S. (2014). Errant Life, Molecular Biology, and Biopower: Canguilhem, Jacob, and Foucault. *History and Philosophy of the Life Sciences, 36*(2), 254–279.

Tirard, S. (2012). Monod, Althusser et le marxisme. In C. Debru, M. Morange & F. Worms (Eds.). *Une nouvelle connaissance du vivant. François Jacob, André Lwoff et Jacques Monod* (pp. 75–88). Éditions Rue d'Ulm.

Turchetto, M. (2009). Althusser and Monod: A 'New Alliance'? *Historical Materialism, 17*(3), 61–79.

Von Foerster, H. (1960). On Self-Organizing Systems and their Environments. In M. Yovits & S. Cameron (Eds.), *Self-Organizing Systems* (pp. 31–50). Pergamon Press.

Massimiliano Simons is an Assistant Professor at Maastricht University. His work specializes in contemporary technoscience (synthetic biology, robotics, data science) and twentieth century French philosophy of science (Gaston Bachelard, Michel Serres, Bruno Latour). He is the author of *Michel Serres and French Philosophy of Science: Materiality Ecology and Quasi-Objects* (Bloomsbury, 2022).

French Philosophy of Technology and Technoscience: A Study on the Mode of Existence of Bio-objects

Jessica Lombard

Abstract This article shows technoscience as an extensively discussed concept in French and Belgian philosophy of technology, with major philosophers such as Bernadette Bensaude Vincent, Xavier Guchet and Gilbert Hottois.

Instead of telling the new narrative of a specific technoscientific object, this article reflects on an ontology of their mode of existence. Technoscientific objects open emerging research fields to indefinite possibilities, while they reorganize epistemic activities and orchestrate the development of technical networks, platforms and structures.

The main question here is to understand what is the ontological mode of existence of those objects and how they relate to the previously existing categories found in the work of Gilbert Simondon, that shaped the French philosophy of technology. To this end, we focus on a submicroscopic technoscientific object; namely the *bio-object*, and put forward an original analysis about the being of (bio)technoscientific objects in their own milieu.

This paper demonstrates that, even if Simondonian notions are critical to study (bio)technoscientific objects, these cannot be considered as proper technical objects. We propose that the mode of existence of bio-objects is closer to that of artificial objects than to technical ones. This peculiarity may be the reason why French philosophers are consistent in studying technoscience as an emerging technical field, with and beyond Simondon's lenses.

Keywords Technoscience · Bio-object · Hypertely · Artificiality · Gilbert Simondon

---

J. Lombard (✉)
Chaire ETH+ – ETHICS EA 7446, Université Catholique de Lille, Lille, France
e-mail: jessica.lombard@univ-catholille.fr

G. Bianco et al. (eds.), *Canguilhem and Continental Philosophy of Biology*, History, Philosophy and Theory of the Life Sciences 31,
https://doi.org/10.1007/978-3-031-20529-3_13

If it is possible to identify some historical trends that underline the connections in a set of objects (such as steam engines during the Industrial Revolution), the convergence phenomenon of our late twentieth and early twenty-first century would probably show the emergence and popularisation of the category of technoscientific objects.

Those objects have a significant place in contemporary debates on technical objects, mainly in French philosophy of technology, with philosophers such as Bensaude-Vincent et al. (2011, 2017, 2018). We observe such a broad spectrum of interactions between these objects and their environments that they are considered both as leading promises regarding ecological or medical issues and as hybrid apparatuses that could weaken our human condition.

The starting point of this paper is to consider that, to embody such expectations and concerns, those technoscientific objects must have some ontological characteristics of their own that distinguish them from other technical objects. This chapter aims at applying to technoscientific objects a categorical analysis that focuses on the specificities of their mode of existence. Indeed, this perspective is rooted in French thought concerning the philosophy of technology since the work of the philosopher Gilbert Simondon.

However, even if we believe that it is possible to read the question concerning technoscientific objects through Simondonian notions, this paper does not purport to be a general ontology of technoscientific objects. By encompassing nanotechnologies, biotechnologies and even information, technoscientific objects are as diverse as technical objects, and each of those subcategories unveils distinctive ontological properties that could not be encapsulated in one universal mode of existence of technoscientific objects, that would gather them all.

To follow Simondon on the idea of convergence phenomenon, the convergence phenomenon of technoscientific objects could be characterised by the submicroscopic scale of some of them. For instance, nanotechnologies at the molecular and atomic level; biotechnologies at the scale of cells and proteins (DNA, RNA and related genetic materials); and information (digital data, IA…) based on digitisation. How are those strange forms of technoscientific objects to be considered, namely those that do not even "appear" as such, for being on a submicroscopic scale? Because they introduce otherwise complex ontological questions, this paper focuses on *bio-objects*. What could be the formal ontology of submicroscopic objects that are not even artifacts but living things that were not produced by human beings, such as molecules, proteins, and genes?

Bio-objects are almost transparent submicroscopic objects that can only be experimented with technological tools or phenotypical reactions. However, rather than telling the narrative of a specific bio-object, this chapter properly examines the ontology of their category by questioning the mode of existence of (bio)technoscientific objects. To this end, I draw heavily on French philosophy of technology, from Gilbert Simondon to contemporary French philosophers, and researchers in the science and technology studies (STS) movement.

1 A New Philosophical Object: What Is a Technoscientific Object?

First of all, this paper needs to answer both straightforward and complex questions; what exactly is a technoscientific object, and why was there any necessity to compound the notion of episteme and technè in one unfamiliar word – although this word is more and more present today? The answer to those questions could make us understand better what is so specific about technoscientific researches that is present at the core of French and Belgian philosophy of technology.

In 1977,[1] the Belgian philosopher Gilbert Hottois coined the word "techno-science" to refer to a situation where technology is the goal of research and the milieu in which it operates. Nowadays, this word can be found in essential works on the philosophy of technology, although sometimes with different meanings, both in France and abroad.

To shed light on this propagation and evolution, Gilbert Hottois explains the current uses of the word "technoscience" and retraces its origins in a critical article (Hottois, 2006). It was first coined and slowly adopted as a central notion of French and Belgian philosophy of technology; before being spread abroad, mainly by Bruno Latour, and being used today by prominent researchers in science and technology studies (STS) movement, such as Donna Haraway. Technoscience is used primarily to qualify "science in the making" (Latour, 1987) in order to elicit a post-humanist criticism of the Enlightenment's belief in the purity and autonomy of universal Science. Gilbert Hottois also remarks that the word "technoscience" is sometimes used to denounce the corruption of scientific research by capitalism, neoliberalism, technocracy politics, postmodern globalization, and so on.

In French philosophy, technoscience is currently the new research field where technological structures and technoscientific objects carry the epistemic mode of production of knowledge.[2] For instance, DNA is not first and foremost a technological object in itself but is constituted as such in a technical net: with biobanks to store biological samples, genomic platforms to produce genetic data, databases to store it, and bioinformatic platforms to generate clinical information from it. As such, technoscience uses models, algorithms, phenomena, diagrams, reactions and experimentation, i.e. technological or technoscientific tools, to produce scientific knowledge. It also deploys new platforms, data structures, digital networks, financial investments to stock, launch, or carry those processes. The introduction of

[1] The word techno-science is already mentioned in the PhD thesis of Gilbert Hottois, which was defended in 1977 and published in 1979 under the title *L'inflation du langage dans la philosophie contemporaine*, by the Editions de l'Université de Bruxelles. Hottois also publicly introduced this concept in the article "Ethique et techno-science" for the Belgian journal *La pensée et les hommes* in 1978.

[2] Sometimes, technoscience is also considered an ideal type based on responsible innovation and sustainable development. The aim is to understand and recognize spontaneous tendencies and collective behavior in things ("self-assembly, self-organisation or self-repair" (Bensaude Vincent et al., 2017, 7)) and how they could organize themselves concerning designed purposes.

Research Objects in their Technological Setting (Bensaude Vincent et al., 2017) also mentions that the emergence of technoscience is not usually considered a historical break because numerous examples of it were found in the past; alchemy, agronomy, chemistry…

However, suppose technoscience is now a predominant topic in French philosophy of technology. In that case, it is not only because it has slowly evolved as a contemporary research field, but mainly because it "denotes a certain mode of existence of research objects" (Bensaude Vincent et al., 2017, 4). Here, the very concept of mode of existence is linked to the important French philosopher Gilbert Simondon and his *Mode of Existence of Technical Objects* (Simondon, 1989). It is not surprising then if French philosophers and STS researchers are deeply impacted today by the ontology of technoscientific objects. I firmly believe that this pursuit of the work of Simondon, even if transposed on technoscientific objects, is one of the main, albeit somehow transparent, arguments on the basis of which technoscience is nowadays a prevalent question in French philosophy of technology.

In my opinion, the strangeness of this word less derives from the idea of gathering making-and-knowing in one field of research ("techno-sciences") than in the fact that technoscience came in reality *after* technoscientific objects. Indeed, the notion of technoscientific objects is meant to recognize that new objects today, with how they perform, integrate new milieus or bear new values, were not easily categorized in what we traditionally call "technical objects". As they adopt new processes of interdependence and organize new internal dynamics, technoscientific objects were to be understood differently than "classical" technical objects – even if they sometimes are the same object, considered from several perspectives. I believe that it was not a new field of research, technoscience, that took shape in new contemporary objects, technoscientific artifacts; but objects as they were newly understood in their inner dynamics and external relations that caused an entirely new understanding of their modes of existence regarding to how they interact. An object is to be called a technoscientific object depending on its way to be structured and understood, and the research field in which this emergent technological setting operates is called technoscience.

As technoscientific objects are now inseparable from technoscience, it seems of paramount importance to understand the main characteristics of a technoscientific object. This question deserves detailed attention, as technoscientific objects are intrinsically different, just as a hammer differs from a longbow. However, the ontological question of their mode of existence can be studied more precisely, as this paper will show with the example of the bio-object.

In general, objects of scientific research are the only bearer of material properties and are left with no signification or value. They are not considered as things themselves and for themselves, but as passages, transitional pathways made for creating, making, and producing something else. They are used for the accidents that their substance bears, i.e. for the dispositional properties or secondary qualities that their primary ontological being manifests.

However, as technoscience entangles the epistemic knowledge and the poïétic bringing-forth, the structure of research behind technoscientific objects shifts from their essential characteristics to their functions in more interdependent processes. Technoscientific objects "propagate, colonise and transform their social and natural surroundings by (re-)shaping them and by imposing their temporal dynamics upon them" (Bensaude-Vincent et al., 2017, 3). A gene is properly a technoscientific object when it is understood as a key component and apparatus in the broader structures of the genome and phenotypical reactions, of genomics platforms and databases, of genomic medicine and biobanks.

Technoscientific objects are less considered through a fixed definition of their substance and properties than from their potential, i.e., depending on the potentialities they open to, or on the horizon of possibilities they can spare. It can be said that matter is paradoxically not what qualifies technoscientific objects nor what import in them for technoscientific research. As such, philosophy needs to rethink the definition of matter, or to never take it for granted. As it overcame the Cartesian notion of "res extensa", philosophy opened the notion of matter to a more plastic reality. Technoscientific objects generate dynamic processes that are never stabilized; thus, they unveil or enhance material potentialities. Bensaude-Vincent et al. (2017, 6) use the term "affordance[3]" to summarize the notion of an object considered in the propensities and relationships it can have; and, I may add, this affordance is brought to light as those propensities and relationships are experimented along.

Nevertheless, if technoscientific objects are objects adequately considered from their environment, this quickly prompts Simondon's notion of "associated milieu". And while technoscientific artifacts are indeed defined as objects that intertwine complex research, stock and production structures (biobanks…) with practical and low-scale technical uses (genetic engineering…), I should yet point out that nothing seems to prevent us from considering that technoscientific objects are "only" technical objects with broadly associated milieus (which also encompass the social fabrics and its organizational and operational elements). Moreover, as no one could pretend that Simondon overlooked the question of the milieu of the technical object, no one could pretend that he considered only secondary qualities and dispositional properties of technical objects instead of their affordances (albeit he does not use this term) or relational uses. Indeed, it is well-known that Simondon worked to overcome Aristotelian hylomorphism[4] and deny the fact that form and information were projected on a passive matter.

Consequently, because the general patterns that distinguish between technoscientific and technical objects are pretty tricky to define, this paper aims to show that a proper study of the mode of existence of technoscientific objects is still relevant as

[3] This term was coined in 1966 by the psychologist James J. Gibson. He defines it as a possibility that emanates first from the environment and that the subject receives and interprets: "The affordances of the environment are what it offers the animal, what it provides or furnishes, either for good or ill. The verb to afford is found in the dictionary, the noun affordance is not. I have made it up. I mean by it something that refers to both the environment and the animal in a way that no existing term does. It implies the complementarity of the animal and the environment" (Gibson, 1979, 127).

[4] Summarily, hylomorphism is the idea that being is the compound of matter and form.

part of the philosophy of technology. Moreover, this study should not only focus on narratives of technoscientific objects, even if this approach has proven quite pertinent.[5] Contrary to what is now one of the main axes covered by French philosophers when it comes to technoscientific objects, this paper takes a group of technoscientific objects – the group of *bio-objects* – and thoroughly questions their mode of existence through ontological categories and characteristics, instead of focusing on the stories that a specific object can deliver.

2 Philosophy of Technology and Biotechnologies: What Technoscientific Object Is the Bio-object?

As previously stated, this essay limits this question to the field of *biotechnology*. It is not surprising if the first concrete example detailed in *The Genesis and Ontology of Technoscientific Objects* is the example of the gene. At first, the gene was this bearer of meaning and imaginaries that led to deterministic inheritable traits, a fate hidden in one's own body that shaped our biological destiny by determining the purity or defectiveness of our entire self. Nowadays, genes are a malleable matter that is both explanatory and manipulable. They conform to our classical idea of a tool, any physical but plastic instrument that could be used in various fields and for several ends and procedures. The gene "became what we call an attractive object of technoscientific research" (Bensaude-Vincent et al., 2017, 2). Because of their hybridity that sways between the living and the non-living, such bio-objects indeed blur the boundaries between artificiality and nature, object and subject.

Biotechnologies are used in several fields of technoscientific research, such as personalized medicine (bioinformation, engineering tools…) or agronomy (GMO…). Are biotechnologies and bio-objects, as a sub-category of technoscientific objects, everyday technical objects; or do they bear a specific mode of existence that should lead us to a more conclusive definition of technoscientific objects? Can they still be understood based on the categories commonly used after Simondon in French philosophy of technology?

To answer those questions, I choose to begin by focusing on the biomarker example, which the French philosopher Xavier Guchet (2016, 353–55) mentions in his book on personalised medicine. This example will make us better understand some characteristics of the mode of existence of (bio)technoscientific objects.

(a) **The Biomarker and its Associated Milieu.**

[5] For instance, the introduction "The genesis and ontology of technoscientific objects" of the book *Research Objects in their Technological Setting* explicitly tell "stories about the genesis and life of a selection of such [technoscientific] objects" (Bensaude Vincent et al., 2017, 8) (cancer stem cells, polar ice core, heroin, nuclear waste…), and the article "Matters of Interest: The Objects of Research in Science and Technoscience" also presents three objects as key examples: carbon, the OncoMouse and STM.

A molecule may prove interesting for researchers as a biomarker if a statistical correlation between its presence and a biological process is proven – presence, over-expression or under-expression. Biomarkers are used for environmental purposes, to detect an old incident (climatic, toxic, etc.), and for medical purposes as part of diagnostics, screening, or monitoring the response to treatment. Therefore, the biomarker is most often a protein that highlights an impact or an effect's biological signature without indicating the cause of it.

At first, I may underline that when the organic molecule becomes a biomarker, we see the true genesis of a (bio)*technical* object. According to Simondon (2017, 26), the technical object is precisely "that of which there is genesis". Yet, the correlation of the word "genesis" with genetics in the genomic sense of the term would suggest that the technical object is precisely that of which there is *no* genesis, but production. We can see that the boundaries are blurred because bio-objects are, or rather were living organisms. Indeed, it seems easier to talk about the genesis of an *organic* technical object than of a more classical (maybe mechanical) technical object. The coming-into-being of biological material such as a molecule seems more understandable than the coming-into-being of the motor. Then, how can it be said that a living organism becomes a technical object? Simondon (1995, 22) speaks of an ontogenesis "that concerns the emergence, and the coming-into-being (*devenir*) of [the technical object's] own being." As such, the term "ontogenesis" refers to all the factors of development of the being, biologically but also ontologically. There is a process of genesis in the molecule as it becomes a biomarker, which makes it as it is.

Secondly, from the perspective of this *coming-into*-biomarker, the molecule undertakes what we could call a specific activity of genesis; the creation of its associated milieu. I follow the French philosopher Victor Petit (2017, 16) when he remarks that "it was Simondon who radicalised the philosophy of the milieu." Simondon does not start from produced and constituted objects, but from their genesis is understood as "mi-lieu or relation of two previously disparate realities" (Petit, 2017, 16). In short, genesis is not a beginning,[6] but a process (movement) that one must constantly see in what comes into being, whose reality of existence each time entails a milieu. The milieu can only be grasped in the midpoint (mi-lieu) of its relationship with the individual. I consider that means, purpose or production category is not enough to address the technical object. Like any object, the technical object must be understood from the question of its milieu.

In my example, the technoscientific object "biomarker", which the molecule is not yet, already bears its associated milieu in the form of specific infrastructures that will respond to its future concretisation. For instance, data networks detect the better organic candidate for the technical role that the biomarker will accomplish and enable the molecule to fulfil its biomarker function. I have said that the biomarker shows a correlation between a molecule and a biological process. This

[6] "All true genesis, whether of being or of thought, has no origin, neither in the individual nor in his milieu, nor their adaptation" (Petit, 2017, 16).

correlation can only be established thanks to a multifactorial sociotechnical system, among which: "very high-speed data acquisition techniques, algorithms that highlight correlations deemed significant, biobanks and tumour libraries, organisational changes affecting research, industry, hospitals and finally regulation" (Guchet, 2016, 353). These elements require the mobilisation of numerous human and non-human resources to be sustained and show the relationship between the technoscientific object and the technical system in which it operates. It is a broad *technical milieu,* so to say, if not even an associated milieu understood as the "condition of existence for the invented technical object" (Simondon, 2017, 59). The technical object's genesis is a convergence of the geographical and technological worlds. It creates an associated milieu (hence the famous example of the Guimbal turbine),[7] without which there can be neither abstract viability nor concrete existence of the technical object. This milieu, both technical and natural, mixes artificial elements with the structural components of its functioning and mediatises them; moreover, it is capable of conditioning itself, i.e. of guaranteeing the possibility of the said functioning.

Consequently, the associated medium "is not fabricated [*fabriqué*], or at least not fabricated in its totality" (Simondon, 2017, 59), which implies that it is not entirely inherent to the human being. A technical milieu is made of a technical ensemble that functions as a stable structure. "To make a good adze, a technical ensemble of a foundry, forge, and quench hardening is required" (Simondon, 2017, 72). As objects depend on each other, they maintain their inner and outer possibility of existence. The nonexistence of a component (be it processors, computers, or genomic platforms for biological samples) will lead to the nonexistence of a compound to which this component would have been necessary. Technical objects depend on inner concretisation to form a proper technical ensemble – even before their material production would be considered in more socio-economic terms. As such, the human supply and demand around them also depend on the possibility of those technical considerations. It must be concluded that technical and technoscientific object create their milieus. On the one hand, they influence the socio-economic and structural environment, i.e. their "geographical world". On the other hand, they also require and cause the development of their "technical world,"[8] the milieu involved in their operations and realisation: the biomarker *calls for* a molecule, a biological organism, a bioinformatics platform, etc.

---

[7] We reproduce this emblematic example for greater clarity. "Such is the case of the ensemble constituted by oil and water moving in and around the Guimbal turbine. This ensemble is concreted and individualized by recurrent thermal exchanges that take place within it: the faster the turbine spins, the more there is an increase in the heat generated by the generator through magnetic losses and the Joule effect; but the fester the turbine spins, the greater the increase in the turbulence of the oil around the rotor and that of the water around the crank-case, thereby activating the thermal exchanges between rotor and water" (Simondon, 2017, 59).

[8] (Simondon 2017, 56) The geographical world is sometimes called a "geographical milieu", and the technical world a "technical milieu".

(b) **The Biomarker and Hypertely.**

Naturally, this technical milieu that the object establishes and demands is also an internal necessity, and therefore a coercive principle on which its genesis depends. The biomarker *cannot be without* a molecule, whereas molecules have an inner coherence. Thus, as Jean-Yves Château (2010, 31) explains, "the genesis of the object overwhelms the object and makes it necessary to create a new milieu for itself", which opens the door to the paradoxical risk of developing an "hypertely" (Simondon, 2017, 53) that ties it too tightly to the said associated milieu.

The notion of hypertely is relevant for this paper in more than one way, mainly because it seems typical of technoscientific objects. As Simondon (2017, 53) shows, an hypertelic milieu is excessively specialised: "there are phenomena of hypertely that manifest themselves in the evolution of technical objects, giving each technical object an exaggerated degree of specialisation and maladapting it to even a slight change in the conditions of its utilisation or fabrication." All technical objects impact and create the milieu that they need, but some of them are "over-adapted" so that they function only under particular conditions. This deficiency prevents them from adapting to any milieu other than their own. Technical objects with the most complex composition and specific purpose are notoriously prone to hypertely. To explain this notion, I will mention the French philosopher Hadi Rizk's example because it does not require prior technical knowledge, unlike the Simondonian comparison between the synchronous single-phase motor and the synchronous three-phase motor (Simondon, 2017, 54). Hadi Rizk (2018, 10) explains the case of a rocket plane whose ramjet can only be started at an altitude of 15,000 m and "which requires that it be initially transported on another plane." The hypertelic constraint resides in the mention of this other plane. Without it, the ramjet roots to the spot the rocket plane.

The notion of hypertely introduces the idea of pathology (in the Canguilhemian sense) in technical objects. Now, the intrinsic considerations of the technical object depend on an adaptability that guarantees its internal viability. But when an hypertelic milieu cancels it out, the adaptability of the technical object is, in fact, over-adapted, over-specialized. In short, the technical object is no longer capable of anything other than its function; the entire milieu set up around it also serves to maintain it (a techno-industrial facility, a particle accelerator, etc.). Thus, hypertely is comparable to a pathological condition of the technical object. It no longer fits in any milieu other than its own; it loses all capacity for adaptation and, consequently, all principles of genesis and coming-into-being.

Consequently, hypertely makes the technical object dependent on its environment. I can directly say that hypertely entails the dependency of the technical object to its *extrinsic* milieu (understood just as human socio-economic conditions are the *extrinsic considerations* of the technical object[9]). Because of the pathological or extreme dependency that it entails, hypertely is the opposite of a normal concretisation of the technical object in its associated milieu.[10] Indeed, hypertely is all the stranger as the technical object is supposed to make circularity on its internal milieu, in the sense that it depends on and benefits from a set of elements making its concretisation viable "like an arch that is stable only once it is finished" (Simondon, 2017, 59). As I just said, hypertely weakens the stability of the technical object in its associated milieu. The rocket is conditioned by its milieu (the plane that makes it start at an altitude of 15,000 m), whereas the Guimbal turbine perpetuates its own movement. When a technical object over-adapts to an external milieu, it becomes unfit for the *adaptation process* itself (the dynamic of its associated milieu).

Consequently, the milieu of the technoscientific object is a determining factor that outlines its ontological specificity. And we know that the extremely high degree of specialisation of this (bio)technical object is impossible to deny. First and above all, the biomarker depends entirely on the correlation "presence of the molecule – biological process" that it is supposed to help clarify. Indeed, if a molecule is not fit for locating the biological process aimed at by technoscientific research, it will not be selected as a potential biomarker. Hence, as the oil is used to lubricate the generator of the Guimbal turbine, the molecule is used to (and because it can) detect a specific biological process. To this end, the biomarker depends on a very complex net of systems that are set up for its proper genesis, stocking and functioning (algorithms, biobanks, tumour libraries…). In this respect, the biomarker must be included in technical objects whose specialisation makes them *hypertelic*.

The said hypertely is already semantically acknowledged in an indirect but explicit way by the linguistic creation of a new category of objects: *bio-objects* (Metzler et Webster, 2011). They are organised along two axes: 1/ organic entities artificially manufactured by the human being in laboratories (e.g. biomarkers); 2/ non-artificial organic entities moved into new artificial spaces (e.g. biological and genetic samples stored and analysed in bio or data banks).

Then, it can be said that (bio)technoscientific objects have a very high tendency to be hypertelic. If most technoscientific objects are intrinsically hypertelic, should we consider then that they are not technical objects? This question needs to be

[9] Extrinsic (or economic) considerations are they object's causes that depend on the human environment. For instance, the availability of the materials that constitute the object, the organization of work and production, the economic interest of the manufacturer, the consumer's demand, etc. These considerations condition the social existence of the technical object in a chain of production, distribution and consumption which make it both feasible and viable. Intrinsic (or technical) considerations, however, are the very possibilities of its inner realization. As the French philosopher Jean-Yves Château writes, they confront the difficulty of the object "to maintain itself in the being" (Château, 2010, 17). Both of them are needed for the technical object to have a viable reality of existence.

[10] The Simondonian associated milieu defines "the recurrence of causality within a milieu that the technical object creates around itself and that conditions it, just as it is conditioned by it" (Simondon, 2017, 59).

considered based on our example; if the bio-object is a technoscientific object, is it not strictly speaking a technical object?

3 A Simondonian Ontology of Technical Objects: How to Redefine Bio-objects?

The bio-object is at first sight what I could call an "*organic technical object*". It is an organic matter that was given the title of object (an organic molecule becomes a biomarker) or that was reified to produce a new one (the tissue or DNA fragment is stored and analysed to create a technical object, such as digital data).

Xavier Guchet (2016, 354) reminds us that bio-objects are "inseparable from virtual *(bio-virtual)* spaces in which they acquire their features." Their hypertely is due to the extreme interdependence between the bio-object and a milieu explicitly created for it. The bio-object (before being a bio-object – i.e. the molecule, before being a biomarker) was externalised from its organic space and moved into the new areas of biomedical research. After what, it was transformed into the bio-object itself (the biomarker) and put back into an organic milieu. As a result, the transductive movement of this genesis, of this coming-into-being, is the *opposite* of the one described by Simondon – and the bio-object is probably not a technical object as he defines it.

(a) **The Technicity of Life: Is the Bio-Object Even a Technical Object?**

For Simondon, the concretisation of the (classic) technical object is made through its social independence. Otherwise, it is a mere utensil, an object separated from its genesis. On the contrary, the bio-object was an organic object that was primarily independent of human practices, and that is now utterly dependent from the technical milieu that the human being has built around it. The technical object follows a dynamic of individuation and concretisation; where a movement of reification or objectification carries out the bio-object. Organic objects, which are essentially caught up in their genesis (take the emblematic case of cancer cells used for medical research purposes, which are constantly multiplying), are at the centre of a process of bio-objectification. They lose their status of living entities to become products (bio-objects) identified in the extrinsic framework of the human economy or ecology.

We could say that an extrinsic cause steers even the intrinsic cause of the coming-into-being of those organic technical objects. Therefore, the bio-object cannot be a technical object from a Simondonian perspective. It is not the case, of course, of proper technical objects derived from bio-objects: mainly digital data. This new specialised entity can be stored and exchanged and comes from reifying the organic object into a bio-object. If the molecule that (bio)marks a medical effect is not a technical object, the data in which translates the biomarker must be considered as such. In short, *the coming-into-being of the bio-object is not the coming-into-being of a technical object*. The coming-into-being of the bio-object is not in its individuation, in the expansion of its associated milieu; those are things already inherent to it as a living thing, which were taken away by a movement of objectification. It must be concluded that the bio-object is not an "organic technical object" as we first thought but an "*organic object (which was) technicalised*". It does not *bear technicity* in itself.

But what does it exactly mean not to bear technicity, and how does it impact the mode of existence of bio-objects? For Simondon (2017, 72), technicity is "more than a quality of [the technical object's] use; it is that which, within it, adds itself to a first determination given by the relation between form and matter." The organization is here the substratum of the technicity of the technical object. The technicity of the nut or the car is what makes these elements technical elements. The nut is not a technical object only because it is a ring with a threaded hole made of metal. It bears or carries technicity: the possibility of screwing the nut on a screw to make a proper "nut and bolt" is technicity. A nut is a technical object because it opens up to the potentiality of technology and therefore contains it by essence. Thus, the technical object is the element that bears technicity and is the object and subject of an inner genesis that creates its associated milieu; hence the fact that technical objects are not only instrumental tools, according to Simondon. The technical concretisation, as Simondon understands it allows us to go beyond the instrumentalization of technology, on the one hand, but also of the milieu, on the other. The concretisation of a technical object is a persistent system which is caused intrinsically, not only extrinsically, i.e. a system that was decentralised from human intervention and whose associated milieu makes it viable.

We understand why it is of paramount importance to consider that bio-objects do not bear technicity in themselves. I said that the coming-into-being of a bio-object, for instance a biomarker which was before that a simple molecule, seemed more easy to picture than the coming-into-being of a technical object such as a motor or the nut. There is a reason for that; the molecule expresses the inner coming-into-being that is inherent to any living thing, and relates to the notions of growth or evolution. However, the coming-into-being of the technical object takes another form; it relates to the concretisation of the technical object in its associated milieu. As already discussed, Simondon considers that the mode of existence of technical objects depends on two types of causality: extrinsic considerations, which are based on human systemic conditions, and intrinsic considerations, understood as the internal conditions of possibility of the technical object. On the contrary, it could be said that the genesis of living things has no extrinsic causes. Of course, the birth of living things is impacted by external circumstances, but they are not human-made contrary to technical objects. In this sense, we must understand that the genesis of living things is not formally based on extrinsic (socio-economic) considerations. Then, living organisms depend first and above all on intrinsic considerations, i.e. the very possibility of their existence, as the French philosopher Georges Canguilhem (1992, 2013) perfectly showed when discussing the viability for a living form to persist in its being on an evolutionary basis.

Then, it is not technicity that organizes biological organisms; the organic object does not bear technicity itself. It was made as a technicalised object. Here, we can see another significant difference between the mode of existence of technical objects and bio-objects – as a category of technoscientific objects. The technical reality of bio-objects is not in their genesis and does not come from an internal necessity, but could be only considered as a result of the human will or purpose: bio-objects do not bear technicity as the basis of their potentiality of individuation and are only caused by extrinsic considerations.

They are not simple "organic" technical objects, but they are reified in the more dense sense of this term. This is why bio-objects are not "organic technical objects," but are in fact "organic objects (which were) technicalised." This is a fundamental difference between technical objects and (bio)technoscientific ones, and I propose that this fact may result from the compound of the notions of "science" and "technology". A technoscientific object is indeed a technical object first and foremost considered from an epistemic perspective; as a mode of production of knowledge that emanates from a broad and hypertelic milieu which is made to grant it technicity. Contrary to technical objects, which are considered as the bearers of material dispositional properties and technicity itself.

As such, as they are based not only on technology but also on sciences, the mode of existence of submicroscopic (bio)technoscientific objects can hardly maintain itself without the human being. To take a simple example, technical objects break or deteriorate without proper human infrastructure. But without appropriate human infrastructure, bio-objects do not exist anymore: the biomarker goes back to being a simple molecule, so to say.

The Simondonian perspective on technical objects deeply impacted French philosophy for its originality; if we were responsible for the extrinsic causes of the technical object, it was completed by so-called intrinsic considerations. These intrinsic considerations showed that something other than human production lied in technical objects. They confronted the standard instrumental view applied to technical objects. But the same cannot be said of the mode of existence of (bio)technoscientific objects.

(b) **The Artificialization of Life: Is the Bio-Object an Artificial Object?**

I explained that the genetic movement of bio-objects seems to reverse one of the technical objects, given the radical hypertely of bio-objects on the one hand and their reification (from organism to object) on the other. As hypertely and reification are two modalities of its mode of existence, I must conclude that the bio-object is not a proper technical object but an "organic object (which was) technicalized".

Consequently, I will draw from the notion of "*artificial object*" an interesting parallel that may help us better understand the ontology of (bio)technoscientific objects.

The artificial object is not necessarily a tool. Nor should the artificial object be entirely reduced to predicates of facticity or illusion, as the word artificial can be understood today. Therefore, contrary to a common idea, one cannot say that the technical object is an artificial object since production does not lead to artificiality.

What leads to artificiality is mostly (if not only) the relation of dependency to the milieu. Thus, if the technical object can be said to be "primitively" artificial since it is manufactured, it finally becomes capable of doing without the artificial milieu through the process of concretisation. "This object needed a regulative external milieu, *at the beginning*, the laboratory, workshop, or sometimes the factory. [...] The concretised object is comparable to the spontaneously produced object; the object *frees itself from the originally associated laboratory*" (Simondon, 2017, 50). Thus, technical objects become self-sufficient or are associated with other objects via a common associated milieu to no longer be isolated. This technical object is not an artificial object as it becomes its principle of coming-into-being in a technical milieu (just as the Guimbal turbine self-regulates itself through thermal plays).

And as the technical object is characterised by its ontological technicity, the artificial object is represented by its artificiality. Artificiality here must be understood in the etymological sense of what is derived from art, hence produced (by the human being).[11] Artificial objects are products of artifice, in the neutral sense of the word.

Consequently, if the technical object is distinguished by its genetic individuation, the artificial object is entirely defined by its *mode of production*. The artificial object bears no technicity and has no inner principle of genesis. It is a thing that was then *made*, i.e. a proper creation. This means that artificial objects were not even technical objects as such before being manufactured; in fact, artificial objects can never be technical objects.

Artificial objects are "living organisms (which were) artificialized". A typical example of this artificialisation process is the glasshouse plant. "[The human being] *diverted* the functions of this plant from their coherent fulfilment, to such an extent that it can no longer reproduce except through procedures such as grafting, requiring human intervention. [...] The artificialized plant *can only exist* in a laboratory for plants, the greenhouse, with its complex system of thermal and hydraulic regulations."[12] As we can see, in this description, I found again the two modalities that were already identified above, namely reification and hypertely, which define bio-objects as technoscientific objects. Although I provisionally named them "organic objects (which were) technicalized", we understand that this category already as a name, and I can now definitely class bio-objects under the group of "artificial objects."

---

[11] One of the main etymological consideration that Heidegger reminded us of, and that shaped our understanding of the word *technè*, is undoubtedly that in Ancient Greek, "art [bore] the modest name technè" (Heidegger, 1977, 34). The discursive relationship between art and technè does not imply that art was reduced to craftsmanship; but that technè embraced a broader domain of experience that the one which is usually implied.

[12] (Simondon, 2017, 49) The coming-into-being of this artificial object can now only be sustained by the human mediation and control. "Its flowering has become a pure flowering, detached, anomic; the plant flowers until it is exhausted, without producing seeds. It loses its initial capacity of resistance against cold, drought, and sun; the regulations of the *primitively* natural object *become* the artificial regulations of the greenhouse" (Simondon, 2017, 49, emphasis mine).

Surprisingly enough, I found a brief mention of this same idea in Yves Deforge's afterword to the *Mode of Existence of Technical Objects* (Simondon, 1989, 301), that shows how French philosophy also opened to the notion of technoscientific objects through Simondon's work on technical object. "Another evolution […] can be considered as an evolution towards artificialization, of which biotechnologies, generally quoted among the new technologies, are an example."

A detail about the notion of artificialization completes this definition, when Simondon (2017, 49) says that: "artificialization is a process of abstraction within the artificialized object." First of all, this strange wording seems to indicate that movement (artificialization) is already in the essence of the object we consider (the artificialized object). Simondon did not write, as would have been expected, that "artificialization is a process of abstraction *of the organic* object"; but *within* the *artificialized* object. This semantic eccentricity nevertheless is a new hint on the question of the mode of existence of technoscientific objects, here biotechnologies.

Indeed, the adverb artificialized differs from artificial, in that the artificial object is the one that has already been turned over by the movement of artificialization. On the contrary, the artificialized object is the one that is caught in the movement of artificialization. It is therefore logical to consider that the movement (artificialization) appears in what is still under its process (i.e. the *artificialized* object).

Furthermore, the full philosophical understanding of such an expression rests in my opinion on the definition that can be given of the word "abstraction". Its clarification will immediately intersect this paper's theory in a way that is, in reality, quite stunning. Abstraction has two meanings that can both shed light on the text. To begin with, "to abstract" means to detach, to withdraw: abstraction is the process of *withdrawal* (of individuation). The concrete independence of the living organism is suppressed by the movement of artificialization, which reifies it. This negativist movement is the reason why the plant becomes the systemic "greenhouse plant". Secondly, "to abstract" means to *isolate*, to *confine*: abstraction is a process of isolation or *virtualization*. When artificialization abstracts the organism, it cloisters it in an artificial and contingent environment. From this virtual environment, the plant becomes a "*greenhouse* plant". By detailing the elements of this expression, I find again my theory according to which artificial objects (bio-objects for instance) are the product of both reification and hypertely.

We can understand here that artificialization refers very precisely to a process where a (living) organism is made into an object by the human hand, and implanted in an artificial milieu. In short, *the artificialization of an organism responds symmetrically to the technical concretisation*. Thus, in a fully antithetical way, the artificial object is extracted from its natural environment by the human being, and cannot trigger the creation of its associated milieu. Its reification and hypertely denote the irremediable deficit of its presence. Whether plant or bio-object, the artificial technoscientific object only dwells in a zone of inauthenticity, for which *glasshouse* could be the symbolic name.

*

This chapter aims to apply distinctions based on French philosophy of technology to new technoscientific objects, in order to question their mode of existence. It would be damaging to argue *a priori* that technoscientific objects confront us, by damaging our nature or condition, without first trying to understand their ontological predicates. French philosophy of technology is a strong standpoint from which to study this question. Furthermore, the specific mode of existence of technoscientific objects led to the creation of technoscience as a research field of major importance in French philosophy, mostly inspired by the impact of Simondon's work in the field of technology.

I put forward this idea by explaining that (bio)technoscientific objects can be carefully thought from the particularities of their own mode of existence.

Any technoscientific research could consider that technoscientific objects have an hypertelic relationship to their associated milieus, even regarding the very vast set of structures and correlations they intertwine. This net is exacerbated by the potential re-internalization of technoscientific objects into organic environments, leading to their inner artificialization, even when they were living organisms themselves. Indeed, I first said that the bio-object was an organic technical object, but the genetic asymmetry with other technical objects led me to speak of an organic object (which was) technicalised. When I compared this notion with the artificialization process, I concluded that biotechnologies are artificial (organic) objects, where the parenthesis already anticipates the loss of this status and the complete reification of those objects in their associated milieu. This process seems to be the opposite of the individuation of technical objects.

As such, with the question of bio-objects, we are confronted to organisms that come to life as artificial objects, that show resilience in an hypertelic milieu and intervene is a very wide set of technical platforms that leads to new epistemic discoveries.

Then, the dynamic of convergence of our late twentieth and early twenty-first century, which leads to technoscientific objects, seems to reverse the process of externalization as conventionally understood since Ernst Kapp. Ernst Kapp considers that hominization is carried out by the deposit or projection of the human gesture, memory and energy in external artefacts. However, nowadays, technoscientific objects lead to a new movement; the internalization of technical artefacts in the human organism. For instance, the biomarker is reintegrated into the organic whole. As technoscientific objects, and bio-objects *a fortiori*, introduce the question of the *internalization* of technical objects, it could be interesting to pursue this line of questioning with the philosophy of technology in German thought.

References

Bensaude-Vincent, B., Loeve, S., Nordmann, A., & Schwarz, A. (2011). Matters of interest: The objects of research in science and technoscience. *Journal for General Philosophy of Science, 42*(2), 365–383. https://doi.org/10.1007/s10838-011-9172-y

Bensaude-Vincent, B., Loeve, S., Nordmann, A., & Schwarz, A. (2017). Introduction: The Genesis and Ontology of Technoscientific Objects. In *Research Objects in their Technological Setting* (pp. 1–12). Routledge.

Bensaude Vincent, B., Guchet, X., & Loeve, S. (2018). *French Philosophy of Technology. Classical Readings and Contemporary Approaches*. Springer.

Canguilhem, G. (1992). *La connaissance de la vie*. Vrin.

Canguilhem, G. (2013). *Le normal et le pathologique*. Presses Universitaires de France.

Château, J.-Y. (2010). La technique. Genèse et concrétisation des objets techniques dans *Du mode d'existence des objets techniques* de Gilbert Simondon. *Philopsis*. https://philopsis.fr/wp-content/uploads/2010/12/pdf_Technique-Simondon-Chateau.pdf

Gibson, J. J. (1979). *The Ecological Approach to Visual Perception*. Houghton Mifflin.

Guchet, X. (2016). *La médecine personnalisée. Un essai philosophique*. Les Belles Lettres.

Heidegger, M. (1977). The Question Concerning Technology. In *The Question Concerning Technology, and Other Essays* (W. Lovitt, Trans.) (pp. 3–35). Garland Publishing.

Hottois, G. (2006). La technoscience : de l'origine du mot à ses usages actuels. *Recherche en soins infirmiers, 86*(3), 24–32. https://doi.org/10.3917/rsi.086.0024

Latour, B. (1987). *Science in Action*. Harvard University Press.

Metzler, I., & Webster, A. (2011). Bio-Objects and their Boundaries: Governing Matter at the Intersection of Society, Politics and Science. *Croatian Medical Journal, 52*(5), 2011.

Petit, V. (2017). Le désir du milieu (dans la philosophie française). *La Deleuziana, 6*, 10–25.

Rizk, H. (2018). *La technique. L'activité technique et ses objets*. Vrin.

Simondon, G. (1989). *Du mode d'existence des objets techniques*. Aubier.

Simondon, G. (1995). *L'individu et sa genèse physico-biologique*. Millon.

Simondon, G. (2017). *On the Mode of Existence of Technical Objects* (C. Malaspina Trans.). Univocal Publishing.

Jessica Lombard graduated from the Special Military School of Saint-Cyr Coëtquidan as a commissioned officer in the French Land Army, and passed her master's degree in philosophy at the École Normale Supérieure d'Ulm and the École des Hautes Études en Sciences Sociales (France). She defended a thesis in philosophy entitled "Towards an Onto-Anthropotechnic of the Human Sphere. The Question of the Technological Ban through a Phenomenological Approach of Transhumanism," in which she specialized in philosophy of technology and phenomenology. She currently works as a University lecturer and researcher at the Chair "Ethics, Technology and Humanities" (ETHICS EA-7446) of the Catholic University of Lille (France).

A Bergsonian Perspective on Causality and Evolution

Mathilde Tahar

Abstract Bergsonian philosophy is not generally regarded as a true philosophy of biology. Bergson's rejection of Darwinism, his silence on incipient genetics, and his unfortunate comparison of the movement of the *élan vital* with the duration of consciousness led Bergson to be considered at best an outdated philosopher, at worst an enemy of science. However, if there is one thing that Bergson's *Creative Evolution* grasped, and offered to biology, it is an understanding of the processual nature of evolution and of its consequences to conceive a causality that considers the efficacy of time in evolution. The aim of this chapter is to reassess Bergson's philosophy of evolution to demonstrate its relevance for contemporary biology. First, I will study the relevance of Bergson's epistemological critique of the theories of evolution under debate in his time. Then, I will provide an analysis of the concept of *élan vital* that will show its topicality to address some of the problems of contemporary biology. Finally, I will indicate the lessons that can be drawn from the *élan vital* for thinking of the causality at work in evolution. My purpose in this chapter is not only to show that the Bergsonian philosophy of life is not completely outdated, but it is to revaluate its importance in the history of biology and the philosophy of biology, while exploring the contemporary extensions of Bergsonian insights.

Keywords Bergson · Theory of evolution · Adaptationism · History of biology · Teleology

In the second half of the twentieth century, Bergson's philosophy of biology which had once been so popular (including among biologists), fell into disrepute. His rejection of Darwinism, his silence on the incipient sciences of genetics, and his unfortunate comparison of the movement of the *élan vital* with the duration of consciousness led Bergson to "almost complete discredit" (Monod, 1970, 44, my translation). Although there has recently been a discreet return to Bergson's philosophy,

---

M. Tahar (✉)
Université Toulouse II – Muséum National d'Histoire Naturelle, Toulouse, France

G. Bianco et al. (eds.), *Canguilhem and Continental Philosophy of Biology*, History, Philosophy and Theory of the Life Sciences 31, https://doi.org/10.1007/978-3-031-20529-3_14

both by philosophers and by some biologists, to address the questions in contemporary biology, his contribution to these debates is often eclipsed by those of more 'frequentable' philosophers – Aristotle's final cause has been used to theorise the function of the organs, and Whitehead is seen as one of the fathers of process-biology.

The aim of this chapter is to reassess Bergson's philosophy of evolution to demonstrate its relevance for contemporary biology, especially for conceiving the link between time and causality in evolution. If there is one thing that Bergson grasped, and offered to biology, it is an understanding of the temporal nature of evolution, and of its consequences to conceive a form of causality that avoids the pitfall of finalism without falling into a reductionist determinism incapable of considering the specificity of the living.

1 The Bergsonian Criticism of the Hidden Finalism of the Theories of Evolution

(a) Against teleology

At the turn of the twentieth century, the evolution of species was well established, but its mechanism remained debated. Besides Darwinian natural selection, several other hypotheses were put forward. In *Creative Evolution,* Bergson reviews the different theories being discussed in his time: Darwinism, mutationism, orthogenesis and neo-lamarckism (the inheritance of acquired characters). I will not go into the detail of his analysis, but rather focus on Bergson's more fundamental objection to all these theories: they subscribe to a form of teleological thinking which is none other than the hidden face of determinism, or as Bergson calls it, "mechanism".

The mechanical explanation, according to Bergson, implies thinking of nature as a closed system on which time has no real effect since only invariable mechanisms are at play: "The essence of mechanical explanation, in fact, is to regard the future and the past as calculable functions of the present, and thus to claim that *all is given*" (Bergson, 2007, 39/1911b, 39–40). Indeed, mechanism assumes, in Bergson's view, a Laplacian vision of the world according to which an omniscient demon knowing all the laws of physics and the initial conditions of the movement of each body at a time *t* would be able to predict the final state of the universe at a time *t2*. Bergson refutes this view as anthropomorphic: the universe is reduced to an object that can only exist if one implicitly accepts the existence of this metaphysical entity: the demon's eye. Furthermore, he deems it inadequate for thinking evolution because it does not seriously take into account the efficacy of duration: the fact that processes, and therefore also their possible outcomes, evolve with time.

> That the appearance of a vegetable or animal species is due to specific causes, nobody will gainsay. But this can only mean that if, after the fact, we could know these causes in detail, we could explain by them the form that has been produced foreseeing the form is out of the question […] how could we know beforehand a situation that is unique of its kind, that has never yet occurred and will never occur again? (Bergson, 2007, 27–28/1911b, 29).

Thus, Bergson rejects both mechanism and finalism *for the same reason*: they are explanations in which time has no effect, and evolution is predetermined. In a finalist framework, the direction of evolution can be likened to the blueprints of an engineer; mechanism on the other hand purports to be blueprint-free but in reality, says Bergson, this intention is only artificially suppressed. Both the "*causalité-impulsion*" (impulse-causality) and the "*causalité-attraction*" (pull-causality) to use Jankélévitch's words, assume that time is ineffective, and that evolution is directed (Jankélévitch, 1959, 133–134). Bergson demonstrates this through a critique of the concept of adaptation, as a mixture of the two explanations. Adaptation is said to "to bring about a precise adjustment of the organism to its circumstances", as though the conditions were a mould in which the organism would mechanically takes its shape. But "this is indeed to be fooled by a metaphor" and not see that the adaptation of an organism "is not repeating but *replying*":

> If there is still adaptation, it will be in the sense in which one may say of the solution of a problem of geometry, for example, that it is adapted to the conditions. I grant indeed that adaptation so understood explains why different evolutionary processes result in similar forms: the same problem, of course, calls for the same solution. But it is necessary then to introduce, as for the solution of a problem of geometry, an intelligent activity, or at least a cause which behaves in the same way. This is to bring in finality again, and a finality this time more than ever charged with anthropomorphic elements (Bergson, 2007, 58–59/1911b, 61–62).

Two contradictory but complementary ideas are expressed here: (1) the idea that organisms are passively shaped by external conditions (mechanical causality); and (2) the idea that the morphological evolution of a species is analogous to the construction of a machine aimed at adapting to external conditions (final causality).

Bergson thus opposes the theories of evolution of his time with a genuine epistemological critique, aiming to thwart the hidden teleological presuppositions of these alleged reductionisms. Darwinian theory does not escape this criticism, although adaptation is an accidental result of the struggle for existence. If selected variations are minute (and Bergson, in a misunderstanding that I have discussed elsewhere – see Tahar, 2022a – goes so far as to say "insensible"), "some good genius must be appealed to – the genius of the future species" (Bergson, 2007, 69/1911b, 72). His main argument is that of structural similarities or evolutionary convergences: "How could the same small variations, incalculable in number, have ever occurred in the same order on two independent lines of evolution, if they were purely accidental?" (2007, 65/1911b, 68).

According to Bergson, the Darwinian theory of evolution fails to account for either the directionality that evolution displays through divergence, or the manifestation of a certain unity that is only expressed through difference. The example on which he relies is what he calls the "analogy of structure" of the Pecten eye and the human eye. But he also brings about examples that are more semantic than structural: the persistence of certain "themes" across different lineages. He invokes the similarity of the instincts in certain insects: "there are manifold variations on one and the same theme. The constancy of the theme is manifest, however, and the variations only fit it to the diversity of the circumstances" (Bergson, 2007, 167–168/1911b, 176).

(b) **Reception**

The idea that the theory of evolution is based on a teleological way of thinking despite an asserted mechanism was echoed by twentieth-century biologists. We find it in the French vitalist zoologist Pierre-Paul Grassé who criticises not so much the finalism of natural selection but rather the fact that it is hidden behind a claim of randomness (Grassé, 1973, 281). More surprisingly, this criticism is also present in the theorists of the modern synthesis. This is particularly true of Sewall Wright, and most strikingly of Theodosius Dobzhansky. Both authors explicitly refer to Bergson, Wright for having opened his eyes to the limits of Laplacian determinism (Wright, 1964, 281), Dobzhansky for Bergson's emphasis on the creativity of evolution, the appearance of the true novelty (Dobzhansky & Boesiger 1968; see also Dobzhansky, 1960, where he develops some ideas very close to those of Bergson). Other authors, such as Julian Huxley and Ronald Fisher, while denying Bergsonian philosophy the status of a scientific theory (which they are not wrong about, and indeed it does not claim to be so), credit Bergson with having brought to light certain aspects of evolution that scientific explanation does not exhaust: the unpredictability of evolution beyond the mechanisms (Fisher, 1950), the unity of evolution that unfolds over time (Huxley, 1923, 33; see Herring, 2018). However, Bergson's misunderstandings of Darwinian theory, his hostility to scientific determinism and his affinity with French spiritualism (which considers spirit as the ontological and explanatory principle of nature) soon made him unpalatable to scientists.

Today, the Bergsonian critique has a new relevance, particularly through the consideration of time in biology. In *Order Out of Chaos*, Prigogine and Stengers write that Bergson would have defined "a program that is beginning to be implemented by the metamorphosis science is now undergoing" (Prigogine & Stengers, 1984, 93), and they develop the criticism of a finality immanent in the technological model of science (see especially *Ibid.*, 174–176). More contemporarily, the neurobiologist Alain Prochiantz refers to Bergson to emphasise the unpredictability of biological phenomena. He goes so far as to say that Bergson's metaphysics is "very close to a scientific theory relevant to biology" (Prochiantz, 2002, 539, my translation).

Nowadays, the teleology implicit in the theory of evolution by natural selection is increasingly put forward, even though it is mainly by authors who do not refer explicitly to Bergson. The criticism of so-called adaptationist explanations that we find in Stephen Jay Gould or Jerry Fodor relies precisely on highlighting their implicitly finalistic nature. Adaptationist explanations consist in asking what a trait is selected for and giving the reason for its appearance based on this purpose, presupposing that it is possible to reconstruct the history of the emergence of an organ on the sole basis of its current function. Thus, seagulls are thought to have their wings *because* these wings are "excellently adapted for flight in a medium having the specific density and viscosity of the atmosphere within a thousand meters or so of the surface of the Earth" (Dennett, 2014, 197). Both Gould and Fodor denounce the teleological presupposition of these explanations: it looks like everything appeared *for* a certain function, and therefore has a structure perfectly suited to that function. Beyond the metaphysical illusion, this postulate is also epistemologically false since it assumes stable living conditions to which the organism is *perfectly* adapted. Against this view, Gould argues for the imperfection of organisms. This is the paradigmatic example of the panda, a vegetarian with a carnivorous gut, whose thumb is in fact an outgrowth of the wrist, which doubles as a similar outgrowth on the foot, that is perfectly useless. Gould also underlines the importance of considering the history of species, which alone can account for the presence of certain organs, with no apparent function – whether they once had one or not (Gould, 1980; see Tahar, 2022b). Fodor and Piattelli-Palmarini also emphasise the fallacy of these explanations: to say, for example, that the dodo became extinct because its living conditions disappeared is a tautology, since the dodo's living conditions are defined in relation to the dodo's very existence: "The extinction of the dodo was the very same event as the extinction of the dodo's way of making a living so neither can serve to explain the other" (Fodor & Piattelli-Palmarini, 2011, 147). Like Bergson, they show that selectionist explanations are adequate to explain what happened *after the fact*, but they do not help define *why one evolutionary path was followed rather than another*. If instead of asking why organisms have similar traits in similar environments, we ask why certain traits do not exist, the selectionist explanation becomes absurd. "Nobody, not even the most ravening of adaptationists, would seek to explain the absence of winged pigs by claiming that, though there used to be some, the wings proved to be a liability, so nature selected against them" (Fodor, 2007, 21). Put into Bergsonian terms, adaptation may explain "the sinuosities of the movement of evolution, but not its general directions, still less the movement itself" (Bergson, 2007, 103/1911b, 107).

But Bergson's interest is not reduced to the persistence of a certain metaphysical and epistemological critique of Darwinism. To understand the scope of Bergson's philosophy, the topicality of the solutions he proposes must also be studied.

2 Understanding the *élan vital*

Against the mechanistic vision of life, Bergson proposed what Huxley called an "intellectual vision of evolution as a fact, as something happening, something whole, to be apprehended in a unitary way" (Huxley, 1923, 33): the *élan vital.* This vision is expressed through images and poetic language that have contributed to the philosopher's discredit in the scientific world. However, poetry does not necessarily mean approximation. Through the *élan vital*, Bergson proposes a list of precise characteristics which he enumerates in *The Sources of Morality and Religion* (Bergson, 2008, 115–120/1935, 101–106). I will only study here those that are especially relevant to addressing the problems that still plague contemporary evolutionary biology. To begin with, I would like to highlight two aspects of the *élan vital*, which reveal ontological proximity, or at least reduce the gap, between Bergson and Darwin.

(a) **Time**

The first is the importance of time. Duration is a fully-fledged Bergsonian concept, the originality of which should not be reduced. It is indeed a concept that is associated in Bergson with the rejection of a substantialist metaphysics, which explains change by laws and properties pre-existing in things. In the framework of such a metaphysics, time is only an unfolding of successive states, and this unfolding can happen more or less quickly without any consequence: time therefore has no real efficacy (Bergson, 2007, 9/1911b, 9–10). On the contrary, for Bergson, time should be understood as a maturation that is also an enrichment: it is a ceaseless production of novelties, and thus of unpredictability. And Bergson goes further: time as the "very stuff" of all things, the 'states' or 'things' being only fixed and transitory views on the movement of time (see among others 2007, 272–273/1911b, 287–288). The difference in being between consciousness, life, and inert matter is thus understood as a difference in tension, *i.e.*, in rhythm (Bergson, 2010, 232–233/1911a, 275; 2010, 279/1911a, 331; 2007, 128–129/1911b, 134–135; 2007, 224/1911b, 235; on this subject see also Tahar, 2022c). Darwin obviously does not develop such a philosophy of duration. But it is noteworthy that in Darwin, as in Bergson, the thickness of time is crucial to understanding evolution, whereas it is very secondary in other evolutionists. Indeed, when Bergson was writing, the biological context, especially in France, was marked by the prevalence of physiology – a biology of short or even present time. At the beginning of the twentieth century, the debate in France was between spiritualist vitalism and materialist mechanism, with both camps arguing from a physiological anchorage inherited from Claude Bernard's determinism, for which history has no depth. For the vitalists (Ravaisson, Caro, Pa. Janet), evolution is the result of a spiritual effort that precedes and explains history; for the mechanists (Perrier, Le Dantec), it is the determined consequence of physico-chemical phenomena that take place during the life of the organism (i.e. the heredity of acquired characters). In this context, a true transformism[1] accounting for the appearance of truly unpredictable novelties in evolution, is almost impossible. It is indeed the consideration of the historical depth of geological time that allows Darwin to say that individual variations can become specific ones and thus explain evolution (Darwin, 1876, 266–270). It is history again that explains the creativity of evolution that can transform the labellum of an orchid into a protection for pollen sacs (Darwin, 2016, 200). Variation is not the result of a pre-determined mechanism, but the very process of life, with stabilisation appearing as the phenomenon to be explained, and which Darwin accounts for through natural selection (remember that the subtitle of *The Origin of the Species* is "the *preservation* of favoured races in the struggle for life", my emphasis). Bergson's starting point is similar: what is

[1]Transformism is a term generally used to refer to pre-Darwinian evolutionism, as opposed to the majority fixism of the time. Here, by true transformism, I refer to the fact that species transform unpredictably: the forms produced in the course of natural history are truly new, not predictable variations from physico-chemical conditions or variations contained from all eternity in a pre-historical spirit.

fundamental is the continuity of change, the "ceaseless upspringing of something new" (Bergson, 2007, 47/1911b, 49). "Evolution implies a real persistence of the past in the present, a duration which is, as it were, a hyphen, a connecting link" (2007, 22/1911b, 24). In relation to this change, stabilisation is always transitory (species are snapshots, taken from the moving process): the form of an organism results from the entire evolutionary history.

(b) **Divergence**

The second aspect of the *élan vital* I wish to highlight is Bergson's emphasis on divergence. Evolution is an upspringing: it does not proceed by *addition* of characters but by *differentiation*. The *élan vital* aims to express this continuity through differentiation, the unity of an impulse that is productive through its division:

> [T]he essence of a tendency is to develop in the form of a sheaf, creating, by its very growth, divergent directions among which its impetus is divided. [...Nature] preserves the different tendencies that have bifurcated in their growth. She creates with them diverging series of species that will evolve separately (Bergson, 2007, 100–101/1911b, 104–105).

Causality is not rectilinear, from cause to effect, but, as with Darwin, causality arises through difference. For Darwin, the divergence of character is in fact the necessary correlate of natural selection: the more varieties differ from one another, the less they compete for ecological space and resources, the more likely they are to survive. From this point of view, Bergson's conception is closer to the Darwinian theory than to that of the heredity of acquired characters: evolution does not (only) happen by accumulation, but above all by differentiation.

However, Bergson's philosophy brings something quite original through the concept of *élan vital*. For, if duration is creative, it is precisely because it is the deployment of an *élan*, that Bergson glosses with several other images. Its productive power is a "tendency", which means that "the evolutionary process [is] splaying out like a sheaf" (Bergson, 2007, 119/1911b, 124); and if its trajectory "would fain go on in a straight line" (2007, 129/1911b, 134), its productions (the living) "like eddies of dust raised by the wind as it passes" (*Ibid.*). These various images aim to constitute what Bergson elsewhere calls a "fluid concept" (Bergson, 2009b, 213/1946, 222): a global but heterogeneous intuition of a certain reality irreducible to the parts that logic discovers. In this way, he proposes a unitary but non-reductionist vision of evolution: he gives an image of what cannot be thought of either on the model of finality or on the model of efficient causality.

(c) **The *élan* as a tendency**

The *élan* is a movement at the origin of movement (see François, 2010, 97), which deploys its potentialities only in its effects, without being identifiable with its effects. From this point of view, unity seems to come from a movement prior to its own realisation, but this does not mean that the *élan* is external to its productions. Bergson adds the image of the tendency, which can only unfurl as a sheaf, *i.e.*, by the deployment of divergent tendencies, which are divergent only in the movement of their actualisation. The tendency does not virtually contain, as prepossibles, the different directions in which it unfolds: these directions are both the result and the surprises of the *élan* as the very efficacy of biological duration. Consequently, life cannot be resolved into physico-chemical elements, despite not being external to matter: it is a certain tendency that is actualised in matter, but whose actualisations do not exhaust the *élan*, which is precisely the impulse that allows this movement of realisation. Therefore, for Bergson, unity is behind us, rather than in front of us: it seems to precede the evolutionary process itself. The tendency is one, but virtual, and its realisation requires differentiation through the living, a differentiation that cannot be deduced from the tendency alone since it depends on the specific conditions of its actualisation.

(d) **A progress with no goal**

« Evolution in general would fain go on in a straight line » (2007, 129/1911b, 134), Bergson adds. The unity of the tendency accounts for the aspect of progress in evolution: it is the progress of a tendency that continues to be actualised. But the directions are not preordained:

> Of course, when once the road has been travelled, we can glance over it, mark its direction, note this in psychological terms and speak as if there had been pursuit of an end. Thus shall we speak ourselves. But, of the road which was going to be travelled, the human mind could have nothing to say, for the road has been created *pari passu* with the act of travelling over it, being nothing but the direction of this act itself (Bergson, 2007, 51/1911b, 54).

The tendency is not goal-oriented: it is an impulse that continues throughout the evolutionary movement, without directing it towards an end. There is an element of contingency, which comes from the conditions of realisation of the *élan* (the same theme can be preserved from one lineage to another but will not be actualised in the same way). The *élan vital* is channelled by the condition of its actualisation: matter. The living appears as a compromise, but also as the centre of two opposing temporalities: that of matter and that of the living, that of entropy and that of organisation. Evolution is "*a reality which is making itself in a reality which is unmaking itself*" (Bergson, 2007, 248/1911b, 261): it is an impulse that is only realised through matter, but which thereby finds its determinations and therefore also its limitation.

(e) **Immanence of the *élan vital* to living beings**

The unpredictability, however, does not only come from the contingency imposed by matter: it also originates from the fact that the creativity of the *élan* is realised through the living beings which, by their action, prolong the vital impetus. Bergson writes that "life can progress only by means of the living, which are its depositaries" (Bergson, 2007, 232/1911b, 243). The *élan vital* is immanent to the evolution of living beings whose power to act derives from the creativity of the evolutionary process, which they singularise and therefore limit by virtue of their individuality... but which they also actualise through their actions. If living beings exhibit a certain stable coherence, this stability is not a closure. This is why Bergson speaks of "eddies" or "circles" (2007, 129/1911b, 134): living beings are a kind of movement that happens 'on the spot' so to speak, yet they are at the same time able to transmit something of this movement: "Every human work in which there is invention […] every movement of an organism that manifests spontaneity, brings something new into the world. […] We are not the vital current itself; we are this current already loaded with matter" (2007, 240/1911b, 252). Living beings actualise the possibilities of evolution, but in the sense that they also create them. This is why Bergson can say that they bring something truly novel into the world: "a living being is a centre of action. It represents a certain sum of contingency entering into the world, that is to say, a certain quantity of possible action" (2007, 262/1911b, 276).

The *élan vital* aims at giving an image of this movement of life, irreducible to efficient causality, irreducible also to finalism, and which implies both harmony and scattering, directionality and radical unpredictability. With this image, Bergson invites us to conceptualise a new form of causality specific to the duration of biological phenomena. But to achieve this conceptualisation, he says, a new biological science would be needed, one capable of integrating the heterogeneous elements included in this image of *élan vital*, a biology that "would become, to the physics and chemistry of organized bodies, what the mathematics of the moderns has proved to be in relation to ancient geometry" (2007, 32/1911b, 34).

3 Topicality of the *élan vital*

(a) The unity is behind

Since Bergson, the reductionist framework of the modern synthesis has been called into question, in favour of a consideration of complexity, of multi-scale causality, which allowed for a third route, beyond neither mechanism and finality. In these more recent theories, some characteristics of Bergson's *élan vital* are rediscovered. Notably, the discovery of genetics, which Bergson does not mention in *Creative Evolution*, gives us a biological, albeit theoretical, resource for thinking the original unity of the living, producing a pulverisation of forms ahead. For instance, the PAX-6 gene, which is found in cephalopods and humans, also has homologues in flies and fish. Its role is to guide the formation of the eye, by controlling the action of other genes which are specific to each of the different species. Consequently, the eyes produced are also specific to each of the species (Nüsslein-Volhard & Neumann, 2000).

(b) Science is rediscovering time

But far beyond this indirect corroboration of certain Bergsonian intuitions, which may appear as mere anecdotal coincidence, Bergson's foresight concerning evolution is manifested above all in the fact that "*science is rediscovering time*" (Prigogine & Stengers 1984, xxviii). The historical dimension of life, and in particular the creativity it implies, has been emphasised by biologists themselves. Genetics and natural selection are not enough to explain the form taken by biodiversity nor, to quote Gould, the "inhomogeneous distributions of organisms across the potential morphospace", *i.e.*, the fact that among the theoretically possible, viable and even "good" morphologies, only some actually exist (see Gould, 2002, 1055–1056). On this point, the theory of evolution has been largely enriched, and the 1500 pages of Gould's *Structure of Evolutionary Theory* reveals both in content and in form the challenge presented by its totalization. In addition, the theory of complex systems strives to integrate the different scales of causality, which often correspond to different time scales. David Kreps (2015), Robin Durie (2002), and Keith Ansell-Pearson (2018, 108), among others, have emphasised the conceptual proximity of complex systems theorists (Brian Goodwin, Stuart Kauffman) to the Bergsonian idea that the living is something making itself in something unmaking itself.

More prominently, in the philosophy of science there is an increasingly marked consideration of the processuality of biological events and a desire to provide new models of causality. The link between the conception of time and models of causality has been highlighted for instance by Nathalie Gontier (2016, 2021) through the reticulate evolution studies. To give a complete theory of evolution, it would be necessary to be able to account for the action of time in evolution, for the way in which different scales of time communicate. More and more attention is being paid to phenomena previously considered accidental consequences of natural selection: the phenomena of exaptations documented by Gould, functional overload (an organ acquiring more and more functions over time), or more recently, the cases of cryptic genetic variation (Paaby & Rockman, 2014). What is observed is that the past changes in meaning as time progresses and may constitute a reserve of innovations.

(c) **Rethinking causality**

These new reflections on the importance of time to conceive biological phenomena are accompanied by the abandonment of a linear causality model in favour of cybernetic or reticular models, and more fundamentally by a new conception of the living world. The unitary vision of nature as governed by eternal laws is replaced by a historical and plural conception of the world, in which biological phenomena really differ from those studied by physics, and where this difference must be accounted for.

Philosophy is increasingly trying to account for the historicity of biological regularities, whose causal power is less "entailing" than "enabling" (Longo et al., 2012). Montévil and Mossio have proposed to speak of historically constituted biological constraints (Montévil & Mossio, 2015) to account for what appear less as causes than as local and transitory conditions, that act on processes (see also Montévil, 2020). Accordingly, they account for the stability of biological systems in terms of a closure of constraints. This closure is explained by the way in which evolutionary history is inscribed in organic matter, therefore channelling the directions taken by evolution, but at the same time creating the conditions for its perpetuation. Thus, these constraints create truly new possibilities, not included in a predefined phase space (Longo et al., 2012; Longo, 2018). It is noteworthy that it is by drawing on the Bergsonian critique of possibility that Montévil develops this idea of new possibilities in evolution, truly unpredictable, even by a Laplacian demon (Montévil, 2019). Bergson criticises the category of possible as a retrospective illusion (Bergson, 2009b, 99–116/1946, 107–125). Once the present event has been observed, it is retrospectively assumed to be possible in the past. The illusion consists in using an unpredictable event of the present to describe an initial situation of the past. It assumes that time has no effect, that everything must be predictable (Bergson 2009b, 19/1946, 26–27): something that happens at a given moment must have already been preordained or at least probable in the previous moment. Montévil relies on this critique to develop the idea of new possibilities brought about by biological processes: when an evolutionary novelty (e.g. articulated jaws) opens up the possibility of truly new, i.e., unpredictable innovations (molars which can crush food) the emergence of the first novelty is a necessary ingredient for the second novelty to play a functional role. This second innovation was not possible before the emergence of the first. Similarly, if it is possible to predefine the space of possibilities of DNA sequences, this space would not take into account the viability of the organism, which is not a generic property, but depends on the conditions of the species, the organism, and even the environment. The possible organism is therefore not preordained but made possible by the particular circumstances of its appearance, which are the result of evolutionary history.

Finally, it is more and more common for living beings to be conceived as actors in the evolutionary process, insofar as they are the ones who live the struggle for existence. Their individual actions are considered as having an impact on the rest of evolution. Even theoretical models increasingly take into account the discrete nature of agents (Zelnik et al., 2015). In addition, there are philosophical contributions on the agency of organisms, which propose to think of a "situated Darwinism" that integrates the way in which organisms experience their conditions and act according to them (Walsh 2015).

All these new theoretical attempts to account for the processuality and historicity of evolution seem to contribute to this new biology that Bergson called for, and which is more committed to the unpredictability of life. Although Bergson may have been considered an outdated vitalist, his intuitions seem to find a certain echo in contemporary biology today, and more explicitly in the philosophy of biology. References to Bergson are generally rare and often discreet. But it seems to us that, more than cultural recognition, they are the recognition of a legacy, and perhaps, from this point of view, a manifestation of the creativity of Bergsonian duration: in which the future comes to open up the potentialities of the past. Furthermore, while Bergson does not provide epistemological tools immediately usable by scientists, he proposes a metaphysics of the living that enables us to think of biological causality in a truly temporal way, through images that continue to nourish our reflection.

4 The Lessons of the *élan vital* on the Causality in Evolution

(a) The critique of the intellect

The transition from a law-based conception of nature to a historical understanding of life also calls into question our means of knowledge. Knowing no longer consists in discovering laws that have existed for all eternity, but it is the meeting of two historical processes – the one that results in the knowing subject, and the one that produces the transient constraints of the studied phenomenon – which lack the stability of objects usually defined by science. As Bergson already pointed out, intelligence itself has a history, it is an "adaptation" "deposited by the evolutionary movement in the course of its way" (Bergson, 2007, v–vi/1911b, ix–x). Consequently, there is a historicity of science, which does not exclude progress. Science is always in the making and can no more be closed than the object it studies can be stabilised. Bergson proposes a metaphysics of life, inseparable from a philosophy of knowledge. Since life is a history open to the future, and intelligence, understood as an *adaptation* is historically situated, our intelligence cannot grasp the very life from which it originated. This implies that "the two ready-made garments that our understanding puts at our disposal, mechanism and finality" (2007, x/1911b, xiv–xv) and more generally our modes of representation must be criticised, in order to think "beyond the human condition" (Ansell-Pearson, 2018), to understand life beyond our own limitations. The *élan vital* serves precisely to open the way.

(b) Philosophy, science, and biology

The *élan vital* is far from being an abstract metaphysical concept. On the one hand, such an interpretation is hardly faithful to Bergson's intention, which was to propose an "idea full of matter, obtained empirically, capable of guiding our investigations, which will broadly sum up what we know of the vital process and will also bring out what is still unknown" (2008, 120/1935, 105). On the other hand, as DiFrisco points out, this interpretation does not capture the organising role of this concept around which a true philosophy of life is articulated, not only based on the data of science, but also aimed at questioning the epistemology of science itself (DiFrisco, 2015). The scope of the *élan vital* can only be understood in reference to the project of making "a truly intuitive philosophy [that] would realize the union so greatly desired, of metaphysics and science", of putting "more of science into metaphysics and more of metaphysics into science" (Bergson, 2009b, 216–217/1946, 227). This union is, according to Bergson, particularly desirable for biology, which lies between the two extremities of science "which have to do with pure intellect" and metaphysics "which calls upon intuition" (2009b, 86/1946, 93). Since its object is both material (and matter is the natural object of scientific intelligence) and living (that is, marked by duration, the object of metaphysical intuition). It is only in reference to this Bergsonian project that we can understand the numerous references to the work of biologists throughout *Creative Evolution.* Philosophers and scientists must collaborate, on the model of the aviator and the diver, proposed by Bergson. If the diver "feels out the wreck on the sea floor", it is because the aviator "has pointed

[it] out from the air" (2009b, 67/1946, 74). The *élan vital* is the indicator that the aviator offers the diver:

> When I relate the phenomena of life and of evolution to an '*élan vital*', it is in no way an ornament of style. [...] The truth is that philosophy only offers philosophers two principles of explanation in this matter: mechanism and finalism. [...] Now [...] the place to be is somewhere in between these two concepts. How should we determine that place? I have to point to it, to indicate it since no concept between mechanism and finality exists. The image of an *élan* is only this indication. (Bergson & Delattre, 1936, 399)

(c) How to "use" the *élan vital*? The *élan vital* as a dynamic scheme

But the *élan vital* remains empty as such if the reader does not make the effort to move from the image to the intuition. From this point of view, it offers what Bergson would call a "dynamic scheme". The dynamic scheme is a concept developed by Bergson to think of intellectual effort. The scheme appears as the horizon of thought that the effort must conquer, that is, filled with images. The scheme is therefore a sort of sketch, but at the same time it guides the process by which the sketch becomes a true image. It is therefore both a sort of representation "[...] which implies reciprocal penetration of all the elements in one another" (Bergson, 2009a, 163/1920, 196) and the "indication of what we must do to reconstruct them" (2009a, 161/1920, 198), of "a certain *direction of effort* to follow" (2009a, 165/1920, 200). This effort consists in "goings and comings, oscillations, struggle and negotiation", between the scheme and the images it calls for (2009a182/1920, 220). In the case of invention, both artistic and scientific, this work also involves a gradual modification of the scheme, a modification "required by the very images which the scheme has aroused" (2009a, 181/1920, 220). It seems to us that the *élan vital* is more a dynamic scheme that a definitive concept that would stand for the ontology of life. It appears as both the starting point of a work and the horizon of the research on biological evolution.

Through the *élan vital*, Bergson does not seek to substitute a metaphysical concept for the explanations of biology, but "to start up a certain effort" (2009b, 185/1946, 195), both on the side of philosophy and on the side of science. One can assume that in the case of the *élan vital*, the comings and goings between the scheme and the images are also comings and goings between scientific discoveries and philosophical intuition: the *élan vital* would thus tend to be enriched and modified as science progresses.

(d) How to Think of the *élan vital* Today

What remains relevant in Bergsonian philosophy is this urge to explore the implications of the consideration of time, and more specifically of the plurality of its rhythms, on our conception of causality in evolution. Enriched and clarified by the discoveries of contemporary biology, and by the new theoretical (philosophical or biological) approaches to evolution, the *élan vital* appears to still convey a vivid and precise vision of evolution as a historical process, weaving together different time scales without contradiction. It invites us to envisage a causality that is both circular and propulsive.

The model of causation is partly circular. Taking time into account invites us to think of the retroaction of evolution at all time scales: that of the organism considered as an open system (see Prigogine & Stengers, 1984; and the work of the geneticist Mae-Wan Ho, *The Rainbow and the worm,* 1998); that of heredity (living beings repeat themselves through generations, although not identically), that of evolution itself considered as a whole (the future emerges from the historical continuity of the past, but the innovations that it carries change the meaning of this past in an unpredictable way). If these circularities are phenomena that differ in rhythm, it must be remembered that duration is one: the difference in rhythm is not a substantial distinction; all these 'levels' that we artificially divide are part of the same evolutionary movement. To conceive their interaction, it must be understood that they are bound together by a causality close to recursion, but a recursion that is *temporal*: the differentiations of rhythms that constitute living beings contribute to the very movement of duration from which they emerge (evolution) and are therefore necessary to the creativity of the process. In this way, life "expands and transcends its own being" (Bergson, 2007, 52/1911b, 55).

It also means that the propulsive aspect of evolution and its apparent directions come from this very circularity. Since, in evolution, the different rhythms are equally relevant, and equally causal, this means that the memory of the entire history, the conditions of life in the present, the particular actions of organisms jointly explain the directions taken by evolution: they produce both the constraints and the future possibilities (on creative circularity, see Hwang, 2017). If evolution is irreversible and unpredictable, if it is a continuity from the past to the future that is less an accumulation than a maturation, if its unpredictability does not exclude it from moving in privileged directions, finally if it is creative, it is because of the interaction of these different rhythms within the very duration of evolution. Circularity cannot mean cyclical causality nor repetition.

Circularity should rather be understood from the image of the spiral that Bergson uses to conceive the movement of history: "the pendulum is endowed with memory and is not the same when it swings back as on the outward swing, since it is then richer by all the intermediate experience. This is why the image of a spiral movement, which has sometimes been used, is perhaps more correct than that of the oscillations of a pendulum" (2008, 311/1935, 281). The present cannot resemble the past in the living because it is enriched with the totality of the movement that preceded it. From this point of view, the future is never encompassed in the present state. Causality is not linear, but it must also be stressed that it is not necessitating, nor even probabilistic. The virtuality of the tendency of the *élan vital* does not mean that evolution is directed towards the actualisation of a predetermined end. For it is not a virtuality of all eternity: it is a *temporal virtuality* whose realisation is really unpredictable. Bergson writes about history: "action on the move creates its own route, creates to a very great extent the conditions under which it is to be fulfilled, and thus baffles all calculation" (2008, 315/1935, 285). This is also true for biological evolution: living beings are only rich in future novelties because of their past history insofar as this history is both an obstacle to be overcome and a reserve of possibilities. This evolutionary duration is therefore both cyclical and propulsive: it recapitulates itself through original situations and makes biological history not repetitive but evolutive.

These Bergsonian suggestions, both highly theoretical and pictorial, are certainly not scientific tools. However, they still aim to reflect on our modes of knowledge, to question our ready-made categories, and they indicate a path to follow. They give rise to a philosophical doubt, rather like that advocated by Claude Bernard (1865): a doubt that does not aim to deny scientific ambition, but to never hold theories as definitive, to always leave open the possibility for new, more comprehensive theories, capable of accounting for more facts. I leave the conclusion to Bergson asking: "Is it not […] dangerous to separate philosophy from science? Should we not rather seek to make our science more philosophical?" (Letter from Bergson to A. Leclère, late July 1902 in Bergson, 2002, 75).

References

Ansell-Pearson, K. (2018). *Thinking Beyond the Human Condition*. Bloomsbury.

Bergson, H. ([1896], 2010). *Matière et mémoire* (Ed. C. Riquier). PUF./1911a. *Matter and Memory* (Trans. N. M. Paul & W. S. Palmer). George Allen and Unwin.

Bergson, H. ([1907], 2007). *L'Évolution créatrice* (Ed. A. François). PUF./1911b. *Creative Evolution* (Trans. A. Mitchell). Macmillan.

Bergson, H. ([1919], 2009a). *L'énergie spirituelle* (Ed. E. During). PUF./1920. *Mind-Energy* (Trans. H. Wildon Carr). Henry Holt.

Bergson, H. ([1932], 2008). *Les deux sources de la morale et de la religion* (Ed. G. Waterlot et F. Keck). PUF./1935. *The Two Sources of Morality and Religion* (Trans. R. Ashley Audra & C. Brereton). Henry Holt.

Bergson, H. ([1934], 2009b). *La pensée et le mouvant* (Ed. A. Bouaniche). PUF./1946. *The Creative Mind* (Trans. M. L. Andison). Philosophical Library.

Bergson, H., & Delattre, F. (1936). Samuel Butler et le bergsonisme. Avec deux lettres inédites d'Henri Bergson. *Revue anglo-américaine., 8*(5), 385–405.

Bergson, H. (2002). *Correspondance* (Ed. A. Robinet). PUF.

Bernard, C. ([1865], 1949). *An Introduction to the Study of Experimental medicine* (Trans. H. C. Greene). Henry Schuman.

Darwin, C. ([1859], 1876). *The Origin of Species by Means of Natural Selection, or the Preservation of Favoured Races in the Struggle for lLfe* (6th ed.), with additions and corrections. Murray.

Darwin, C. ([1862], 2016). *The Works of Charles Darwin:* Vol 17: *The Various Contrivances by which Orchids are Fertilised by Insects* (Ed. P. H. Barrett & R. B. Freeman). Routledge.

DiFrisco, J. (2015). *Elan vital* Revisited: Bergson and the Thermodynamic Paradigm. *The Southern Journal of Philosophy, 53*(1), 54–73.

Dobzhansky, T. (1960). Evolution and Environment. In S. Tax (Ed.), *Evolution after Darwin* (3 vols., pp. 403–428). The University of Chicago Press.

Dobzhansky, T., & Boesiger, E. (1968). L'Évolution créatrice. In *Essais sur l'évolution* (pp. 145–165). Masson & Cie.

Dennett, D. C. (2014). *Darwin's Dangerous Idea: Evolution and the Meanings of Life*. Simon and Schuster.

Durie, R. (2002). Creativity and Life. *The Review of Metaphysics, 56*(2), 357–383.

Fisher, R. A. (1950). Creative aspects of Natural Law. In *The Eddington Memorial Lecture* (pp. 179–184). Cambridge University Press.
Fodor, J. (2007). Why Pigs don't have Wings? *London Review of Books, 29*(20), 19–22.
Fodor, J., & Piattelli-Palmarini, M. (2011). *What Darwin Got Wrong*. Profile Books.
François, A. (2010). L'évolution de la vie. Mécanisme et Finalisme. In A. François (Ed.), *L'évolution créatrice de Bergson* (pp. 17–110). Vrin.
Gontier, N. (2016). Time: The Biggest Pattern in Natural History Research. *Evolutionary Biology, 4*(43), 604–637.
Gontier, N. (2021). *Teleonomy as a Problem of Causation, and Causation as a Problem of Hierarchy and Time*. In Conference: Evolution 'on purpose': Teleonomy in living systems, by the Linnean Society. https://www.youtube.com/watch?v=h6b-p3-P2h0
Grassé, P.-P. (1973). *L'évolution du vivant. Matériaux pour une nouvelle théorie transformiste*. Albin Michel.
Gould, S. J. (1980). *The Panda's Thumb: More Reflections in Natural History*. W.W. Norton.
Gould, S. J. (2002). *The Structure of Evolutionary Theory*. Harvard University Press.
Herring, E. (2018). 'Great is Darwin and Bergson his Poet:' Julian Huxley's Other Evolutionary Synthesis. *Annals of Science, 75*(1), 40–54.
Ho, M.-W. (1998). *The Rainbow and the Worm. The Physics of Organisms* (2nd ed.). World Scientific Publishing.
Huxley, J. (1923). Progress, biological and other. In *Essays of a Biologist* (pp. 3–68). Alfred A. Knopf.
Hwang, S.-Y. (2017). La figure du cercle vertueux comme aspect essentiel de la causalité créatrice. Une lecture critique de Bergson. *Dialogue, Revue canadienne de philosophie, 56*(2), 317–335.
Jankélévitch, V. (1959). *Henri Bergson*. PUF.
Kreps, D. (2015). *Bergson, Complexity and Creative Emergence*. Palgrave Macmillan.
Longo, G., Montévil, M., & Kauffman, S. (2012). "No Entailing Laws, But Enablement in the Evolution of the Biosphere." In *GEECO'12: Proceedings of the 14th annual conference companion on genetics and evolutionary computation*, pp. 1379–1392.
Longo, G. (2018). How Future Depends on Past and on Rare Events in Systems of Life. *Foundations of Science, 23*(3), 443–474.
Monod, J. (1970). *Le hasard et la nécessité. Essai sur la philosophie naturelle de la biologie moderne*. Seuil.
Montévil, M., & Mossio, M. (2015). Biological Organisation as Closure of Constraints. *Journal of Theoretical Biology, 372*, 179–191.
Montévil, M. (2019). Possibility Spaces and the Notion of Novelty: From Music to Biology. *Synthese, 196*, 4555–4581.
Montévil, M. (2020). Historicity at the Heart of Biology. *Theory in Biosciences, 141*(2), 165–173.
Nüsslein-Volhard, C., & Neumann, C. J. (2000). Patterning of the Zebrafish by a Wave of *Sonic Hedgehog* Activity. *Science, 289*, 2137–2139.
Paaby, A. B., & Rockman, M. V. (2014). Cryptic Genetic Variation, Evolution's Hidden Substrate. *Nature Reviews Genetics, 15*(4), 247–258.
Prigogine, I., & Stengers, I. (1984). *Order out of Chaos. Man's New Dialogue with Nature*. Bantam Books.
Prochiantz, Alain. 2002. À propos de Henri Bergson, être et ne pas être un animal. *Critique*, 68(661–662), 542–551.
Tahar, M. (2022a). The History of the Bergsonian Interpretation of Charles Darwin. *Bergsoniana*, 2. Online since 01 July 2022, connection on 10 July 2022. http://journals.openedition.org/bergsoniana/740
Tahar, M. (2022b). Bergson as Visionary in Evolutionary Biology. In M. Sinclair & Y. Wolf (Eds.), *Bergsonian Mind* (pp. 446–460). Routledge.
Tahar, M. (2022c, in press). Bergson's Vitalisms. *Bergson and Vitalisms.* Special Issue of *Parrhesia* (Ed. M. Tahar).
Walsh, D. (2015). *Organisms, Agency and Evolution*. Cambridge University Press.

Wright, S. (1964). Biology and the Philosophy of Science. *The Monist, 48*(2), 265–290.
Zelnik, Y. R., Solomon, S., & Yaari, G. (2015). Species Survival Emerges From Rare Events of Individual Migration. *Scientific Reports, 5*, 7877.

Mathilde Tahar received her Ph.D in philosophy of biology from the University of Toulouse II and the Muséum National d'Histoire Naturelle. Her dissertation topic was 'Evolution beyond all finalism: Bergson's critique of evolutionists' and aimed at confronting Bergsonian evolutionary thought with contemporary biology. She has contributed to the collective volume *The Bergsonian Mind*, edited by Mark Sinclair and Yaron Wolf (2021), and published several articles on Bergson and biology.

The manufacturer's authorised representative in the EU is Springer Nature Customer Service Centre GmbH, Europaplatz 3, 69115 Heidelberg, Germany. If you have any concerns regarding our products, please contact ProductSafety@springernature.com

Printed and bound by CPI Group (UK) Ltd, Croydon, CR0 4YY
28/09/2026
02224075-0009